1953-1999

C. Lee Campbell was a scholar in the finest sense of the word. His reputation as a world-class plant pathologist is well known. But what is not as well known is that Lee had broad interests, and a desire to understand and support research in other disciplines that would enrich his own. He had a passionate desire to understand the complex historical forces that shaped plant pathology and continue to influence his science today. Whether in plant pathology or history, Lee's research and writing were characterized by enthusiasm, clarity, and a breadth of vision. Those traits fill the pages of this book. As such, it is a fitting tribute to a life of scholarship.

Bookplate

The Formative Years of
PLANT PATHOLOGY
in the
United States

The Formative Years of
❧ PLANT PATHOLOGY ❧

in the
United States

C. Lee Campbell
Paul D. Peterson
Clay S. Griffith

APS PRESS

The American Phytopathological Society
St. Paul, Minnesota

*To our families, whose love has sustained us and made
the completion of this book possible*

CONTENTS

Scientists generally concern themselves with the current and future state of knowledge. In a science like plant pathology, such concern is directed to the application of new knowledge to real-world problems. Yet from time to time we pause to wonder about those who came before us, how our science came to be what it is today, and what factors catalyzed its development. An occasional glance back into the history of our science does not impede the quest for new knowledge. In fact, such a retrospective glance benefits the discipline as practitioners develop a more complete view of the formation and growth of their science.

The Formative Years of Plant Pathology in the United States fills a gap in the history of science. No previous book has examined and synthesized the events, both biological and social, that fostered the development of plant pathology in the United States. Herbert Hice Whetzel's 1918 book, *An Outline of the History of Phytopathology*, provides a tantalizing but all too brief treatment of some of the significant individuals and key events in the science. Andrew Denny Rodger's 1952 book, *Erwin Frink Smith: A Story of North American Plant Pathology*, offers a scientific biography of only one of the founders of American plant pathology. Geoffrey Clough Ainsworth's 1981 book, *An Introduction to the History of Plant Pathology*, though a worthy contribution to the history of the science, takes a much larger temporal and international perspective than we have chosen to pursue here. His examination is a useful collection of events, names, places, and dates that outlines the essentials of what transpired, rather than a critical probe of the factors, people, and institutions that catalyzed and led to the formation of the science of plant pathology. Other useful, shorter papers and chronologies cover specific aspects of the history of plant pathology in the United States, but again these lack the context or "big picture" approach that we felt was needed. Our book draws on the strengths of the histories that came before us to provide not an encyclopedic chronicle, but rather a readable and accurate examination that weaves many elements—biology, individuals, institutions—into a unified view of the formative years of plant pathology in the United States.

Although we briefly mention and acknowledge key European contributions, our focus is almost exclusively on events in the United States of America. This limited geographic perspective allows us to explore a largely untapped wealth of primary sources available from those who participated in the formation of American plant pathology. We focus on events key to the discipline, but we also attempt to place our subject within the larger context of the history of American agriculture and science. However, attention to well-worked historiographical arguments about "pure" versus "applied" and professionalization in science are beyond the scope of this book. We realize that we have only scratched the surface of the context of American history, but we hope that we have provided a good jumping-off point for future scholars.

The Formative Years of Plant Pathology in the United States is intended for three specific audiences. The primary audience is graduate students entering careers in plant pathology or closely related disciplines. We have worked to provide an accurate and readable story of the formation of the science they are considering as a profession. We wish to provide a background and context from which they can continue during their own careers to advance knowledge about plant diseases and their control. A second audience is the current cadre of plant pathologists and other plant biologists who are creating the future history of the science. We seek to entice these busy professionals to pause and reflect upon the accomplishments of those who preceded them and to wonder how those in future generations of plant pathologists will view what we do today. A third audience is historians of agriculture, biology, and science as a whole. We believe that these historians will be intrigued by the events occurring in the plant sciences and plant pathology in concert with and sometimes in advance of similar events in animal and medical sciences.

The preparation of this book has offered many challenges, but one in particular has been choosing the scientific names of pathogens. Many of the pathogens we discuss have undergone name changes over time. For example, the pathogen that causes late blight of potato was known as *Botrytis infestans* and *Peronospora infestans* before receiving its current name *Phytophthora infestans*. We have chosen to use pathogen names as they would have been used by the original, contemporary authors. We have provided a listing of the older and current names in the appendix. Also, we acknowledge that the oomycetes such as *Phytophthora infestans* and *Plasmopora viticola* are no longer classified as fungi. However, we refer to them as fungi in this book because that was the accepted practice in the contemporary literature.

The writing of this book has been a stimulating academic endeavor and a labor of love. We have benefited from the professionalism and kindness of many individuals in many places throughout the United States. Without the assistance of these people, many of whom are listed in the acknowledgments, the completion of our task would not have been possible. We owe these many individuals our heartfelt thanks.

We wish to convey our special thanks to several individuals whose efforts and support have helped measurably in making this book a reality: O. W. Barnett, Jr., gave continuing strong support and encouragement for our research and publication efforts; William C. Kimler provided historical insights and assistance in maintaining focus in our research and writing; the late Kenneth F. Baker, in early conversations, conveyed to us his infectious enthusiasm for the history of our science; Marvin Williams worked his magic on the photos that so richly illustrate our history narrative; and Kurt J. Leonard, of APS Press, served ably as our patient, understanding, and supportive senior editor.

Finally, our thanks to our families, who have provided us with encouragement when we needed it, patience and understanding to see us through this project, and unquestioning love, which continues to sustain us.

❧ ACKNOWLEDGMENTS ❧

Many individuals assisted us in the preparation of this book by reviewing what we had written, answering our questions, and offering helpful suggestions and leads in our search for information about early plant disease researchers in the United States. We are indebted to them for their efforts and appreciate all that they have done in helping us bring this book to completion.

We sincerely appreciate the comments, suggestions, and encouragement provided by those who read the earlier drafts of one or more sections of this book: Robert Aycock, Department of Plant Pathology, North Carolina State University; Kenneth F. Baker, formerly of the University of California, Berkeley; Deborah R. Fravel, USDA ARS, Biocontrol of Plant Diseases Laboratory, Beltsville, Maryland; Douglas Helms, USDA NRCS; R. Douglas Hurt, Department of History, Iowa State University; William C. Kimler, Department of History, North Carolina State University; Thor Kommedahl, Department of Plant Pathology, University of Minnesota; Patrick E. Lipps, Department of Plant Pathology, Ohio State University; Jeanne D. Mihail, Department of Plant Pathology, University of Missouri; Richard A. Overfield, Department of History, University of Nebraska, Omaha; Wayne D. Rasmussen, retired chief historian, USDA; John M. Riddle, Department of History, North Carolina State University; Amy Y. Rossman, USDA ARS Systematic Botany and Mycological Laboratory, Beltsville, Maryland; G. Terry Sharrer, Smithsonian Institution; Turner B. Sutton, Department of Plant Pathology, North Carolina State University; Carol E. Windels, Department of Plant Pathology, University of Minnesota; Tom van der Zwet, USDA ARS, Appalachian Fruit Research Station, Kearneysville, West Virginia.

The assistance, support, and encouragement of a number of individuals at our home institution, North Carolina State University, made the day-to-day progress of the research and writing effort for this book possible. Our thanks to the following individuals who helped in special ways: Robert Aycock, Samia El-Allaf, John Riddle, Charles R. Harper, and Patricia Sullivan. Also, our sincere thanks to the undergraduate students who assisted us in many phases of preparing this book: Annette Huetter, David Norden, Amy Polen, Courtney Reuter, and Beth Sykes.

Many individuals took the time to locate material for us and assisted us in many ways during our research trips to their universities. We are most grateful for all of their efforts. Special mention is due to the following colleagues: Sandra Anagnostakis, Connecticut Agricultural Experiment Station; Neil A. Anderson, Department of Plant Pathology, University of Minnesota; Deane C. Arny, Department of Plant Pathology, University of Wisconsin; Myron K. Brakke, Department of Plant Pathology, University of Nebraska; Helene R. Dillard, Department of Plant Pathology, New York State Agricultural Experiment Station and Cornell University; Craig R. Grau, Department of Plant Pathology, University of Wisconsin; Dennis C. Gross, Department of Plant Pathology, Washington State University; John Haines, New York State Museum, Albany; Joseph G. Hancock, Department of Environmental Science, Policy and Management, University of California, Berkeley; James W. Hilty, Department of Entomology and Plant Pathology, University of Tennessee; Thor Kommedahl, Department of Plant Pathology, University of Minnesota; Philip O. Larsen, Department of Plant Pathology, University of Minnesota; Kurt J. Leonard, USDA ARS Cereal Disease Laboratory, St. Paul, Minnesota; Kent Loeffler, Department of Plant Pathology, Cornell University; Jeanne D. Mihail, Department of Plant Pathology, University of Missouri; Dewey Moore, Department of Plant Pathology, University of Wisconsin; Forrest W. Nutter, Department of Plant Pathology, Iowa State University; Donald H. Pfister, Farlow Herbarium, Harvard University; Amy Y. Rossman, USDA ARS Systematic Botany and Mycological Laboratory, Beltsville, Maryland; Gregory E. Shaner, Department of Botany and Plant Pathology, Purdue University; Steven A. Slack, Department of Plant Pathology, Cornell University; James B. Sinclair, Urbana, Illinois; Sue A. Tolin, Department of Plant Pathology, Physiology and Weed Science, Virginia Polytechnic Institute and State University; Roy C. Wilcoxson, Department of Plant Pathology, University of Minnesota

The faculty, staff, and students of a number of current and former departments of plant pathology made our research trips more enjoyable and productive and were willing to assist by providing materials and helping us find answers to a wide range of questions about the history of our science. We especially appreciate the hospitality and assistance from members of the departments at University of California, Berkeley; Cornell University; Iowa State University; University of Minnesota; Arthur Herbarium, Purdue University; University of Wisconsin.

We also received invaluable assistance from archivists and staff members at the following repositories where records concerning the earlier years of plant pathology in the United States are found: Academy of Natural Sciences; American Philosophical Society; Hunt Institute for Botanical Documentation, Carnegie-Mellon; Department of Manuscripts and University Archives, Cornell University Libraries; Forest History Society; The Botany Libraries (Farlow Reference Library and Herbarium of Cryptogamic Botany, Library of the Gray Herbarium, Library of the Arnold Arboretum), University Archives, Harvard University; University Archives, University of Illinois; Department of Special Collections and University Archives, Iowa State University; Library of Congress; University Archives, University of Minnesota; Missouri Botanical Garden; National Academy of Sciences; National Agricultural Library; National Archives; University Archives, University of Nebraska; University Archives, Purdue University; Systematic Botany and Mycological Laboratory; University Archives, University of Vermont; Division of Archives, University of Wisconsin.

We are also indebted to Rhone-Poulenc, Inc., for their early financial support of the research associated with this book.

Part One

PLANT DISEASES AND AGRICULTURE IN EARLY AMERICA

As the ingenuity and invention of man may increase to an unknown and inconsiderable degree, so may the improvements and arrangements of husbandry keep pace therewith, until the most fruitful spot that now exists, may produce a tenfold quantity, and the land which now supports a hundred men, give equal enjoyment to a thousand.
—Samuel L. Mitchill

Cultivators of the earth are the most virtuous and independent citizens.
—Thomas Jefferson

Although science historians have referred to the American colonial period to the mid-nineteenth century as the "Dark Ages" of plant pathology, Americans have always been interested in plant disease—even before the scientific elaboration of the germ theory of disease and the rise of plant pathology in the middle years of the nineteenth century. European colonization brought a movement of agricultural crops and pathogens that led inevitably to troubling plant diseases.[1] Among the many diseases observed and considered by informed agriculturists were outbreaks of wheat rusts and smuts, fire blight of pear and apple, peach yellows, potato late blight, and cotton root rot.

The responses to plant disease reflected the main currents of agricultural and scientific knowledge in early America. An influential movement for agricultural improvement occurred in Europe and America in the late eighteenth and early nineteenth centuries. This movement, led by some of the most respected European agricultural writers of the time, such as the Englishmen Arthur Young, Sir Humphrey Davy, and William Forsyth, championed Enlightenment ideas of agricultural practices based on careful observation and reason. Informed agriculturists applied their powers of reason to rational laws for agricultural improvement. They advocated a number of dependable cultural practices such as manuring, fertilizing, tillage, drainage, and crop rotation, with the primary goal of increasing yields.[2] Understanding and controlling plant diseases fit well into this rational model for improved agriculture.

In the late eighteenth and early nineteenth centuries Americans began to merge their interests in agricultural improvement with a focus on understanding plant diseases and mitigating their effects. Colonists seldom did more than record scattered observations. After the Revolution, however, confronted by escalating plant disease problems, Americans took advantage of new outlets of communication to routinely discuss their ideas. The vanguard of this movement was an elite group of gentleman horticulturists. By the 1840s the movement expanded to include a larger assembly of agriculturists, amateur scientists, and agricultural chemists concerned with practical agriculture. These enthusiasts actively organized and joined societies, participated in meetings, collected and distributed European agricultural and scientific information, and published their views on causes and controls. Such careful observations on America's earliest and most important plant diseases sometimes resulted in useful cultural practices for disease control. Moreover, controlling plant disease became an important topic in America's expanding agricultural press.

But as these early responses reveal, the nature of the contemporary understanding of plant disease was precarious. Early plant disease theorists in the United States worked by and large from suppositions reflecting agricultural practice. The tools they had to determine the causes of diseases and to suggest cures or preventives were, for the most part, the use of their eyes, their powers of reasoning, possibly European information, and years of agricultural experience. Their causal theories reflected observation and reasoning based on a scattering of disease incidents, not on experimental method. The generally accepted causes of plant diseases encompassed a wide spectrum, variously indicting insects, the environment, poor soil, incorrect cultural practices, inappropriate plant nutrients, divine intervention, and evil spirits. Naturally, prescribed methods of treatment mirrored agricultural practices and the one thing over which they had the most control—cultivation. Practical methods of battling plant diseases centered, for example, on planting and harvesting time, drainage, and the removal and destruction of infected plant parts or debris.

Some Americans did participate in scientific discourse on plant disease etiology, if only in a limited way. A number of scientific agriculturists, amateur naturalists, and medical men in the early nineteenth century, dissatisfied with other explanations of disease, articulated the concept that fungi could cause plant diseases. America had a mycological tradition, but early American mycologists were largely uninterested in practical economic work. In the United States, advocates of fungal pathogenicity were in a minority, even within the circles of amateur naturalists and scientists, who more often looked to theories of disease causation like atmospheric influences or chemical explanations than to germ theory.

One reason for the dearth of American plant disease knowledge was the lack of public or private support for fundamental science.[3] Early scientific and agricultural societies had only a limited effect on plant disease problems. Another impediment was the lack of government assistance in matters related to plant disease, because the responsibilities of state and federal governments to agriculture had not yet been clearly defined. One particular disease outbreak, the potato late blight epidemic of the 1840s, engendered an institutional response at local, state, and federal levels, particularly in the United States Patent Office. The ultimate failure to solve the problem exposed the limitations of both the science and the institutions. The experience with potato late blight demonstrated that it would take more than words to answer questions of cause and control of devastating plant diseases.[4]

By the 1850s American agriculturists found themselves adjusting to rapidly changing conditions. Expanding population, the growth of cities, westward migration, innovations in transportation and communications all brought tremendous economic change to the United States. Largely as a result of these forces, agriculture was transformed into a market economy. Farmers adapted to new market conditions with increased specialization, diversification, and mechanization.[5] In turn, these innovations encouraged noticeable and disruptive plant disease problems. In a market economy, however, plant diseases were no longer just sources of irritation for horticultural enthusiasts or checks to surpluses. The ability to market a crop and to realize a profit was no longer of incidental importance but rather determined an entire standard of living.

In this changing environment, the impact of crop losses emphasized the limits of knowledge and the need for an organized, institutional approach to dealing with agricultural matters. Agricultural societies, combined with a vibrant agricultural press, began to function as a "self-interested lobbyist group" for agriculture.[6] An expanded and democratized movement for agricultural improvement, bolstered by the promise of the developing fields of agricultural science like chemistry, demanded a more substantive government response by the mid-1800s. After the creation of the land-grant college system and the United States Department of Agriculture (USDA) in 1862, agricultural science and the study of plant diseases began to look more promising.

Fig. 1.1. Samuel L. Mitchill, courtesy of the National Archives.

Fig. 1.2. Richard Peters, courtesy of the Pennsylvania Academy of the Fine Arts, Philadelphia.

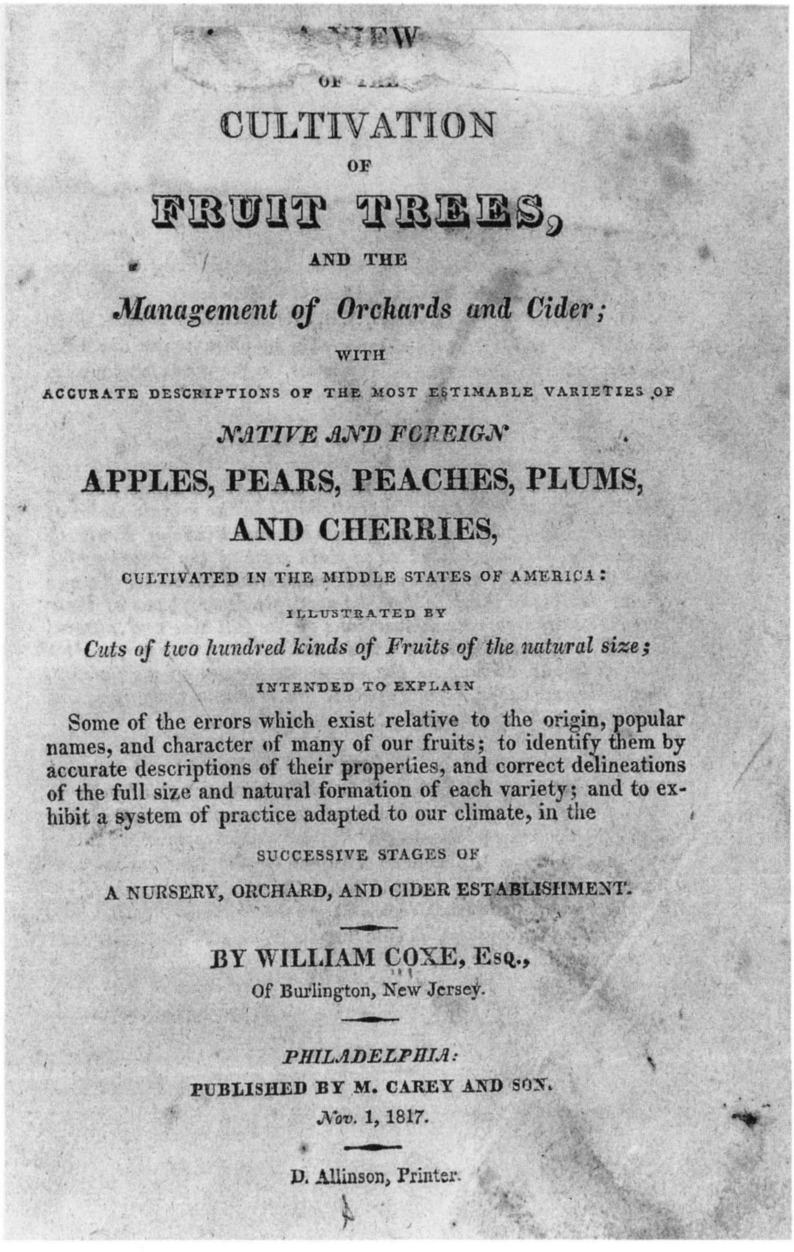

A VIEW

OF THE

CULTIVATION

OF

FRUIT TREES,

AND THE

Management of Orchards and Cider;

WITH

ACCURATE DESCRIPTIONS OF THE MOST ESTIMABLE VARIETIES OF

NATIVE AND FOREIGN

APPLES, PEARS, PEACHES, PLUMS,

AND CHERRIES,

CULTIVATED IN THE MIDDLE STATES OF AMERICA:

ILLUSTRATED BY

Cuts of two hundred kinds of Fruits of the natural size;

INTENDED TO EXPLAIN

Some of the errors which exist relative to the origin, popular names, and character of many of our fruits; to identify them by accurate descriptions of their properties, and correct delineations of the full size and natural formation of each variety; and to exhibit a system of practice adapted to our climate, in the

SUCCESSIVE STAGES OF

A NURSERY, ORCHARD, AND CIDER ESTABLISHMENT.

———

BY WILLIAM COXE, Esq.,

Of Burlington, New Jersey.

———

PHILADELPHIA:

PUBLISHED BY M. CAREY AND SON.

Nov. 1, 1817.

D. Allinson, Printer.

Fig. 1.3.   Title page from William Coxe's *A View of the Cultivation of Fruit Trees*, 1817.

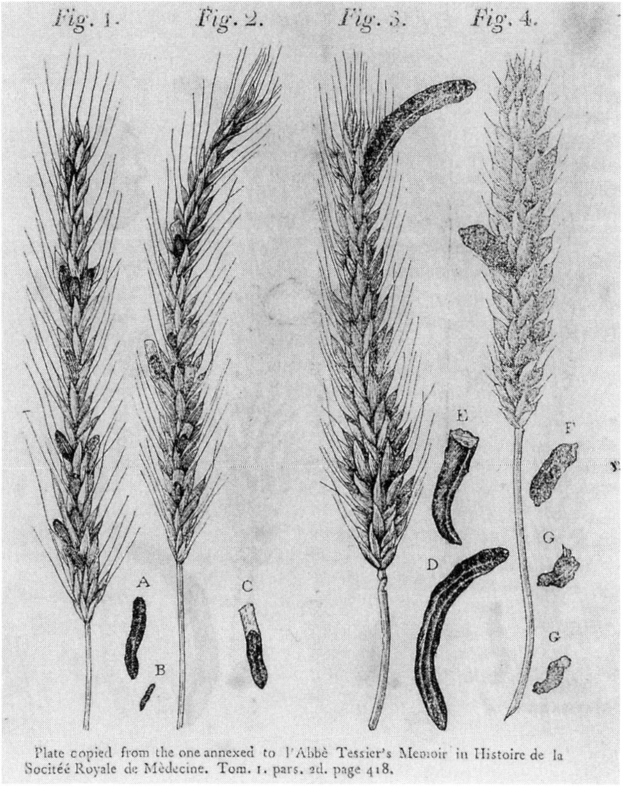

Plate copied from the one annexed to l'Abbè Tessier's Memoir in Histoire de la Socitéé Royale de Mèdecine. Tom. 1. pars. 2d. page 418.

Fig. 1.4   Illustration from Oliver Prescott's *A Dissertation on the Natural History and Medicinal Effects of the Secale Cornutum, or ergot,* 1813, courtesy of Duke University, Rare Books.

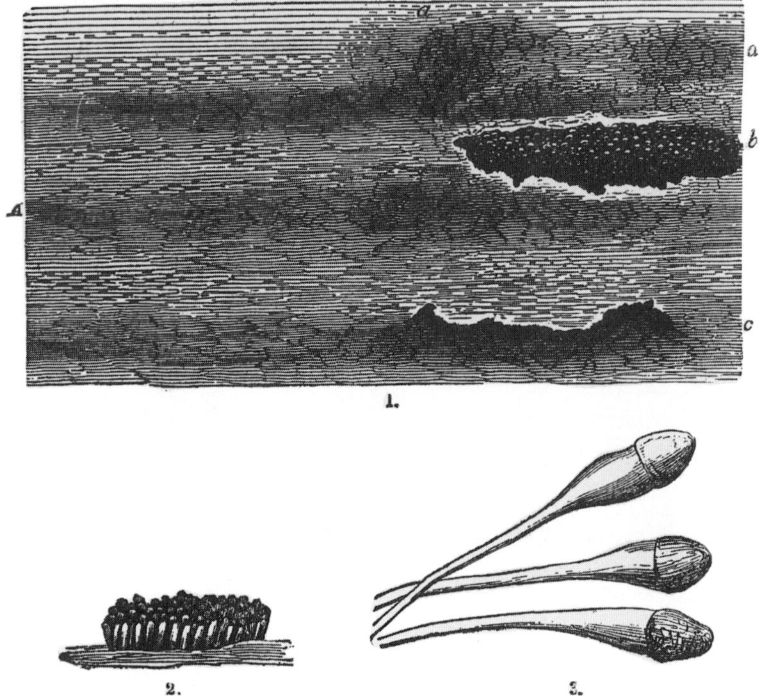

Fig. 1.5.   Illustration from John J. Thomas's *The Diseases and Insects Injurious to the Wheat Crop*, 1843.

Fig. 1.6.   Jared P. Kirtland, courtesy of the Cleveland
Museum of Natural History.

Fig. 1.7.   Henry L. Ellsworth, courtesy
of the National Archives.

Fig. 1.8.   Justin Morrill, courtesy of
the National Archives.

CHAPTER 1

Cultivating the Art and Science of Agricultural Improvement

Plant disease troubled farmers even in early colonial days. Two-thirds of American agricultural crop plants had come from Europe with the colonists, fostering ecological conditions in the new land that favored regular and serious outbreaks of disease.[1] Crops that had not evolved in the New World were susceptible to native pathogens; many crop plants of the cool temperate zone had been transferred to a warmer, more humid environment, where pathogens were more numerous and survived the milder winter with greater success. As early as the seventeenth century, colonial agriculturists and foreign travelers made sporadic observations of diseases on a variety of crops such as apples, grapes, and wheat, suggesting remedies for the resulting damage.

The English traveler John Josselyn visited the American colonies in 1638 and later recorded his observations on diseases of New England apple trees. Under the title *An Account of Two Voyages to New-England,* Josselyn described a visit to the orchard of Governor Endicott on Governor's Island in Boston Harbor. There he saw fruit trees that "are subject to two diseases, the measles which is when they are burned and scorched with the sun, and lowsiness, when the woodpeckers jab holes in the bark." Sharing some of his European experience, Josselyn advised apple growers that "the way to cure them when they are lowsie [*sic*] is to bore a hole in the main root with an auger, and pour in a quantity of Brandie or Rhum, and then stop it up with a pin made of the same tree."[2] This "cure," not uncommon for its day, most likely had little if any effect. It was perhaps connected to the humoral theory of disease associated with human medicine, with rum being prescribed as a tonic for balancing the vital plant humors.

From the scattered observations available, diseases of fruit were probably relatively infrequent during the first century and a half of the colonization of North America. The lack of fruit disease most likely reflected cultivation patterns. Orchards were generally small and isolated, and the soil was not heavily cultivated. On the other hand, colonists had difficulty growing grapes. The European *Vitis vinifera* was susceptible to damage during cold winters, to attacks by the native phylloxera, and to disease-causing fungi. But the settlers were slow to realize that native grapes fared much better, and they continued to advocate planting the European cultivars.[3] Essays on vine cultivation and management proclaimed the virtues of importing foreign stocks and contained little discussion of disease.[4] According to one such tract, grape growers had more to fear from young boys stealing their fruit than from losses due to disease.[5]

Colonists were more concerned with the age-old problems of rusts, smuts, and mildews of cereals. The English visitor Josselyn told of the blasting and mildewing of the New England wheat crop during a visit from 1663 to 1671.[6] Thomas Hutchinson, the last royal governor of the Massachusetts Bay colony, wrote about the blasting of wheat in New England in 1664 and how it discouraged farmers.[7] Wheat pests were a constant source of frustration for colonial farmers throughout the second half of the seventeenth century. Conditions worsened to such an extent in certain areas of New England that days of prayer were mandated in 1680 and 1685 for relief from the ravages of "blastings, mildews, worms, floods, and drought."[8]

But not all colonial farmers were resigned to accept Providence. Although prayer often seemed the most comforting response to crop failures, and a logical one given the limited understanding of diseases, years of agricultural experience had led farmers to certain empirical solutions. To reduce the damage caused by grain rusts, American agriculturists applied Old World control ideas such as plant removal.

Both in Europe and in the American colonies, farmers often associated the appearance of rust on wheat stems with the presence of a rust on barberry plants. Although the biological relationship was not understood, the perceived connection was strong. Thus, the concept that diseases were contagious had a grounding in the popular mind even before germ theory was born. While the discipline of medicine was debating theories about microbes and disease, common sense and repeated experience told many farmers that barberry rust must be related to wheat rust.[9] This practical "cause and effect" understanding influenced legislative action to eradicate barberry bushes in Connecticut in 1726, in Massachusetts in 1754, and in Rhode Island in 1766.[10] A portion of the Connecticut law, titled "An Act Concerning Barberry Bushes," read:

Whereas the abounding of barberry bushes is thought to be very hurtful, it being by plentiful experience found that, where they are in large quantities, they do occasion, or at least increase, the blast on all sorts of English grain,

Be it therefore enacted by the Governour, Council and Representatives, in General Court assembled, and by the authority of the same, That the inhabitants of the several towns within this Colony may, and they are hereby fully impowred [*sic*], at their annual town meetings, to determine and agree upon the utter destroying of the said bushes within their respective townships, and the time when and manner how.

But plant disease, though certainly irritating, was most likely not economically devastating in colonial American agriculture. Although disease outbreaks occasionally caused sufficient concern to influence government action, large-scale epidemics were rare. Two major factors limited their development. First, early American agricultural practice was characterized by small, noncontinuous plantings, with farmers often producing their own seed. These kinds of practices mitigated against the possibility of larger, widespread crop losses. Second, abundant land and other suitable crops made it possible for many Americans to simply move away from disease problems. Incidences of plant disease became more frequent and more devastating in the late eighteenth century, when social, economic, and agricultural conditions in the United States began to change.

SCIENTIFIC AND AGRICULTURAL SOCIETIES

Plant disease began to receive more attention in North America when the revolutionary period, a time of increased nationalism and cultural spirit, brought about the rise of scientific and agricultural societies. Supporting and encouraging the dissemination of "useful" scientific and agricultural knowledge, these societies became meeting places, forums for discussion, and clearinghouses of information for those Americans interested in plant disease. Much of this scientific activity was centered in the urban areas of the Northeast, mainly around Philadelphia and Boston, where gathering and communication were feasible.[11]

The formation of America's learned societies began in 1768 with the creation of the American Philosophical Society in Philadelphia. Boston soon followed with the founding of the American Academy of Arts and Sciences in 1780.[12]

The societies were populated by a curious assortment of merchants, agriculturists, lawyer-statesmen, physicians, and ministers, some already members of London's Royal Society. The members of the new societies were not "full-time" scientists; there were no institutions for advanced training, nor was there financial support for careers in science. They were amateurs, with an education that enabled them to read and understand the current scientific literature and the wealth and leisure to indulge their scientific curiosities. These enlightened, prominent men kept an eye on scientific events in Europe and exercised their scientific approach to gain new knowledge through a passion for observation and experimentation.[13]

The first American learned societies were broad in mission, encompassing practically all known fields of learning. In the absence of specialization, to study science meant investigating, most broadly, all of the natural world. Men of science were "natural philosophers," interested in the systematic study of a wide range of scientific phenomena.[14] Members of the American Philosophical Society wrote that their discussions ranged from natural philosophy and moral science to history and politics, while their investigations extended to botany, medicine to mineralogy and mining, mathematics, chemistry, mechanics, arts, trades and manufacturers, geography and topography, and agriculture.[15]

Though generalists, some of the early society members expressed interest in agricultural improvement. Often these men resided on estates just outside the cities. Many of these gentleman farmers had traveled to Europe and had become fascinated with the new interest in improved agriculture, particularly as it was occurring in England. They viewed American agriculture as undeveloped, and this led them to communicate ideas to their societies about crop rotation, manure, tillage, fencing, and drainage.[16]

With the zeal for agricultural improvement on the rise, a recognition of common purposes inevitably led some groups away from the general learned societies and into specialized associations.[17] In the late eighteenth century, a movement began that promoted the creation of agricultural societies and clubs in many states. To some extent, the enthusiasm for agricultural science had grown beyond the scope of the general learned societies. In addition, some members believed that general science societies had failed to address practical needs. They looked to separate organizations to fulfill practical agricultural objectives.[18] By the early decades of the nineteenth century agricultural associations had appeared in Pennsylvania, South Carolina, Maine, New Jersey, New York, Massachusetts, Connecticut, New Hampshire, and Virginia.[19]

In some ways, these new agricultural organizations closely resembled the learned societies that had preceded them. They were located in economic and cultural centers like Philadelphia and Boston, and their often overlapping membership was largely "local, amateur, and elite."[20] But whereas the general learned societies were interested in nature in the broadest, philosophical sense and were primarily self-interested, the new agricultural societies were more utilitarian and outward-looking.[21]

The common threads that ran through all of the new agricultural associations were the "promotion" and "improvement" of American agriculture and the diffusion of "useful knowledge" to farmers for reasons of "national betterment," with the goal of increasing yields and preserving the soil.[22] The Philadelphia Society for Promoting Agriculture, founded in March 1785, claimed as its object the promotion of "a greater increase of the products of land within the American States" by publishing memoirs and agricultural essays, offering prizes for experiments and improvements, and encouraging the establishment of other societies across the country.[23] The gentleman farmers and novice horticulturists who joined the agricultural societies were driven by a vision of agricultural progress.[24] Understanding and controlling plant disease became a part of a general movement for agricultural improvement.

A PROFESSORSHIP AT COLUMBIA COLLEGE

The early agricultural societies took various routes to reach their objective of increasing agricultural knowledge. Their results, at best, were mixed. From its foundation in 1791, the Society for the Promotion of Agriculture, Arts, and Manufacturers of New York City moved to apply science to agriculture. One of its first activities was a survey of local cultivation and husbandry practices. The survey was conducted by means of circular letters, and the response rate was minimal. To promote the value of experimentation to agriculture, Robert R. Livingston, noted Revolutionary statesman, farmer, and president of the society, carried out a number of experiments. It would be some time, however, before a significant number of agriculturists accepted the merits of experimentation.[25]

Probably the most significant effort to promote the concept of improved agriculture, with a direct relation to plant disease, occurred when the New York society championed the establishment of a professorship of natural history, chemistry, and agriculture at Columbia College. As one historian later wrote, "The natural sciences were . . . taking more definite shape, their relations to agriculture were becoming more apparent, and there was a growing hope that their future developments would greatly promote agricultural advancement."[26] The New York legislature agreed to furnish a five-year grant for support. Thus, in 1792 one of the first American college appointments in natural history was made at Columbia College.[27]

The scientist chosen to fill this position was Samuel Latham Mitchill. Like most scientists of his day, Mitchill had a diverse background and European training. Born in North Hempstead, Long Island, in 1764, he had studied medicine at the University of Edinburgh in 1786 before returning to the United States to study law. He farmed and served in the New York legislature for two years before accepting the professorship at Columbia in 1792. For the next nineteen years, Mitchill was an influential figure in education. Both during and after his years at Columbia, he was a driving force in support of agriculture in New York.[28]

The establishment of the professorship in natural history, chemistry, and agriculture marked a substantial broadening of the curriculum at Columbia. As in most American colleges of the period, science at Columbia was a small part of the existing classical curriculum.[29] Mitchill, however, was well suited to introducing natural science. As one observer later wrote, he was "a versatile, progressive, and widely respected scientist," who "ranged over such topics as earthquakes, lightning, mineral springs, limestone, manure, and milk as he lectured on geology, meteorology, hydrography, mineralogy, botany, and zoology."[30] An advocate of Antoine-Laurent Lavoisier, Mitchill has been credited with pioneering the teaching of modern chemistry at Columbia.[31]

Mitchill's instruction in botany also was forward-looking. In a course outline for his first year of classes, Mitchill included a discussion of plant disease. Under the heading "II. Agriculture or Cultivation of Plants. IV. Disease," he listed as topics for study "vermin—blast—smut—mildew—rust—coalgrain—drought—winter-kill."[32] Although we do not know the specifics of Mitchill's plant disease instruction, a description of his work two years later testifies to his having included plant disease in his courses. Clearly, Mitchill's views on plant disease were informed by his background in human medicine. Describing his botany course, Mitchill reported, "In this course, besides the discussion of the Linnaean or sexual system, the explanation of terms and phrases, and the arrangement or classification of the vegetable species, an attempt is made by the Professor, who is a practical farmer, to elucidate and explain the economy of plants, their affinity to animals, and the organization, excitability, stimuli, life, *diseases*, and death of both classes of beings. The physiology of plants, including their food, nourishment, growth, respiration, perspiration, germination, &c. is therefore particularly enlarged upon, as connected with Gardening and Farming."[33]

Unfortunately, little else is known of Professor Mitchill's interests in plant disease. Low enrollments led college officials to drop the course from the curriculum in 1797. The loss of the Columbia appointment did not, however, hamper Mitchill's career. He remained active in both the civic and scientific life of New York for many years, maintaining his close connections to agriculture. Continuing to support an alliance between botany and agriculture, he endorsed the creation of agricultural societies. New York's Society for the Promotion of Agriculture, Arts, and Manufacturers made him their secretary in 1799. He also contributed many publications and lectures on agricultural subjects to this and other societies until his death in 1831.[34]

HORTICULTURISTS AND FRUIT DISEASES

The most immediate effect of the early agricultural societies in the United States was in stimulating attention on agricultural concerns through shared information. Agricultural societies diffused "useful" knowledge through the publication of society proceedings, transactions, and memoirs. Society members, particularly horticulturists, often published separate treatises on specific subjects of interest.[35] This activity was the first stirring of an agricultural press in America. Agricultural societies also played an important role in providing information to privately owned agricultural periodicals, starting with the *American Farmer* of Baltimore, Maryland, founded in 1819. Agricultural periodicals remained largely dependent on societies for much of their information throughout the first half of the nineteenth century.[36]

As members of these societies and through these agricultural publications, various Americans with interests in plant disease began to share their observations and ideas. In the late eighteenth and early nineteenth centuries, reports of plant problems surfaced from time to time in the agricultural literature. But the first Americans to enter into a protracted discourse on plant disease were amateur horticulturists. As orchards enlarged and as entrepreneurial investment rose, fruit growers experienced a sharp increase in disease problems. Concerned gentleman orchardists had the time and money to perform experiments and to record observations of disease conditions.

Tree fruit diseases were soon in the spotlight of the movement for agricultural improvement. Wealthy horticulturists were eager to hear and read information on agricultural subjects, to test new crop cultivars, to examine methods for improving soil fertility, and to investigate innovative designs for better tools.[37] Horticulturists were active and vocal in desiring practical solutions for disease problems, which meant first determining the causes and then devising preventive measures. Their ideas on the causes of the diseases affecting their fruit trees were both imaginative and eclectic.

The two most influential plant diseases—both in terms of human impact and the early development of plant pathology in the United States—were fire blight of pome fruits and peach yellows. These two diseases threatened fruit trees from the late eighteenth century on, a time when horticulturists were trying to popularize fruit culture.

There is perhaps no disease of fruit trees so utterly devastating as fire blight, caused by the pathogen *Erwinia amylovora.* This bacterial disease plagues pear, apple, and other pome fruits. The term *fire blight* is a fitting description. Flowers, twigs, and branches that have been invaded by the bacterium appear to be scorched.[38] Indigenous to North America, fire blight had long coexisted in an ecological balance with native, rosaceous host plants such as hawthorn, mountain ash, service berry, and *Malus* species. But this quiescent situation began to change when European pomaceous fruit cultivars were brought to North America, beginning in the seventeenth century.[39] The first pear, apple, and quince trees introduced to colonial America probably escaped the ravages of fire blight. Scattered plantings and cultivation practices did little to promote vigorous growth, and propagation mostly as seedlings limited susceptibility.[40] But as cultural practices changed in the late eighteenth century, with more widespread planting of fruit trees, increased focus on the soil and fertilizers, and more use of European cultivars, the disease-causing bacterium spread rapidly.

One of the earliest reports of an outbreak of fire blight in America was made by William Denning in the *Transactions of the New York Society for the Promotion of Agriculture, Arts, and Manufacturers*. Denning recorded his observations of a "disorder" on apples, pears, and quinces in his orchards in the Hudson River valley of New York, beginning in 1780. After continued inspection of his trees and other orchards "further south," Denning concluded that the disease was "spreading rapidly." He observed that the ailment struck "pear trees and quince trees, to the total destruction of them in a few years." "This disorder," Denning continued, seemed to be "of a more serious nature than any thing that has ever infested orchards." Discounting the general opinion of farmers that "the trees were blasted by lightning," he suggested that a worm, or borer, found in the trunk of the tree was responsible for the malady.[41]

After Denning's reports, there was little discussion of fire blight in the agricultural literature until William Coxe of Burlington, New Jersey (twenty miles northeast of Philadelphia, Pennsylvania), published a comprehensive description of the disease in 1817. Coxe, a successful merchant, was typical of the men who joined scientific and agricultural societies and wrote on agricultural topics during this period. He had the wealth to pursue his pomological interests and the progressive spirit to experiment with methods of planting and fertilizing. Coxe was no stranger to the local scientific and literary scene; he was active in several societies in Burlington and Philadelphia, and even became a member of the Horticultural Society of London, a membership that few Americans could claim. A strong advocate for fruit growing in America, his orchards were extensive. In 1807 he had more than 2,000 apple trees on about seventy-five acres.[42] A North Carolina farmer claimed that Coxe had "paid more attention to the raising of orchards, than perhaps any other person in our country."[43] A self-described "scientifick [*sic*] cultivator," Coxe found his orchards "inexhaustible sources of intellectual occupation."[44]

In 1817 Coxe wrote one of the first comprehensive American books on pomology, *A View of the Cultivation of Fruit Trees, and the Management of Orchards and Cider*. In this book he described in detail a disease affecting certain pear cultivars, identifying the condition as "fire blight." He wrote, "Whether the climate of the United States is so well adapted to the cultivation of the pear as the apple, is doubtful, in the opinion of some experienced cultivators—that species of blight, which is sometimes called the fire blight, frequently destroys trees in the fullest apparent vigour and health, in a few hours, turning the leaves suddenly brown, as if they had passed through a hot flame, and causing a morbid matter to exude from the pores of the bark, of a black ferruginous appearance."[45]

Coxe observed that the disease came more often with "weather both hot and moist." This led him to conclude that the malady was caused by the sun's rays in combination with humid conditions. He believed that the repeated plantings, or as he termed it, the "long duration of the variety," exhausted stock, and he made the astute observation that overpruning of the pear trees made them susceptible to the blight.[46]

Fire blight was particularly destructive from the mid-1820s to the mid-1840s. From the initial observations in New York in 1780 and New Jersey in 1817, reports of the "extensive ravages" from fire blight came from a widening geographical area. By 1844 the celebrated horticultural writer and clergyman Henry Ward Beecher spoke of its movement westward from the Atlantic coast to Pennsylvania, Maryland, Virginia, Ohio, Kentucky, Indiana, and Tennessee. Beecher added that already in Indiana and Ohio, the "*blight* has prevailed to such an extent as to spread dismay among cultivators."[47]

During and immediately following severe outbreaks in 1826, 1832, and 1844, the disease aroused a growing flurry of attention, thanks in part to the debate over its cause in the expanding agricultural press. Among the popular causal theories were lightning damage, insects, overheating of the pear tree by the rays of the sun, and debilitation due to cultivation.[48] The insect theory, which became popular in the 1830s, found the most favor. Many growers quite naturally made the logical association between the fire blight symptoms and the repeated presence of minute insects, particularly the aphid *Scolytus pyri*.[49]

Around 1840, a new theory on the cause of fire blight appeared. Called the frozen-sap theory, this theory asserted that autumn or winter freezing of unripened wood manufactured a poison that was transported by the sap through the tree in the spring and summer.[50] From the start, the frozen-sap theory collided with the insect theory. Henry Ward Beecher's article, published in the *Magazine of Horticulture* in 1844, went to great lengths to dismiss the theory that "small, red . . . insects [*Scolytus pyri*], briskly moving from place to place on the branches" of pear trees, caused fire blight. Beecher considered the ravages of the disease in several states, examined different weather patterns and cultivation practices, and concluded that poisoned sap was the culprit. He noted that serious blight had followed early and harsh autumn freezing conditions, and that orchards had fared well when the wood of the fruit trees had ripened early in the year.[51]

About as numerous as causal theories of fire blight were remedies to control it. Many home cures reflected "tried and true" agricultural practices, but some had their origins in human medicine and the theory that health required a balance of bodily humors. Doctors often treated human ailments by prescribing therapeutic substances taken internally or applied externally to correct imbalances. Similarly, horticulturists battled plant disease with a curious assortment of liquid substances. These concoctions usually were applied to the trees by syringes, brushes, or by washing (painting) the plants with rags or sponges.[52]

In 1837 the Pennsylvania Horticultural Society offered a premium of $500 "to the person who shall discover and make public an effective means of preventing the attack of the disease usually termed pear blight." Suggestions presented to the society ranged from soaking the soil and trees with soapsuds to wrapping the limbs with rags sprinkled with brimstone (sulfur).[53] The assumption seemed to be that the more offensive the substance, the more effective the remedy. Moreover, because foliage always appeared brighter and healthier when wet, horticulturists thought they saw immediate benefits due to their liquid compounds. Many late-eighteenth- and early-nineteenth-century horticulturists, for example, recommended the adoption of the "composition," developed by William Forsyth, the gardener to the king of England. Forsyth's "composition" was a mixture of cow dung, lime, wood ashes, river sand, and urine.[54] An American horticulturist who believed that insects were responsible for fire blight advised applying chloride of lime to destroy the insect and as "an antidote to the poison" left in the tree.[55] Another insect theorist advocated the use of spirits of turpentine "on and about the diseased part."[56] Some of these liquid applications actually worked against insects, but few of them had any effect on plant pathogens.[57]

Some fruit growers combated tree fruit diseases with dry or solid substances. Some placed iron around diseased trees, presumably to revive the vital spirits or balance the nutritional needs of the plant.[58] Others endorsed the use of sulfur in various forms. One grower gave the following sulfur remedy, which he claimed "never failed to revive" the pear tree from the blight: "I make an incision in tree with an auger, about two inches deep, and in a slanting position, so that the bottom or farther point of the hole shall be inclined towards the roots and generally about two feet above them; the cavity is then filled with sulphur and plugged up perfectly tight: I then cut off the plug smoothly even with the bark. This operation I perform in the month of March or April, or any time when the sap is circulating in the tree."[59] On the other hand, frozen-sap theorists, who blamed the disease largely on cultivation practices, advocated the use of certain soil types, the selection of naturally early growing and ripening cultivars, and root pruning.[60]

Because many orchardists shared a vague sense that the disease was contagious, an often-repeated control suggestion was to remove and burn diseased branches before the fire blight attacked the entire tree or to remove infected trees altogether.[61] Even with this sound advice, growers felt generally helpless to prevent or retard the injurious force of fire blight in their orchards. "It is so destructive to the pear tree in some sections of the country," one horticulturist lamented in 1840, "that its culture is almost given up."[62]

Fire blight was not the only disease of fruit trees to draw attention from concerned horticulturists in early America. Peach yellows, now known to be caused by a phytoplasma, was also a recurring source of economic frustration beginning in the late eighteenth century. Recognized by the premature ripening of red-spotted, useless fruit and the yellowing and curling of foliage, peach yellows often affected whole trees and destroyed entire orchards.

Peaches were probably introduced into North America around 1630.[63] Peach trees grew well in the colonies and plantings increased throughout the next century. By the time of the American Revolution and in the years immediately following, interest in peach orchards rose noticeably. Orchardists took advantage of the conducive climate in the vicinity of Philadelphia, Pennsylvania, and along the Delaware to devote considerable attention to its cultivation.[64] Unfortunately, this cheerful picture was not to last. By the 1790s reports began to circulate that the peach, long favored for its health and vigor, was becoming increasingly prone to "degeneracy." Samuel Deane, vice president of Bowdoin College, was one of the first to voice his concern. Dating the decay of the peach back to 1760, Deane suggested, "We have room for making great improvements, it seems, in the culture of this fruit." The "degenerated" state of current peach cultivars led Deane to conclude that the time had come for restoring orchards "by bringing the trees or stones from some other country."[65]

Around Philadelphia, difficulties with peach cultivation reached such threatening proportions that in May 1796 the American Philosophical Society found it necessary to offer a premium of $60 "for the best method, verified by experiment, of preventing the premature decay of Peach-Trees."[66] The society received several meritorious replies and decided to split the award between two submissions that were believed to be of equal worth. Both contributors came directly or indirectly to the conclusion that the cause of the peach problem was insects. John Ellis of New Jersey attributed the peach decay to the assault of worms and recommended encircling peach trees with straw bands to prevent the worm from depositing its eggs at the base of the trunk.[67] Thomas Coulter of Bedford County, Pennsylvania, however, considered the worm to be only an ancillary cause. He blamed the decay on excessive cultivation "owing to planting, transplanting, and pruning the same stock, which occasions it to be open and tender," allowing birds and insects to make wounds in the tree.[68]

Because their cures failed to mention symptoms associated with yellows, it is impossible to say for sure whether Deane, Ellis, or Coulter were referring precisely to peach yellows. However, later reports from the same vicinity suggest that they were. Richard Peters of Philadelphia, one of the driving forces behind the creation of the Philadelphia Society for the Promotion of Agriculture, was probably the first to describe and name peach yellows. In a paper read before the Society on February 11, 1806, "On Peach Trees," Peters recorded his extended observations and experiences with peach culture: "Fifteen or sixteen years ago, I lost one hundred and fifty peach trees in full bearing in the course of two summers; by a disease engendered in the first summer. I attribute its origin, to some morbid affection in the air, which has the most to do with all vegetation, as well in its food and sustenance, as in its decay and dissolution. The disorder being generally prevalent, would, among animals have been called an epidemic. From perfect verdure, the leaves turned yellow in a few days, and the bodies blackened in spots. Those distant from the point of original infection, gradually caught the disease. I procured young trees from a distance, in high health, and planted them among those the least diseased. In a few weeks they became sickly, and never recovered."[69]

Although Peters was sparse in his comments on symptoms, he did mention the appearance of "sickly shoots" and the death of "extremities," both characteristic of yellows. It was also significant that Peters knew that the disease he called the "yellows" was contagious. He recommended the one control measure most likely to be effective in this event: the quick and total removal and destruction of all infected trees.[70]

Having "recruited" his "peaches from distant nurseries, and promptly, on the first symptoms appearing, removed the subjects of it," Peters claimed to have more success after the initial outbreak. But in an addendum to his February 11, 1806, communication, Peters hinted at future troubles: He mentioned seeing yellows in the orchard of his neighbor, Edward Heston.[71] In September 1807 Peters reported that not only were "the yellows . . . making destructive ravages in Mr. Heston's peach plantation," but that he himself had "lost a great proportion of . . . trees, by the same malady, this year." Peters added that he could "not recollect ever to have seen more general destruction among peach trees." "Two successive rainy seasons," he claimed, were behind "this irresistible disease."[72]

On November 17, 1807, Peters wrote again "on peach trees," insisting "that the disease, so generally fatal (more so this year, than any other in my memory), called the *yellows*, is *atmospherical*." He did not elaborate. Instead, he included a letter he had received from James Tilton, a medical doctor from Bellevue, Delaware, dated November 6, 1807, that provided additional remarks on the peach malady. Tilton attributed "all diseases of the peach tree" that he had observed, "even that sickly appearance of the tree, called the yellows, attended by numerous weakly shoots on the limbs generally," to the destructive properties of insects. He acknowledged, however, that "the diseases and early death of our peach trees is a fertile source of observation, far from being exhausted."[73]

From the initial sightings near Philadelphia, the yellows disease gradually spread into other peach-growing areas. William Coxe wrote to Peters in 1807, asserting that he was ignorant of the disease.[74] By 1817, however, Coxe was writing about peach yellows in his book *A View of the Cultivation of Fruit Trees, and the Management of Orchards and Cider*. Evidence suggests that by 1817 the yellows had already devastated much of the peach crop in New Jersey.[75] This probably explains why William Coxe did not hold much hope for the future of peach culture in the United States. In 1817 he wrote with regret, "It is, when in perfection, the finest fruit of our country, for beauty and flavour: it is deeply to be regretted that its duration is so short, and that it is subject to a malady which no remedy can cure, nor cultivation avert. Of the numberless modes of mitigating or preventing the diseases of the peach tree, with which our publick [*sic*] prints are daily teeming, none have yet been found effectual—the ravages of the worm, which destroys the roots and trunk of this tree, may be sometimes prevented, and with care may be at all times rendered less destructive, but the malady which destroys much the largest portion of the trees, has hitherto baffled every effort to subdue it; neither its source, or the precise character of the disease, appear to be perfectly understood."[76]

By the 1820s the yellows disease was well on a course that in the next two decades would take it north, west, and south from the vicinity of Philadelphia.[77] As its range increased, new reports multiplied. In 1824 David Thomas planted trees on the shore of Lake Cayuga in western New York and reportedly lost peaches to yellows.[78] Samuel Mitchill of New York, writing in 1826, remarked that the "malady under which the peach trees linger and eventually die, in our neighbourhood, is a calamity whose visitation has not yet passed away."[79] Two years later, William Prince, a distinguished nurseryman from Flushing, Queens, in New York, wrote on the yellows in *A Short Treatise on Horticulture*. Prince not only corroborated the symptoms of yellows based on "the leaves putting on a sickly, yellow appearance," but he added "premature ripening of the fruit" as further evidence of the disease. His ardent belief that the yellows was contagious led him to urge the swift removal of infected trees.[80]

Just as it had for fire blight, the increasing prevalence of peach yellows brought an expansion in theories for the causes of the disease. Explanations were often similar to those offered for fire blight, including the effects of weather, attacks from insects, and poor cultivation practices.[81] Nevertheless horticulturists felt generally helpless to prevent this "most insidious and fatal disease to the peach tree."[82] Even widespread agreement that the disease was contagious offered little solution. Although better cultivation practices and removal of infected limbs and whole trees seemed to alleviate the immediate severity, the disease nonetheless destroyed entire orchards. As a result, many nineteenth-century horticulturists gradually gave up on peaches, as they had on pears in the areas ravaged most frequently by epidemics.[83]

Answers not yet possible

The mysteries of peach yellows and fire blight called for answers not yet possible in the worldview of these eighteenth- and early-nineteenth-century horticulturists and enthusiasts of agricultural improvement. Appreciable change in the approach to plant-pathological problems would have to wait for the demonstration of the principles of infection made possible by advances in biology later in the 1800s. Nevertheless, most Americans could take comfort in the fact that plant disease, though devastating enough, had not seriously threatened American agriculture. In the case of tree fruits, where fruit was grown largely for home consumption, to make beverages, or as a substitute for animal feed, diseases were not important enough to cause much concern outside of a small group of entrepreneurial horticulturists.[84]

For now, horticulturists and agricultural improvers were learning about the vulnerability of plants to disease. By searching for causal agents and practical solutions, discussing their ideas in society meetings, and publishing the results of their observations and experiments in the agricultural literature, these gentleman farmers took a sincere first step in the direction of understanding plant disease and plant health. Horticulturists offered accurate descriptions of disease symptoms, gave credible causal explanations, and suggested plausible control strategies such as the removal and destruction of infected plants. Most of all, they kindled a spirit receptive to new ideas. Among horticulturists these interests would intensify, particularly as the commercial fruit industry developed in the 1800s. So far, serious attention to plant disease had involved a relatively localized and elite circle of gentleman farmers aspiring to the promises of agricultural improvement. Many more Americans would realize the potential of crop losses and the human costs associated with plant disease during the potato late blight epidemic of the 1840s.

CHAPTER 2

The Potato Epidemic: Causes and Cures

In the first half of the nineteenth century, plant disease was generally attributed to poor cultivation, insects, bad weather, or an inexplicable act of God's will. Although some close observers noticed that tiny fungi often appeared in association with diseased or dead tissues, they explained this phenomenon quite easily by the theory of spontaneous generation. This popular doctrine argued that microbes were the result rather than the cause of disease.[1]

Long-held concepts of plant disease were shaken to their foundations in the 1840s, however, with the appearance of a new disease—potato late blight. Until potato late blight appeared, no recorded disease had been so destructive as to threaten large populations with starvation and death. The blight created famine conditions in parts of Europe, and, although it was less devastating in the United States, it nevertheless was widespread and widely publicized. The magnitude of damage stimulated a burst of activity in the study of plant disease.

THE DEVELOPMENT OF FUNGAL THEORY

Although farmers had recorded scattered observations of crop problems for years, botanists and naturalists did not begin to demonstrate an appreciable interest in plant disease until the eighteenth century. As was the practice in natural history, most of this activity was devoted to identifying and classifying plant disease into a variety of systems of botanical nomenclature. Classification of plant life added a wealth of descriptive data to botanical science, but it contributed little toward the actual understanding of plant disease.

During the late eighteenth and early nineteenth centuries, a handful of naturalists pushed the study of plant disease in new directions. Stimulated by the development of a subclass of botany, the study of fungi, which heralded the beginnings of mycology, they used magnifying lenses and simple microscopes to examine the tiny fungi that mysteriously appeared on diseased plants.[2] In 1728, for example, Henri Louis Duhamel du Monceau demonstrated clearly that the disease of fall-blooming Saffron crocus *(Crocus sativus)* occurred only with the presence of a fungus and that it was contagious. A year later, Pietro Antonio Micheli showed through his experiments with melons that fungi reproduced by "seeds." In 1766, responding to a severe epidemic of wheat rust, Felice Fontana and Giovanni Targioni-Tozzetti, described both teliospores and urediniospores of the wheat rust fungus.[3]

By far the most significant experimental work of the eighteenth and early nineteenth centuries regarding pathogenic fungi, however, was that of Mathieu Tillet and Isaac-Bénédict Prévost. Tillet did experiments in the 1750s that convinced him that wheat bunt or smut was contagious and that chemical treatment could prevent the occurrence of the disease. In 1807, Prévost published the results of ten years of research and experimentation in which he carefully observed the germination of bunt spores under a microscope, concluding that these spores were the cause of bunt. Prévost also was the first to publish on the effectiveness of copper as a fungicide.

On the whole, however, the so-called fungal theory fell on deaf ears. Although excellent descriptive and experimental studies occurred over a nearly 100-year span from Duhamel du Monceau to Prévost, the fungal theory lacked the widespread acceptance necessary to refute the more "sensible" claim of spontaneous generation. Observation had established irrefutably that fungi were often associated with decay, but the majority of scientists continued to hold the opinion that fungi on plants were the result of disease—not the cause. Thus, the prevailing biological precepts of the autogenesis school, which did not allow for pathogenic fungi, continued to dominate scientific thought.

While America had not produced a Duhamel, Tillet, or Prévost, the plant disease problems that occurred in America during the eighteenth and early nineteenth centuries were serious enough to attract attention from numerous agriculturists. For the most part, however, these horticulturists were overwhelmingly amateurs and were absorbed primarily with the general concepts of agricultural improvement. Taking their ideas from Great Britain, Americans were concerned with cultural practices—that is, crop rotation, manuring, and drainage. Although connections between agriculture and biology were emerging in Europe, particularly on the Continent, such scientific links were yet to be made in America.

The disparity between America and Europe in the amount of attention to plant disease may lie in the maturity of European scientific institutions. Long before American science was established in societies, universities, governments, or experiment stations, Europeans interested in plant disease took advantage of scientific, institutional connections. Tillet was encouraged in his work by the Académie Royale des Belles-Lettres, Science & Arts de Bordeaux, and he was sponsored by the French government. Prévost was a university professor.[4] Such institutional support encouraged Europeans to develop ideas about plant disease, and it exposed them to the newest, circulating biological concepts. As one historian has written, "Scientists above all other men need the stimulation provided by interaction within a group."[5] In contrast, in late-eighteenth- and early-nineteenth-century America, plant disease information was just beginning to be disseminated by way of societies and the agricultural press. University connections, with few exceptions, were still years away.

When the germ theory did appear in America, it battled against traditional agricultural opinions that plant disease was caused by insects, weather patterns, or cultural practices. Even the connection of fungal parasitism with wheat rusts and smuts, which was generally accepted in Europe by the early 1800s, received a much weaker, if not timid, endorsement in the American agricultural press. Samuel Deane of Bowdoin College wrote in 1797 that he was familiar with the work of Duhamel du Monceau, Tillet, and other Europeans on smut of grain. But Deane's discussion that microscopic fungi were implicated went only so far as to grant that it seemed "rather probable" that a "mould" similar to a "minute moss" might be the source of the smut.[6] Several years later, James Mease, another influential agricultural improver, suggested that fungal microbes were involved in wheat smuts and mildews (rust). Mease, of the Philadelphia Society for Promoting Agriculture, offered few details to support his argument, however, other than to add that Prévost had associated the origins of wheat mildew to "an intestine *parasitical plant*."[7]

Other American agriculturists were aware of European ideas about parasitic fungi but were unwilling to accept them. Timothy Pickering, a former member of George Washington's cabinet and a prominent member of the Massachusetts Society for Promoting Agriculture, wrote about mildews of grain in 1810, acknowledging his familiarity with the suggestion from English botanist Sir Joseph Banks that "invisible seeds of fungi" may be responsible for mildew. But Pickering rejected Banks's theory, choosing to look back to an explanation offered forty-two years earlier by "a New-England Man," which blamed the disease on "a *super abundance* of sap."[8]

Empiricism was firmly entrenched in America, even among proponents of agricultural improvement. Farmers hoped to control disease and insects by altering planting time, draining the land, plowing more deeply, paying attention to weeds, and rotating their crops.[9] To battle the rusts and smuts of grains, many American farmers accepted and advocated European recommendations for seed treatments.

Some of the agricultural improvers who communicated control ideas were aware of the developing fungal theory of disease. Deane, like Tillet, recommended the practice of brining and liming seed, and Mease asserted that Prévost's "blue vitriol (sulphate of copper) was also a preventive."[10] But neither of these enlightened agriculturists went so far as to advance the concept that microscopic fungi could cause plant disease. More important, there is little evidence to suggest that any of the control recommendations offered in the American agricultural press had any conceptual connection to the developing fungal theory. The early seed treatments, based almost exclusively on empirical conclusions, were usually prescribed because of the stimulative effect on the host plant. Thus, seed treatments were customary long before the scientific aspects of disease were understood.

Nevertheless, some Americans were convinced that microscopic fungi caused certain reoccurring plant diseases. One of the most impressive early discussions of the smuts and rusts of grains came by way of a recent "transplant" to America, Anthony Fothergill, of Sedbergh, England. Fothergill had retired from his medical practice in Bath and moved overseas to Philadelphia in 1803.[11] He had a respectable knowledge of the current work taking place in England on the fungal theory and desired to share this information, together with his own observations in England and America, with his new countrymen. In a paper read before the Philadelphia Society for Promoting Agriculture on November 11, 1806, he laid out a well-reasoned argument for a fungal cause of the diseases of some grains.

First addressing the question of the origin of grain rust, Fothergill insisted that the mystery of the cause of this crop disease had been resolved. Microscopic examinations had revealed that the true cause of the disease resided in a microscopic fungus that "infests the stem and leaves with yellow and dark brown stains, and forms an orange coloured dust." He further explained that the "clusters of a fungus or parasitical plant, . . . insinuate themselves into the pores of the absorbent vessels of the stem, and deprive the grain of the sap destined for its nourishment."[12]

Addressing next the origin of smut disease, Fothergill intimated that fungi were probably also responsible for this plant condition, but the evidence, he concluded, was not as overwhelming as it was for rust. He urged his fellow "writers on husbandry" to explore the true cause of smut and methods for its prevention "by attentive observation and accurate experiments." He proposed a number of experiments to test causal theories for smut and to disclose the true benefits of current control practices, particularly the efficacy of different seed treatments. But Fothergill left little doubt that he believed that microscopic fungi would be found to be associated with the disease. He was convinced that "the inquisitive naturalist, possessed of a penetrating eye, and a powerful micro-scope" would "prove that many of the diseases termed blights, hitherto attrib-uted to other causes, will, on a more close inspection, be found to originate from a parasitical vegetation."[13]

During the next two decades of the nineteenth century several Americans, all with backgrounds in medicine, turned their efforts to understanding ergot of rye and came away with a belief in fungal pathogenicity. In a paper read before the Massachusetts Medical Society on June 2, 1813, Oliver Prescott, a doctor, dis-cussed ergot both as a medical phenomenon among humans and as a plant dis-ease. Prescott expressed his awareness of the disease in Europe and the United States and indicated that he was familiar with the work of Tessier, among others, in France. He attributed "the peculiar disease" that attacks the grains of rye and produces "convulsions and spasmodic affections" in man and animals to the properties of the "four diseases enumerated by Linnaeus, and by him denomi-nated *clavus*."[14] During the next decade, two other Americans, also influenced by medical training but reflecting an interest in mycology, probed the mysteries of ergot. William Tully, a physician and teacher in Connecticut, considered the ergot of rye as *Sclerotium clavus* and gave eleven reasons why he believed "that the *Clavus* is a parasitic *Fungus*, like the different sorts of blight, smut, etc." Five years later, M. Field repeated these causal explanations in his own paper on ergot.[15]

Prescott, Tully, and Field came to an interest in parasitic fungi because of their medical backgrounds, but in early-nineteenth-century America there was also developing a separate tradition of descriptive mycology. Among the most recognized mycologists was L. D. de Schweinitz, an official in the Moravian church in Salem, North Carolina, and Bethlehem, Pennsylvania. De Schweinitz is generally considered the first American mycologist. His major work, *Synopsis fungorum Carolinae superioris*, was published in 1822 and was quickly accepted by the European mycological community. In 1832 de Schweinitz was the first to describe the fungus later found to be responsible for sooty blotch of apples, but his description occurred merely as an unexpected and unintended result of his mycological studies.[16] Other naturalists with an interest in fungi, such as Edward Hitchcock of Massachusetts and M. A. Curtis of North Carolina, published the findings of their collecting efforts.[17] Although there was little interest in diseases caused by fungi among the practitioners of descriptive taxonomy, the descriptive scientific tradition established by these and other American mycologists would create much of the taxonomic knowledge base that merged later in the century.

Between the time of the discussions of ergot in the first quarter of the century and the publication in 1843 of John J. Thomas's prize-winning paper on wheat diseases, agricultural and scientific writers made only occasional reference to the fungal theory. Thomas's paper, "The Diseases and Insects Injurious to the Wheat Crop," published in the *Transactions of the New York State Agricultural Society*, was therefore a major breakthrough. Thomas, a well-respected pomologist and agricultural editor, sought to advance the theory of fungal parasitism in America by drawing on the work of European authorities. In his paper, he referred to Duhamel du Monceau and the English agricultural chemist Sir Humphrey Davy, among others, to support his argument that the mysteries of wheat smuts and rusts rested with microscopic fungi.[18] He demonstrated his firm allegiance to the fungal theory with a remarkable willingness to verify this principle by microscopic examination. "Hundreds of observations by means of a compound microscope," Thomas wrote, "satisfied beyond scarcely a doubt" his opinion that the fungus associated with smuts and rusts "is a small plant, of as regular and uniform a growth as the wheat itself."[19]

Thomas also was aware that these tiny plants were contagious. He asserted that proof of infection came from "rubbing the seed in smut before sowing" to cause "smut in the ear."[20] According to Thomas, seed treatments, particularly methods practiced in Europe but with some American modifications, were the most effective preventives against smut.

Turning his attention to rust, Thomas considered this enemy of wheat to be "a much more formidable disease than smut, and far less under the control of any remedy which human ingenuity has yet devised." In the Northeast alone, he insisted, rusts brought more crop failures "than all other diseases and all insects put together."[21] His examinations convinced him not only "that rust spreads by a contagion," but also that it was reinforced by certain environmental factors.[22]

Thomas's work was derivative and added little that was new to the developing fungal theory. But his paper on wheat disease was important in keeping the idea of fungal pathogenicity alive in America at a time when support for this new concept was expanding in Europe. His work also proves that there were those in America who were receptive to new theories about the causes of plant disease, able to grasp their biological significance, and willing to share them. If Americans were to alter their understanding of disease, novel ideas, particularly from Europe, had to reach a responsive audience. Thus, Thomas's prize essay was part of the contemporary scientific discourse breaking down the barriers to innovative plant disease theory.

Nevertheless, fungal theory was far from the prevalent view in late-eighteenth- and early-nineteenth-century America, and fledgling American plant pathologists were far from matching European advances. Other than scattered and isolated investigations of ergot and wheat diseases, Americans, by and large, concentrated their efforts on practical solutions to the problems caused by plant disease. Popular, "farm-tested" preventive measures circulated by the agricultural press generally did little to control plant disease, but thus far the ramifications of mistaken views had not been disastrous. That was about to change.

POTATO ROT

Of all the recorded outbreaks of plant disease, the potato late blight epidemic that struck parts of North America and Europe in the 1840s was the most severe and the most devastating.[23] No one could have predicted the extent and seriousness of the disease and its far-reaching impact, even if they had been more aware of the early warning signs. But by the time the potato late blight was noticed, most of its damage was done.

Potato diseases had occurred previously without devastating results. The late blight, known colloquially as potato rot or potato murrain and caused by the fungus *Phytophthora infestans,* was the third major potato malady to occur. At the end of the eighteenth century, the curl, which is now known to be caused by a virus, became a problem for farmers. The 1830s saw the advent of dry rot, now attributed to fungi that are species of *Fusarium.* Given the limited extent of scientific knowledge about plant disease, there was some early misinterpretation of late blight as a reappearance of the other two potato ailments, which could explain the lack of alarm.[24] In any case, late blight struck so rapidly and with such devastation that the argument over whether it was a "new disease" took second place to attempts to alleviate its effects.

When the potato was introduced into Western society, few could have foreseen the importance it was to have. Native to South America, the plant was introduced to Spain by the conquistadors around 1570.[25] Europeans viewed the new plant with a jaundiced eye. Many believed it to be poisonous because it belonged to the nightshade family. However, by the eighteenth century, after a considerable period of selection that produced better cultivars, the potato became a staple garden and field crop for much of Europe and North America. The hardy potato could be grown from the highlands of Spain to the wetlands of Ireland and throughout much of the settled area of North America. Problems with smutted grains in the late eighteenth century encouraged people in more areas of Europe and North America to turn to the potato as an important source of food.[26] During the eighteenth century, a dependence on the potato spread over the Western world. This dependence made late blight a serious threat to human survival in some areas of Europe. Ireland, particularly, depended on the potato for most of its nutrition.

Cultivation of the potato was not without problems. In addition to curl and dry rot, farmers experienced the usual problems from too little moisture during dry years, an overabundance of water in wet years, and frost damage in cold years. The potato was also prone to maladies such as scab, but such maladies were usually little more than irritations.[27]

The late blight was different. The most obvious characteristic of the blight was its quick, thorough destruction of the leaves and stems of the potato plant. On these parts, the blight caused dark lesions of different sizes and shapes that, under moist conditions, produced a whitish growth of visible sporangiophores on the underside of leaves. The number of lesions on potato leaves and stems increased rapidly, and eventually the tubers became infected. Under favorable weather conditions, wet or cool, the pathogen spread widely. In its most severe form the entire disease cycle could transpire repeatedly within three or four weeks, destroying an entire potato crop.[28]

Although the most famous epidemic of this disease was in Ireland in 1845, potato late blight struck earlier in North America. Due to poor communication or misinterpretation, the center of origin for the American disease is unknown. In 1843, however, there were specific accounts of the disease in a five-state area surrounding the port cities of Philadelphia and New York. During 1844, potato late blight spread to other parts of the United States and to Canada.[29]

The sudden and devastating eruption of potato late blight stirred widespread interest in plant disease in the United States. In 1843, potato late blight was the first plant disease to receive extensive attention from the federal government's nascent agricultural arm in the Patent Office. In 1839 Henry Ellsworth, commissioner of patents, had convinced Congress to earmark a portion of his office's yearly budget for agricultural purposes. Ellsworth's primary agricultural job was to collect and distribute seeds of new plant species and cultivars that had been sent back to the United States by Americans overseas. Another duty was to collect statistics and issue a yearly report on the state of agriculture in America.[30]

There had been a small yet active movement for government support for agriculture dating back to the founding of the nation. A handful of senior statesmen and prominent agriculturists concerned primarily with increasing production called for federal coordination. George Washington, a practical farmer who made experiments at Mount Vernon and read European farming literature, proposed to Congress in 1796 the idea of a federal board of agriculture with state societies as part of it. The following year, at the urging of the president, a special House committee recommended the creation of an "American Society of Agriculture," to be headquartered in Washington, D.C., under a secretary to be paid by the government. But this plan failed to become law when opponents like Thomas Jefferson, though an experimental agriculturist himself, argued against the legislation on constitutional grounds. When the question of a national board came up again in 1817, it was again defeated by forces sympathetic to constitutional limitations.[31]

But the idea of governmental involvement in agriculture did not die. In 1819,

William H. Crawford, secretary of the treasury, instructed American consuls abroad to collect and forward rare plants and seeds to the United States for distribution. A year later, Congress granted five acres of land to the Columbian Institute for a botanical garden.[32] In the 1820s, both houses of Congress established committees of agriculture. But the struggles to establish these congressional committees demonstrated that serious doubts remained about the need for government involvement.[33]

Although the federal government had previously played a small role in American agriculture, circumstances in the 1830s favored increased government involvement in promoting agricultural improvement and farming interests. The hopes of "progressive" agriculturists were heightened in 1835, when Henry L. Ellsworth was appointed commissioner of the Patent Office.

A graduate of Yale in 1810, Ellsworth had become involved in successful farming and commercial ventures in his native Connecticut. He brought to his post in the Patent Office strong support for agricultural improvement. The Patent Office historically had concentrated on patents on technology, but Ellsworth believed that the time had come to expand its reach. In 1837, without authorization, he began free distribution of seeds and cuttings sent from Europe and elsewhere. After repeated attempts to elicit government support for this activity, in 1839 he persuaded Congress to attach a rider to the Census Bill for an appropriation of $1,000 to the Patent Office for the collection of agricultural statistics and other purposes. Almost single-handedly, since his arrival in 1835, Commissioner Ellsworth had redesigned the functions of the Patent Office and, in doing so, had begun a federal agricultural service.[34]

Not long after the landmark legislation of 1839, Ellsworth began to publish yearly reports on crop statistics and methods of cultivation. Culled largely from the regional agricultural press, material for these annual reports also came from Ellsworth's extensive correspondence with many agricultural authorities. Though derivative, the commissioner's yearly reports had value. They brought together scattered agricultural information and provided improved visibility for American farming problems.

Beginning in 1843, the commissioner of patents devoted space in his annual reports to Congress to summary discussions of the potato late blight. Based primarily on information from the agricultural press, the material in these discussions demonstrated the expanding concern over large-scale crop losses due to the disease. The first references to the potato rot in the 1843 report detailed the range and extent of damage and featured observations from various locations. Unnamed "authorities" also put forward causal explanations for the disease, generally blaming a combination of wet weather and late harvesting. "Had these potatoes been dug just at the point of time at which they were ripe," one writer argued, "there can be no question that they could have been saved from rotting." The correspondent went on to suggest that the best remedy was to plant later the next year, since "the late planted potatoes are said to have fared much better than the early planted."[35]

Although the consensus was that the blame for the potato rot lay with the weather and poor cultivation, Patent Office officials also mentioned a fungal theory in the 1843 report. Without crediting anyone specific, the report stated that "an eminent agriculturist" had declared the disease to be caused by "a fungus belonging to the vegetable growth, as rust and smut in wheat and corn, and mould and mildew." The writer recommended further "that all diseased potatoes be carefully taken out, and thrown away; and that, finally, pulverized *lime*, either slaked or unslaked, be sprinkled among the healthy potatoes, just enough to whiten their surfaces lightly." The report added, however, that no one had attempted experiments to discover if this control method worked.[36]

If the 1843 Patent Office report contained only brief attention to the "new disease," the commissioner's summary to Congress the following two years devoted voluminous concern to the condition of the American potato crop. In his 1844 report Commissioner Ellsworth confessed, "The great anxiety felt in the United States respecting the disease in the potato, by which whole sections of our country have been seriously affected, has induced me to devote much time to investigate this subject." Ending on a somewhat pessimistic note, Ellsworth asserted that "if no satisfactory reasons are assigned for the disease, it is hoped some partial preventives, at least, are suggested."[37]

By 1844 scientific and agricultural journals, as well as newspapers, were publishing scores of articles discussing the new potato disease. Pamphlets describing a vast array of causes and remedies proliferated. Scientific organizations, agricultural societies, and state and federal governments promoted efforts to obtain all the facts about this mysterious affliction. Some offered monetary rewards to anyone who could explain and control the current malady.[38] Officials at the Patent Office wrote that "the country could well afford to pay a handsome sum to any one who could devise the way to insure the agriculturists of our land from the destructive progress of this threatening calamity."[39]

Although it is impossible to determine how much response the promise of monetary compensation attracted, both the seriousness of the disease and a lack of information stimulated a large group of American farmers, "progressive" agriculturists, amateur scientists, and agricultural chemists to investigate its cause and to seek a cure. Scientific interests focused on causal explanations for the mysterious disease. These different and sometimes sharply conflicting interpretations touched off a lively discussion in the agricultural press, and in turn in the Patent Office, and revealed the most forward-thinking ideas about plant disease among Americans in the 1840s.

One of the most widely held theories was that an atmospheric agent was responsible for the potato problem. In 1844 Edward Hitchcock, professor of chemistry and natural history at Amherst College, claimed that the potato, like all living organisms, relied on a balanced relationship between the external and internal environment. "Plants, being like living beings, like animals, requiring a certain state of air, moisture, light, heat, and electricity, and proper food," he wrote, "will of course be liable to disease, like animals, from an excess or deficiency of any of the agents that affect them." Hitchcock maintained that the absence or surplus of any one or more of these aspects could explain the current epidemic much better than other explanations like parasitic plants or insects.[40]

Patent Office officials reported, however, that "the advocates of this theory" often poorly defined those atmospheric influences that were "the supposed cause."[41] Hitchcock offered no evidence to suggest which of these influences might be responsible for the potato's condition. Instead, he asserted that the rot owed its existence to "some atmospheric agency, too subtle for the cognizance of our senses, like those which bring such epidemics as the influenza and the cholera over particular districts or continents." Arguing that "modern science has shown us that many of the most powerful agencies of nature are concealed from common, and even acute observation," Hitchcock assumed that "others," like those attacking the potato, were "yet undiscovered."[42]

Although Hitchcock declined to define the exact cause of the potato disease, others holding this view were ready with somewhat more precise descriptions of what atmospheric influences they thought precipitated the potato rot. In 1845, the New York Agricultural Society awarded a twenty-dollar prize to Andrew Bush, a physician from Pennsylvania, for his paper on the potato disease. Echoing widespread opinion blaming the rot on excessive moisture and high temperatures, Bush found the epidemic to be the "condition of the atmosphere, brought into active influence by heat and moisture, and producing the 'rot' in the more tender varieties of the potato, or those raised from seed, or badly cultivated, or under any circumstance unfavorable to their growth or preservation."[43] Bush recommended that farmers adopt the following field practices: the planting of only healthy potatoes from areas free from the disease; the application of lime, soda, and potash; the planting of different cultivars from separate geographical locations; inducing germination; the planting of whole potatoes to propagate hardy offshoots; cultivating the potato within 120 days of planting; and the gathering of potatoes as quickly as possible, storing them in a cool, dry environment.[44]

Because of their interest in agricultural science, some "progressive" farmers and amateur scientists began to espouse the view that the origins of the potato disease were chemical. When the work of German chemist Justus Liebig stimulated interest in the chemical composition of plants and soils, American attention turned to the effects of manure on the potato's growth, and, among farmers, interest spread to the appropriate use of fertilizers.[45] Some farmers, finding the high ammonia concentration in some organic manure, linked the application of manure to the source of the potato problem.[46]

Liebig's work became a veritable fad in America in the 1840s, and agriculturists who were inclined to embrace science as a way to advance farming thought that Liebig's theories might well be a panacea, the perfect tool with which to define and solve agricultural problems. Liebig's chemistry reinforced efforts by agricultural journals to urge farmers to employ new practices such as drainage, the use of new crops, and the introduction of soil amendments such as gypsum. Writing to Commissioner Ellsworth in 1845, Thomas Croft of Wilkes-Barre, Pennsylvania, expressed his faith that a chemical explanation, based on the balance between organic acids and bases, would resolve the issue of the origin of the potato disease. "All plants, for a healthy growth, need certain elements *in certain proportions*," he wrote. "If more of one of these elements be given, by man or by nature, to the plant, than is proper to form a compound with the other elements, then the elements furnished are useless, if not injurious." Giving "Dr. Liebig" the credit for his theory, Croft concluded that a deficiency of "alkalies" had upset the harmonious chemical relationship with the organic acids in the potato plant. "The perfect development of a plant is dependent on the presence of alkalies or alkaline earths," Croft quoted Liebig.[47] Diseased potatoes, Croft argued, had been rendered susceptible to the destructive effects of carbonic acid, which produced decay. To counteract the properties of the carbonic acid, he followed Liebig in recommending the application of potash, soda, lime, or magnesia to the soil. "By supplying the alkalines in sufficient quantities," Croft asserted, "we would at least save our potatoes from the 'rot.'"[48]

Although Liebig's chemical theories were gaining acceptance as a key to a solution to the potato rot, one man championed an idea that was not so widely accepted. Writing in October 1844 in the *New England Farmer and Horticultural Register*, James Teschemacher of Boston said that his observations of potato rot led him to conclude that a fungus similar to the common mushroom was responsible for the misfortune. He based his belief that a fungus was involved on the peculiar odor of the diseased plants and the accounts of animals dying after eating them. His initial opinion was strengthened when he microscopically recognized among the "greyish slimy mass" taken from the potato "the spores or reproducing bodies of the fungus." These spores were not unlike "the seeds of other vegetables," forming and spreading in the air with "inconceivable rapidity."[49]

Teschemacher carried out his studies with a simple, homemade microscope. Typical of most amateur scientists of the day, he had a lifestyle that provided him with time to devote to his avocation of natural history. Born in Nottingham, England, in 1790, he came to the United States in 1832, settled in Boston, and went into the mercantile trade. He joined the Boston Society of Natural History, one of the premier scientific institutions in America, and the society published many of Teschemacher's papers on botany, mineralogy, geology, and chemistry. He also delivered numerous addresses before the Massachusetts Horticultural Society and the Harvard Natural History Society and served as coeditor of Boston's *Horticultural Register and Gardener's Magazine* in 1835.[50]

In a follow-up to an earlier article on potato rot, Teschemacher noted that because objections to his views on the cause of potato rot were not forthcoming, he had continued his investigations of the fungal theory. In his second report, he declared that it was impossible for insects to be responsible for the rot because potato decay was present before their appearance. To address the prevailing theory that the fungus associated with the rot occurred as the result rather than the cause of the disease, he observed parts of the plant first affected and then watched the disease spread. Observing that the fungus originated on the skin of the potato, he traced its penetration of the healthy interior by the discoloration of the tuber. If the fungus were the result instead of the cause, Teschemacher concluded, he would have found parts of the rotten potato without the fungus, and this, he asserted, had not happened.[51]

Teschemacher went on to describe to the editor of the *New England Farmer and Horticultural Register* experiments he did to determine the contagious nature of the disease. He cut a diseased potato in half and placed each half on an intact, healthy potato under a bell jar in a dark, cool environment. He observed the potatoes after five days and found the sound potatoes "uncontaminated." He also buried a "much diseased potato" five centimeters in the soil and six centimeters away from a healthy potato, at a temperature between 13.9°C and 16.6°C. Again, after five days, he observed that the disease had not affected the sound potato.[52]

Teschemacher closed his letter to the editor by explaining his intention to leave the potatoes in their locations for a longer period of time, and he promised to report on his findings later. He revealed the results of his lengthened analysis at a meeting in Boston convened to discuss potato rot in January 1845. Potatoes that showed no contamination after five days were diseased after two weeks.[53]

In another series of letters to the editor of the *New England Farmer and Horticultural Register* beginning in October 1845, Teschemacher restated his ideas, and to strengthen his case, he mentioned a fungal theory held by Charles Morren of Liege, Belgium. Although Teschemacher claimed that Morren's "opinion had been generally received as true by the best informed circles in Europe," critics of the fungal theory were unconvinced.[54] They continued to argue against Teschemacher's ideas in the agricultural press, maintaining that the fungus was a result of a disease, not the cause.[55]

Teschemacher returned fire on his critics, particularly those who favored the vague idea that atmospheric influences such as cold, wet weather were the key to the rot. Insisting that weather patterns in America had been too irregular to draw such a conclusion, Teschemacher added that this type of weather, common in Europe, did not always coincide with "the appearance of this peculiar disease in the potato." He judged that "it is clear, therefore, that this so called atmospheric influence is merely a name, without a distinct tangible meaning."[56]

In addition to his etiological work, Teschemacher proposed control measures for late blight. To do so, he turned to the growing field of agricultural chemistry. In his October 1844 article in the *New England Farmer and Horticultural Register*, he advocated the use of common salt, "as it destroys the fungus vegetation." Teschemacher also asked that anyone with a better microscope than his study the "action of sulphate of iron, sulphate of soda, or of ammonia, or any other substance which can be cheaply applied to the soil as a preventative."[57]

Although Teschemacher's opinions continued to find opposition, he continued along the same path. In November 1845 he added to his chemical arsenal against potato blight, suggesting the application not only of salt, but also of lime and several other chemical compounds. He preferred salt because "when mixed in the soil, it may get into the juices and circulate through the whole plant." Lime or limewater would perform the same function, but "it is far less soluble than salt."[58]

Teschemacher advised farmers to apply chemicals with some forethought and scientific guidance to be effective. There were several instances of "chlorine gas" killing the blight fungus, but large-scale application could cause "fatal accidents." Lime, although effective, was also problematic and could lead to failures. Teschemacher recognized that the potato crop of 1845 was a near-total loss, and he suggested specific measures to prepare the soil and the seed for the next year. He advised farmers to combine chemical treatment with the proper choice of soil and drainage area and the proper storage of seed tubers. The key to a healthy crop the next year would be plowing lime and salt into the soil and soaking the seed potatoes in brine, and he added a new compound to his formula— bluestone, a common name for copper sulfate.[59]

Teschemacher realized that nothing could be done to save potatoes already infected with the fungus. He also lamented that "my microscope, being made by myself, is of course very inferior to those now manufactured in London and Paris," and thus, it did not allow him to do more extensive work into the nature of the disease and possible remedies. Teschemacher believed that, given better equipment, he could have made more of a contribution to the understanding of the potato rot.[60]

In fact, better work was occurring in Europe. As early as 1842 in Germany, Von Martius may have recognized and drawn, from his microscopic examinations, the fungus responsible for late blight. In 1845 the French mycologist Montagne collected the fungus and named it *Botrytis infestans*. He sent samples to

Miles Joseph Berkeley in Great Britain, who, after some hesitation, became the British spearhead for the fungal theory. And, as previously mentioned, Morren did experiments on cause and control and widely communicated his results.[61] Although these European investigations were transmitted to a small number of American amateur naturalists like Teschemacher, who were capable of appreciating them, the American naturalists were not yet able to replicate or transmit them successfully to the American agricultural audience.

THE LIMITS OF AGRICULTURAL IMPROVEMENT

At the end of 1844 Commissioner Ellsworth wrote, expressing his feelings as well as those of most others, that he hoped that 1845 would bring relief "from the ravages" of the potato epidemic. When this did not occur, Ellsworth had faith that all the attention devoted to the disease in the United States would "at least (throw) so much light . . . on the probable cause of the evil as would enable us to remedy its destructive effects." But Ellsworth was to be sadly disappointed: At the end of the 1845 growing season, a solution to the origins of the disease still seemed remote. Although Americans had proposed many theories as well as numerous cures, no consensus had emerged, and Patent Office officials began to take a pessimistic view. "While the public journals and agricultural papers have abounded with numerous articles embodying recommendations and suggestions as to the cause or remedy," Ellsworth complained, "we scarcely seem to be any nearer the desired result."[62]

In all, the Patent Office devoted over 200 pages of its 1845 annual report to Congress to the potato rot. As a whole, the report cost the federal government nearly $20,000 to publish.[63] This attention to the potato late blight was the first substantial involvement by the federal government in an occurrence of plant disease. But the government's interest did not last. The agricultural arm of the Patent Office was disrupted in 1846 by the resignation of Commissioner Ellsworth and the withdrawal of appropriations by Congress, which decided that the Patent Office's budget could better be used in other ways. No report was issued in 1846.

In 1847 Congress once again appropriated money for agricultural services, and the Patent Office resumed its annual reports. The potato disease, however, received less attention than in previous years. The restricted space devoted to the disease in 1847 cannot be explained by any lessening of the severity of the epidemic. From the evidence, potato late blight seems to have continued to pose an agricultural problem for the United States as late as the 1860s. Indeed, while they did not explain why they had "neither time nor space for even an abstract, and far less for a condensed survey of opinion and facts" in their 1847 report, patent officials expressed the view that the potato rot continued to be a "most important subject," and the Patent Office's reports to Congress in 1848, 1852, and 1856 all referred to the disease.[64] Journals such as *Scientific American, Country Gentleman,* and the *New England Farmer and Horticultural Register* ran articles about the disease throughout the late 1840s, the 1850s, and into the 1860s.

A survey of the literature in the agricultural press makes obvious two points: First, the early efforts of Americans to understand the nature of the potato disease produced few tangible results. Second, this lack of progress was evident to those following the literature about the epidemic. Commissioner of Patents Edmund Burke echoed the mood of frustration in his report of 1847, bemoaning the unsuccessful investigations of the disease. He wrote:

> Since our report of 1845, which contained many pages in reference to this evil, renewed attempts have been made to discover, if possible, the cause, and to devise a remedy. Numerous experiments have been tried and the disease has been investigated in almost every possible shape and form. Theories have been propounded and renounced. Vegetable physiology, chemistry, and every kindred science, has lent its aid to trace the monster to his lurking place; but hydra-like, if for a time one head seemed to be destroyed, another would start-up, and as yet science and practice have alike been baffled. Men of learning, and the unlearned, are as yet equally at fault.[65]

The 1847 report offered little beyond a review of what had been proposed in 1845. No new theories had been proposed since the earlier report, nor had contemporary investigations of existing ideas advanced anywhere. What Patent Office officials did choose to discuss, though, gives a clear picture of the state of American science with regard to plant disease in the mid-nineteenth century. Recognizing the "patient investigation[s] which give so high a value to not a few of the . . . men of science and practical knowledge in Europe," the 1847 report turned its attention to examinations of the disease by European scientists.[66]

Patent Office officials were not the only observers who noticed that Americans had made little progress toward unlocking the secrets of the potato disease.

The Farmer's Library and Monthly Journal of Agriculture labeled the theories summarized in the 1845 Patent Office report as mostly "philosophical and erudite speculations" and "puerile balderdash." According to the editors of this farm journal, the Patent Office had wasted public money and time by gathering and republishing so many misleading theories. It would have been more prudent to have offered awards of "$500 or $1,000 to practical farmers and vegetable chemists and physiologists, at home and abroad," for intelligent studies on the disease. The editors of the journal proceeded to criticize the federal government for its refusal to create a department of agriculture separate from the Patent Office.[67]

For the most part, American and European observations of the potato disease, the causal theories proposed, and the suggested remedies were about the same. The popular American view of the disease, however, lagged behind the European view. This lag reflected the state of general scientific knowledge in America. Except for a small minority of amateur scientists like Teschemacher, Americans were not participants in the important mycological work taking place in Europe that would soon disclose the parasitic nature of fungal microorganisms. In fact, Patent Office officials, briefly surveying the American setting in 1847, remarked that the fungal theory, "which was perhaps the prevailing theory at the time of our last report," had become "less confided in at the present time."[68]

Research and debate on potato late blight, its cause and management, marked James Teschemacher's last significant public foray into science. His investigations seem to have gone no further than 1845. As the potato blight faded from a public-policy crisis to an endemic irritation, his name became absent from the agricultural press. In their critical piece on the Patent Office in 1847, the editors of *The Farmer's Library and Monthly Journal of Agriculture* complained that Teschemacher's ideas had not received enough attention in the United States. Interestingly, they stated that Teschemacher was, in their view, "the first to announce the opinion founded on observation, that the cause of the rot was a *fungus*."[69] Teschemacher died in 1853, never having transcended the role of amateur dabbler to become one of the small but growing number of professional scientists. Europeans such as Anton De Bary and Miles Joseph Berkeley received the credit for developing and promoting the theory of fungal pathogenicity. But both Berkeley and De Bary gave Teschemacher his due. De Bary's 1853 landmark studies with fungi linked Teschemacher's name with the discovery of fungal pathogenicity. And Berkeley cited Teschemacher as one of the earliest scientists to recognize that a fungus caused late blight of potato.[70]

In Europe, scientific entrepreneurs cast aside the empiricism of agricultural tradition for the guiding hand of Newtonian science and, increasingly, applied themselves to the study of plant disease, which meant, more specifically, to the study of fungi. The early mycological work of Tillet, Prévost, and others had laid the groundwork for the scientific response to the crisis in Europe by scientists such as Morren, Berkeley, and De Bary. The potato blight was a catalyst for much of the modern plant pathology that followed. The blight focused and solidified concepts of plant disease that opened new vistas in the sciences. L. R. Jones wrote in 1914 that "it required the plague of the potato disease and the example of the Irish famine finally to focus attention upon the fundamental problem—the relation of the mildew to the sick potato plant, of the smut and rust fungi to the infected grain—the problem of parasitism."[71]

It would be a mistake to assign too much blame to America's amateur scientists for failing to match Europe's success in plant disease investigations during and immediately following the potato epidemic. The potato epidemic was much less of a hardship for Americans than it was for Europeans, particularly the Irish. The potato was the major source of food for many people in Europe, but the American food supply was more diversified. In America, where the potato was an important cash crop, the loss of the potato crop was somewhat mitigated by the law of supply and demand. As agricultural historians P. W. Bidwell and J. I. Falconer noted in 1925, "To the farmer the severity of the loss was considerably mitigated by the increased price received."[72]

Even in America, however, the potato late blight epidemic stirred an unusual degree of interest, for a disease of this magnitude had important ramifications. Americans in Maine, for example, where it has been estimated that in 1844 "nine tenths of the crop rotted in the ground or in the cellars," were made painfully aware of the catastrophic potential of plant disease epidemics. And it may be that the economic impact of the potato late blight in America has been underestimated. In Maine the epidemic came at a time when farmers there had only recently turned from wheat to potatoes as a leading cash crop. The onset of the disease also coincided with the formation of a nascent starch industry in that state. The effect of the disease was to halt efforts to set up factories designed to use potato starch in sizing for cloth.[73]

The limited economic impact of the devastation of potato late blight in the United States was not the only reason America lagged behind Europe in potato disease research during the mid-1850s. Another important reason was a general lack of understanding in American society about the connections between science and agriculture. Without such an understanding, Americans did not recognize scientific education or experimentation as a way of aiding the average farmer. This led to political decisions that failed to nurture an institutional base that would help agricultural science develop. In contrast, such an institutional base was developing rapidly in Europe in the 1840s and 1850s; in fact, scientific education in relation to agriculture and the creation of research stations was expanding rapidly.

Americans' faulty perceptions of plant disease in the first half of the nineteenth century mirrored a general lack of scientific knowledge and training in the United States at that time.[74] The American response to the potato epidemic revealed that American farmers and amateur scientists who were interested in plant disease had little comprehension of the nature of disease or the relationship between the fungus and the host plant. Although some, like Teschemacher, made contributions, they were limited by their lack of knowledge and equipment. During the early 1800s American science was still the province of the amateur. By the 1840s an indigenous, professional scientific group was beginning to emerge, but, despite exceptions like Amos Eaton, John Torrey, and Asa Gray, most professional scientists specialized in the physical, not the natural, sciences. Amateur naturalists and scientific agriculturists in America, with their eclectic interests, had not divorced themselves from the inductive tendencies of the Baconian tradition, nor had they been much exposed to the new theoretical trends in nineteenth-century biology.[75]

❧ CHAPTER 3 ❧

The Movement for Agricultural Science

By the middle decades of the nineteenth century the agricultural press was publishing increasing numbers of reports of crop losses from all of the cultivated regions of the United States. The clear implication was that America's problems with plant disease were escalating. Disease struck every sector of American agriculture, afflicting both crop plants that had long been grown in the oldest cultivated regions of the country and crops that were gaining in significance. The devastation of the potato crop due to potato late blight was the most obvious example, but crops of all kinds were prey to a variety of diseases, and these diseases attracted increasing attention from agriculturists and amateur scientists. The perceived magnitude of disease problems intensified the search for solutions through the new movement for agricultural science.

PLANT DISEASE PROBLEMS OF THE MID-NINETEENTH CENTURY
Except for potato rot, most of the coverage of plant disease in the agricultural press during the middle years of the nineteenth century was devoted to diseases of fruits and fruit trees. Some of this interest no doubt reflected a lingering curiosity on the part of affluent horticultural enthusiasts for avocational fruit culture. But it also showed that more and more Americans were finding fruits to be an attractive commodity. By the late 1840s, transportation links to urban markets had facilitated the rise of a commercial fruit industry in the United States. Fruit growing for market increased during the late 1840s and 1850s in the older orchard areas of the East and the recently settled lands of the western states and territories.[1] As one perceptive horticulturist reported to Patent Office officials in 1849, "The cultivation of fruit trees has of late received a far greater and more general attention in this country, than at any former period in our history."[2]

The interest in fruit growing was also evident in the enormous volume of literature on fruit trees published in America in the late 1840s and in the creation of specialized horticultural societies and fruit growers' associations. The American Pomological Society, the most prominent of these organizations, had its origins in Buffalo, New York, in 1848. The literature reflected the changing status of fruit culture. When fruit was used primarily for on-the-farm consumption, beverages, or livestock feed, issues such as yield, quality, and storage environments were not significant. But when marketing became the incentive, production levels, appearance, and taste became important considerations.[3]

Throughout the late 1840s and 1850s, peach growers continued to suffer from damage wrought by their old nemesis, yellows. By the middle decades of the nineteenth century, peach yellows had been reported from parts of Maryland and Delaware, north through New York, Connecticut, and Massachusetts, and west into Ohio, Indiana, Illinois, and Michigan. One eastern observer wrote in 1845 that because of peach yellows, "no tree in the United States, or elsewhere, has become so universally fruitless and short-lived."[4] The next year, William R. Prince, of Flushing, New York, observing the destruction from yellows in New Jersey and the surrounding region, complained that an "almost universal extermination" of the peach had occurred in 1846 and that "any one who will visit the once splendid peach orchards in various parts of New Jersey will be struck by the desolate aspect of innumerable plantations of dead trees."[5]

Nor did the yellows strike only old plantings. In time, yellows spread even into areas where the fruit industry was in its infancy. In 1853, the yellows was blamed for peach failures in Indiana.[6] Four years later, a Michigan fruit grower wrote regrettably that "the peach tree is subject to a disease called 'the yellows' which has ruined many fine orchards, and in some localities entirely blasted the hopes of the orchardist."[7]

The mood of dismay was echoed by other orchardists who were growing other crops. Apple, quince, and especially pear growers continued to suffer the effects of fire blight. One disheartened grower wrote in 1847 that the blight is the "only one drawback to the extensive cultivation of the pear tree. . . . To see a fine healthy tree, apparently in the condition of the utmost vigor, suddenly— sometimes in a single day—turning black in its branches, and dying as if struck by lightning; this is indeed discouraging to the cultivator."[8] The following year, John J. Thomas of Ohio complained that the "past season has been unusually destructive to the pear tree. In some gardens, scarcely a tree has wholly escaped; and in many instances entire trees have been destroyed." Thomas added, "The frequency and extent of this disaster is likely to prove a serious drawback on the general cultivation of this delicious fruit, which otherwise would rank, perhaps, second to none in importance."[9]

Such discouraged declarations continued throughout the 1850s. In 1854 an orchardist in Illinois questioned the future of pear culture in America, writing that "whilst we justly extol the pear, and assign it the highest place in the list of *exquisite* fruits, our enthusiasm is tempered by the recollection of the . . . maladies which threaten its utter extinction in this country."[10]

Apples, probably the most popular orchard fruit in America, were also affected by fire blight, but apple growers also had to contend with a long list of other maladies, including bitter rot, rust, leaf blight, and scab.[11] Orchardists in the South complained of apple diseases they referred to as "rot" or "black rot."[12] One apple grower challenged pomologists in 1849 to consider apple bitter rot as closely as agriculturists had studied potato rot during the previous few years.[13]

Another fruit disease, which had troubled orchardists since the early nineteenth century, was black knot of plum.[14] Black knot became a greater problem as plum cultivation expanded.[15] In 1858 a correspondent to the *Valley Farmer* claimed that "in many parts of the country, the disease known as the black knot on plum trees, is more formidable than even the mischievous little curculio." This writer went on to say that "in many . . . places, so destructive has this disease become, that many fruit growers have abandoned the culture of the plum altogether."[16]

Also prominent on the list of the most harmful enemies to fruit culture were rots and mildews that struck grapes and grapevines. Like plum black knot, these diseases had been a thorn in the side of growers for some time.[17] In 1833, the editors of the *Southern Agriculturist* reported that grape culture was acquiring great popularity. But while the activity was spreading, "the great check to this has been, and is, the rot, which sometimes destroys nearly the whole crop."[18] During the 1850s accounts in the agricultural press of the mischievous effects of grape rot intensified. One viticulturist disclosed in the *Southern Cultivator* for July 1851 that "every person of my acquaintance engaged in vine raising, after about the second crop of grapes, reported a total failure from 'the rot.'"[19] Grape rot proved baneful to growers in other regions as well. An Ohioan complained that the disease was widespread and destructive in the Midwest.[20] And grape rot was not the only threat. Another often-reported disease of the vine during the period was grape mildew. As one grower complained in the *Ohio Cultivator*, "The great obstacle in the way of growing the finest kinds of foreign grapes in this country is the *mildew* or blight which attacks and destroys the fruit before it is half grown." He added that "this disease has of late years so generally destroyed the fruit, after the first or second year, that few persons now think of planting foreign grapes, except in houses constructed for their culture; and even in these much care is often requisite to guard against this evil."[21]

As American fruit growers expanded the range of their crops, they encountered more unfamiliar diseases. The agricultural press abounded with reports of previously uncommon diseases such as pear leaf blight, gooseberry mildew, cherry black wart, peach mildew, and peach leaf curl.[22] A group of orchardists promoting fruit culture in Pennsylvania in 1853 felt compelled to admit that "every kind of fruit is affected . . . by insects and diseases, and none flourish without care and culture."[23]

Fruits and fruit trees were not the only cultivated plants susceptible to plant disease. Farmers also continued to record in the agricultural press their experiences with diseases of field crops. Wheat diseases had taken a toll on cultivation in the eastern and Atlantic states, and by the mid-1840s, wheat production in the United States had shifted west into the fertile soils of Ohio, Indiana, Illinois, and Michigan. The movement westward was promoted by improved transportation, better agricultural implements, and new markets in growing cities.[24]

The older growing areas of New England, New York, and the Middle Atlantic states continued to grow wheat, but wheat growing was an upward struggle. In New York, a decline in wheat production began in the 1840s, and during the 1850s, production reached a new low, with a 44 percent drop from 13,121,000 to 8,681,000 bushels.[25] The threats to the wheat crop in the eastern and Atlantic states seemed never ending: insects, winter-killing, soil depletion, rust, and smuts.[26]

Wheat diseases troubled farmers across the nation.[27] One writer remarked in 1856 that "there is no important crop produced by the citizens of Indiana so suggestive of fear and so suspending of hope, from seed time till harvest, as the wheat crop."[28] In 1855 the editor of the *California Farmer*, J. F. Morse, reported that in his state "in relation to grain, very much has been injured by smut. . . . Some farmers have not made more than half crops."[29]

The expanding catalog of diseases of field crops at this time related directly to the pace at which American agriculture was diversifying and expanding. New diseases were often reported in areas where farmers were attempting to grow new crops after others had failed. The failure of the wheat crop and the growing market for grains, according to Patent Office officials, led farmers to expand corn acreage, particularly in the eastern states hit hardest by wheat pests. Before long, however, farmers reported attacks of smut on corn.[30] Other cereals, like oats, were affected as well. The editors of the *Valley Farmer*, startled by an outbreak of oat rust in 1858, declared that "the very general destruction of the oat crop by rust in many portions of the West, the present season, has caused no little surprise among the farmers." Furthermore, they insisted, "this disease has never been known . . . to attack this crop, and is considered by many as something 'new under the sun.'"[31] A contributor to the *Southern Cultivator* noted that 1858 was the worst year for oat rust that he had seen in a quarter of a century.[32] Besides rust, the oat crop was also susceptible to smut, causing some growers to lament that information on oat smut was rare compared to that of wheat.[33]

In the South, reports of disease problems proliferated as that region intensified the cultivation of important cash crops such as tobacco and cotton. Because cotton was the South's chief agricultural product, its diseases attracted a great deal of attention in the regional agricultural press. Cotton suffered from a number of diseases popularly known as rust, rot, and blight.[34] In Georgia, one grower noted that in 1828 cotton rust destroyed "one entire field of forty acres" in one month.[35] In 1835 Thomas Spalding, a cotton farmer from Darien, Georgia, noted that rust, "that extraordinary and increasing evil, has, apparently, taken hold of all of our high lands."[36] Although diseases of tobacco received less attention than cotton, tobacco suffered from a disease often referred to as "spot." It also fell victim to such colloquial maladies as "firing" and "hollow stalk" and a condition of stunted growth that produced "walloon" tobacco that "is good for nothing."[37]

In the 1850s diseases of vegetable crops associated with truck farming began to be reported across the country. Writing in the *Cultivator* in 1852, one farmer described "a disease . . . now very prevalent in turneps [*sic*]" whereby "a flag in the leaves is discovered some sunny day, and by and bye the plants wither and die."[38] The following year another vegetable grower reported on what he referred to as "clump-root" of cabbage that attacked the plants' "roots, which become distorted, knobby, and monstrously swollen." This observer added that the disease was revealed by the "wilting and flagging" of leaves and that it "affects broccoli, cauliflower, and all kinds of cabbage."[39] And in 1857, a horticulturist in Massachusetts recorded one of the first references to what he considered a new disease of the onion. "A serious obstacle has lately come in the way of the culture of the onion," this grower insisted, "in the form of *smut* or *rust* upon the young plant."[40] The rise in attention to vegetable diseases was another sign of economic shifts in American agriculture.

Mid-century investigations

Some mid-nineteenth-century American farmers took a fatalistic view of plant disease. As one frustrated Pennsylvania farmer wrote in 1854, plant pests are "the price we must pay for everything most necessary to our existence."[41] But Americans in greater numbers began to propose causal theories and remedies that indicated a growing unwillingness to accept the plant disease losses as inevitable. One Illinoisan "hoped that cultivators will not relax their efforts in searching out the cause and cure of these terrible maladies."[42]

Unfortunately, enthusiasm for finding causes and cures largely failed to translate into a scientific understanding of crop losses. To a great extent, the ideas and information offered on plant disease during the middle decades of the nineteenth century differed little from those that had come before. Agriculturists and amateur scientists generally looked to tradition to formulate their theories and remedies, and as a result made little progress in determining the true causes of disease. For the most part, the lineup of usual suspects never changed: insects, poor cultivation practices, bad weather, atmospheric influences, exhausted soil, and the degeneration of the plant itself. Enlightened farmers, horticulturists, and amateur scientists continued to recommend disease-control methods such as plant removal, seed selection and treatments, soil amendments, and the search for less susceptible cultivars.

But plant disease investigations in the late 1840s and 1850s contrasted with earlier efforts in ways that were subtle yet important. The changes stemmed from America's rapidly changing agriculture in the mid-nineteenth century, which led progressive agriculturists to increase their demands for better farming through science.[43] Although the details of what science meant to agriculture were unclear, the spirit of this movement was obvious: a break from tradition and a receptiveness to new ideas.

In 1845, for example, Noyes Darling, a close observer of peach culture in Connecticut, was led to investigate peach yellows because he was unsatisfied with current theories that attributed the disease to inconsistent factors such as soil exhaustion and climate. Persuaded by the notion that the malady was contagious, he attempted "to inoculate artificially by the pollen" the agent of the contagion from a diseased tree to a young, unaffected tree in his garden.[44] At the end of his experiment, Darling reported that the inoculated tree "showed no mark of disease." When he was able, however, to induce symptoms of the yellows by bud transfer from a diseased tree, he became convinced, as were some other investigators, that the yellows was an infectious disease not unlike cholera.[45]

The contagion theory was gaining ground in America. In 1845 Reuben Ragan carried out inoculation tests with fire blight. Samuel Gookins recounted a visit to Ragan's Indiana farm where Ragan told him that he had inoculated "a thrifty young pear tree in his nursery with the sap of a blighted tree." Gookins went on to describe the experiment: "He had made an incision about three feet from the ground, lifted the bark as in the process of budding, and injected a small quantity of the diseased sap. We found the leaves of the patient changing color, and emitting that peculiar odor which indicates the incipient state of decay, which is always present in cases of blight; and upon applying the knife, the inner bark was found to be black from the root to the top, while nothing of the kind appeared elsewhere in the nursery."[46]

These experiments with peach yellows and fire blight, however, were isolated events. Most studies of fruit tree diseases continued to be largely observational and descriptive rather than based on scientific experimentation. Although a tenuous link was made to infection and disease in a few instances, fruit tree diseases would remain mysteries to nurserymen for some time to come.

Agriculturists and amateur scientists in the 1840s and 1850s came a little closer to understanding the secrets of other plant diseases. Although most Americans were not prepared to accept the fungal theory in relation to potato late blight, many were willing to concede that parasitic fungi were somehow responsible for the rust and smuts of small grains and for some other diseases. From the time of John J. Thomas's important work on wheat diseases in 1843, the fungal theory was receiving regular coverage in the American agricultural press.

Some Americans, of course, had believed in the fungal theory since the beginning of the nineteenth century. They based this belief largely on European sources. Some enlightened agriculturists and amateur scientists received books and other publications directly from overseas. Many agricultural journals and the Patent Office reports carried foreign news stories on the developing fungal theory. One such report in the *American Agriculturist* in 1845 declared that European experts agreed that wheat smut "is produced by the minute seeds of a parasitic fungus, which, being carried by the wind, enter the plant by the fine openings of the epidermis."[47] Two years later, Patent Office officials, looking to Europe, wrote that "it appears as abundantly proved by Corda and various writers" that the cause of wheat rust "is a parasitic fungus which insinuates itself into the ears of the plant or changes the character of its tissues, scatters abroad its spores, and ruins the prospects of the husbandman."[48]

One important source of European information on plant disease was James F. W. Johnston of the University of Durham, England, an eminent champion of agricultural science and a disciple of Justus Liebig. Johnston's views on fungal pathogenicity circulated widely in America during the 1840s and 1850s. His opinion that the smut of wheat was a "disease, which is a species of fungus," and his description of the disease agent and how it enters the wheat plant were published in the United States as early as 1844.[49] In 1850 Johnston visited the United States, where he addressed the New York State Agricultural Society on the state of agricultural science and the fungal theory. "By examining [fungi] closely through the microscope," Johnston explained, "botanists have discovered how they grow—what they are—how they propagate—how they get into the plant and seed—and how they may be exterminated." Johnston was convinced that botany had much to offer agriculture. "There are none of you who may not see," he told listeners, "that the application of the results of this branch of study, has a direct bearing on the practical, pocket interests of the farmer."[50]

Some informed American agriculturists were ready to embrace the fungal theory, and by the 1850s, they routinely associated parasitic fungi with the rusts and smuts of wheat, oat, and corn, as well as with ergot of rye.[51] Others connected fungi with leaf blight of the pear and apple, cedar-apple rust, peach curl, plum decay, and mildew and rot of grape.[52] And because fire blight symptoms resembled those of potato late blight, there were some who believed that fungi might be the source of this tree fruit malady.[53]

Americans were generally good observers, and with their eyes, experience, and sometimes with the aid of simple microscopes, American proponents of the fungal theory performed intelligent investigations. But American fungal theorists derived most of their ideas from overseas and failed to contribute much to the experimental work occurring in Europe.

The high water mark for American plant disease studies at mid-century was the work of two Ohioans, Jared P. Kirtland and John H. Klippart. Kirtland, a prominent nurseryman and horticulturist in the Western Reserve, typified many plant disease enthusiasts of his day. A career in medicine provided both financial support and the scientific training to spark an interest in fruits and flowers. He became an influential figure in Ohio horticulture and on the national scene in the second quarter of the nineteenth century.[54] For thirty years, Kirtland observed brown rot of plum, "a fatal disease . . . insidiously progressing" among plum orchards in Ohio.[55]

Pomologists generally attributed the plum's decline to the curculio or to age. Kirtland was convinced, however, that observers had failed to differentiate between these enemies of the plum and the prevalence of a fungus disease causing most of the fruit's troubles. In 1855 he set out to correct this misjudgment by publishing a description and causal theory for the disease in *The Florist and Horticultural Journal.* After the fruit attained full size, but just before ripening, he wrote, the symptoms of disease first appeared "as a soft or discolored spot on the surface of one or more of the fruits." He continued, "This rapidly extends, soon reaching all the adjoining clusters and involving them in one common and agglutinated mass of corruption. The surface is studded with a mould or parasitic fungus. The surface of the mass soon desiccates and assumes a black color. For months, and even years, it will adhere tenaciously to the limb of the tree. This malignant and cankery action will, likewise, extend to the adjoining bark, wood and fruit spurs; and often either entirely destroys their vitality or induces a sickly condition; and little or no fruit will again set under two or three years."[56]

Having described the disease symptoms, Kirtland took "the liberty to suggest [his] theory in regard to it." He was certain that the fruit of the plum had been deprived of "the elements of saccharine matter" prerequisite for essential formation by the parasitic feeding of a fungus belonging to the genus *Torula*. He was, however, much less certain of a remedy. Kirtland knew that orchardists applied sulfur and soap suds to similar diseases of other fruit trees, but he was not prepared to recommend their use against the plum fungus. Cautiously, he suggested that "a solution of lime, slaked with brine, and suffered to remain a year before applied," might be adopted with success.[57]

Fellow Ohioan John H. Klippart's "An Essay on the Origin, Growth, Diseases, Varieties, etc., of the Wheat Plant," published in the *Twelfth Annual Report of the Ohio State Board of Agriculture for 1857*, ranks as one of the best informed discussions on wheat diseases up to this time. In this essay Klippart, a self-taught agricultural improver, demonstrated not only a thorough knowledge of American views on plant disease, but also a firm grasp of European theories.[58] A steadfast supporter of the fungal theory, he explained articulately why some important diseases of wheat were caused by "vegetable parasites."

Klippart provided a convincing case for fungal pathogenicity. "By nature wheat is exposed to the attacks of diseases whose ravages are very great," he argued, "which are caused by agents which are independent existencies . . . which affect the plant by feeding upon its nutricious [*sic*] juices and destroying . . . its vitality." Going to some length to prove his point, he described "vegetable parasites" as "minute plants of the cryptogamic class, . . . so minute as to escape the scrutiny of the unassisted eye," and provided illustrations of his own observations of parasitic wheat fungi through the microscope. His discussion of rust, smut, and bunt also included detailed explanations of parasitic transmission and a brief mention of physiology in relation to fungal reproduction. Influenced by European work, Klippart asserted that "since the invention of convex glasses, and the attentive studies of the learned physiologist, Benedict Prévost, it can no longer be doubted that the moulds, the rusts of plants, etc., are real vegetables, which . . . although they do not conform entirely like others, . . . follow the same general rules of birth, growth, and death, and of reproduction by seeds."[59]

The work of Klippart and Kirtland demonstrated that some Americans in the first half of the nineteenth century were contributing in at least small ways to the developing fungal theory, but unfortunately, careful observations alone did not provide many practical solutions for American agriculturists whose crops were troubled by plant disease. For over half a century, plant disease had been observed on the farm, discussed in society meetings, and reported in the agricultural press. These approaches had done little to slow the steady progress of crop failures. The careful observations of Klippart and Kirtland represented the best that America had to offer during the period, but observational skills had taken them about as far as was possible at that time. "By the 1850s," as one student of the period wrote, "the fund of common experience had been exhausted, and the bankruptcy of the program was becoming increasingly clear."[60] Americans had genuinely sought to improve their agriculture, and in some cases had been successful. But by the 1850s, their means of sharing information and their methods had served out their usefulness.

When Kirtland wrote in 1855, he noted wisely that "a knowledge of the pathology of a disease . . . is frequently an advancement at least one half of the way towards the discovery of a successful mode of treatment."[61] But if Americans were to really understand and control plant diseases, a more scientific and organized approach was required than that presented in the first half of the nineteenth century. To close the case on the fungal theory, Americans were going to have to collect not only observational, but also experimental, evidence.

It was obvious to many Americans that their methods of dealing with diseases were failing. One correspondent to *The Country Gentleman* wrote in 1858 that "there was little hope of success" in finding solutions to fruit tree diseases "unless care and perseverance, with a sufficient amount of scientific knowledge, be brought to bear on the subject." This writer added that "it is enough for us to know, that annually a great number of valuable fruit trees are destroyed by an undefined malady, . . . without any decision being arrived at as to their nature, development, or cure."[62] Another disappointed agriculturist, writing to the same journal, scorned the lack of advances that were being made to treat diseases while these "maladies . . . are becoming more numerous and malignant in character and fatal in effect. . . . Take for instance, the blight of the pear tree, the yellows of the peach, the black knot of the plum and the potato disease," he insisted, "and tell us what beneficial discoveries in the treatment or cure have been made." This correspondent declared that Americans were "yet in the dark as to the cause and treatment as much as we were twenty or thirty years since."[63]

Neither individuals, societies, nor the press seemed capable of meeting America's agricultural challenges. The efforts of the Massachusetts State Board of Agriculture to investigate plant disease in 1859 are particularly revealing. When the Massachusetts legislature decided to offer a $10,000 prize to solicit a remedy for potato rot, the Board of Agriculture appointed a committee to "consider and report upon the diseases of vegetation." This committee sent circulars out across Massachusetts asking farmers to respond to twelve general questions about plant disease and eight additional queries on potato rot.[64]

To the disappointment of the Committee on the Diseases of Vegetation, as it became known, the response to its landmark attempt to cast light on plant disease problems in Massachusetts was dismal. Although it claimed to have distributed the circulars widely to farmers, it received fewer than thirty responses. Among those, the committee found "that . . . a more degrading record of ignorance of the first principles of natural science" could scarcely be found, and wrote that it considered "this want of interest on the part of the agricultural portion of society as somewhat disgraceful." Its only explanation was "the necessity of greatly increased exertion on the part of the friends of agricultural education."[65]

There was wisdom in the Committee on the Diseases of Vegetation recommendation for increased education. Other than a small number of enlightened agriculturists and amateur scientists with an eye on events in Europe, most American farmers of the 1850s understood little about plant disease or agricultural science. Under such circumstances, the poor response rate was understandable, and many believed that the committee had raised its expectations too high. As one critic asked, why should the committee have "expected that our farmers . . . be scientific men?"[66]

Perhaps, some argued, answers to plant disease problems should be sought not from farmers, but from scientists who understood the revolutionary ideas in biology and could perform the experiments requisite for the explanations they sought in agriculture. But agricultural scientists were difficult to find. For over half a century, the subject of plant disease had been the province of loosely organized, enlightened amateurs working through agricultural societies and journals toward the goals of agricultural improvement. This establishment had yet to move far beyond enthusiasts and empiricism. As the experience of the Massachusetts Committee on the Diseases of Vegetation proved, this was largely a dead-end path to a true understanding of plant disease. By the end of the 1850s, the agricultural societies and state committees had reached the limits of their capabilities, both organizationally and financially, and the difficulties of securing substantial and unwavering state support led the advocates to turn to the federal government.

INSTITUTIONS FOR AGRICULTURAL EDUCATION AND RESEARCH
Advocates of agricultural improvement in the mid-nineteenth century generally agreed that their goals would best be achieved by rational organization and improved methods, beliefs born of the Enlightenment and the Industrial Revolution. Nevertheless, debate continued to rage over how best to apply the various advanced forms of knowledge to farming, an avocation bound by tradition and often averse to new ideas. Creating and exchanging new ideas for agricultural improvement and dispersing these ideas to the practical farmer would require new institutions. Such institutions for education and research, often government supported, already existed in Europe, and they frequently held the improvement of agriculture as one of their primary goals. In the United States, however, there were no such institutions, and many agricultural improvers thought they were desperately needed.

Most advocates of agricultural improvement in the United States agreed that the most important step in improving agriculture, particularly in a rapidly changing arena, was education of the farmer. New principles were worthless if they were not communicated and used. Alarmed by the failure of lands in the eastern United States and the surge of disease problems, alert agriculturists and journal editors recommended immediate action. Education in the new sciences, they reasoned, could improve cultivation practices, renew failing eastern soils, increase crop yields, give farmers assurances against plant pests, and prevent the deterioration of the newly settled and fertile western lands.[67]

But the advocates of agricultural education failed to agree on how it should be accomplished. Some thought it should consist primarily of vocational training in subjects such as cultivation and drainage practices, accompanied by perhaps a little farm management, called "rural economy." Others recommended a curriculum more in line with a traditional, classical education with elements of agriculture added. Such a curriculum increasingly included science as a proper and necessary component. The list of sciences most often considered vital to agricultural education was led by chemistry, but other recommended fields included botany, mineralogy, geology, entomology, and the physiology of plants and animals.[68]

American advocates of agricultural improvement through science education looked to Europe for institutional models. The 1850 New England lecture tour of James F. W. Johnston heightened interest in European activities, including the investigation of "mildew, smuts and rust," which Johnston said was "deserving of all possible encouragement."[69] American agricultural journals reprinted articles and speeches from European sources, particularly from Great Britain. German universities were highly praised for their scientific achievements.[70]

The promise of agricultural science and the European institutional models found a ready audience in the eastern states of New York, Massachusetts, Connecticut, and Maryland. Here, where soil exhaustion and disease problems were a reality, farmers were deserting the land for the cities or moving to the fertile Ohio and Mississippi valleys to the west. As production expanded in the western United States, forward-thinking leaders expressed an interest in agricultural education as a preventive against the failure of their lands and crops due to poor farming practices and disease.[71] Thus landowners, politicians, and scientists led the movement for agricultural education, first within their own states and then by pushing for federal involvement in the late 1840s and 1850s.[72]

The agricultural press began to call for a federal commitment to support agricultural education. Editors complained that Congress was squandering tax money primarily raised from farmers on nonfarm expenditures, while ignoring the growing needs of agriculture.[73] "If a pirate on the high seas," Daniel Lee, agricultural chemist and editor of *The Genesee Farmer*, wrote for the Patent Office, "injures the property of a citizen to the amount of a few dollars, millions are expended . . . to punish the offender." Why then, Lee queried, could Congress appropriate no money to protect crops from "parasitic plants, such as rust on wheat"?[74]

Although proposals for federally funded agricultural colleges had been made for years, the most important plan came in 1851 from Jonathan Baldwin Turner of Illinois College. Turner's "Plan for an Industrial University," which received widespread attention in the agricultural press, spoke specifically to plant disease, anticipating the need to increase research in that area. One of the important obligations that he intended for his "industrial university" was the study of "blights, blasts, rots, rusts, and mildews which so often destroy the choicest products of industry."[75]

In 1857 Vermont congressman Justin Morrill introduced a land-grant college bill that reflected Turner's concept, but it was vetoed by President Buchanan. Then, in 1862, in the absence of Southern congressional opposition due to the

Civil War, and after the bill had been fine-tuned, the law passed. President Abraham Lincoln signed the "Morrill Land Grant College Act" on July 2, 1862.[76]

While the advocates of agricultural improvement were arguing for land-grant colleges, they were also fighting for a second institutional base—a centralized federal department in Washington, D.C. The Patent Office's agricultural bureau had fulfilled this role through the 1840s, collecting statistics, distributing seeds, and publishing a bulletin. However, the potato late blight epidemic contributed to a feeling that these limited activities were insufficient to meet the daunting task of coordinating the issues facing the nation's changing agriculture. The editors of the *American Agriculturist* wrote that they considered "the agricultural operations at Washington . . . a sham."[77]

Through the late 1840s and into the 1850s, there was a strong movement for expansion of the agricultural activities of the federal government, including scientific research. From this movement came specific suggestions that an expanded bureau of agriculture could prove its value by dealing effectively with plant disease. "If, under the judicious administration of the proposed department," Congressman Holloway wrote, "investigations should result in securing a prevention of the potato rot . . . the expense attending these investigations alone will be but as a drop in the bucket in comparison with the resulting benefits."[78] Claiming that "diseases of vegetation" were the worst problems facing farmers, the editors of *The Magazine of Horticulture* asserted that their study "deserves encouragement by legislative action."[79]

Against this background, in January 1862 a bill to establish a department of agriculture was introduced into Congress. The bill introduced in the Senate emphasized the scientific role of the new department, but because the endeavor had its critics, the bill that finally passed was conservative, concentrating on the same economic and political duties that had been successful in the Patent Office. The new department certainly had scientific goals and duties, but the extent and specifics of those duties were to be worked out later. President Lincoln signed the bill into law on May 15, 1862.[80]

The creation of the United States Department of Agriculture and the land-grant college system was the culmination of much effort, all aimed at the improvement of agriculture and, with it, the control of plant disease. The exact means to achieve this goal were often subject to strong debate. But it was clear that institutional support was necessary at local, state, and federal levels. In 1862, with the watershed legislation that created both land-grant agricultural colleges and the United States Department of Agriculture, the essential elements for agricultural research in the United States were in place. It was within these two institutions that plant disease research was to rise to national and international prominence in the second half of the nineteenth century.

Part Two

THE ORIGINS OF U.S. PLANT PATHOLOGY

That botany which hopes to satisfy the demands of the advanced agriculture of today must include a knowledge of pathology.
—Charles E. Bessey

In spite of the fact that our cereal and other crops of the farm and orchard are damaged to the extent of many millions of dollars annually by the attacks of fungi, there has been very little done towards instituting experiments to find means to guard against them.
—Norman J. Colman

During the second half of the nineteenth century, the era of the amateur naturalist, the casual plant disease observer, drew to a close. The naturalists' dream of improving plant health, however, still flourished in a small group of professional scientists eager to find the causes and cures for plant disease. Supported and encouraged by a matrix of new institutions and new knowledge, scientists at American universities and land-grant colleges, agricultural experiment stations, and the United States Department of Agriculture entered into a period of unparalleled creativity and growth.

The new era of professionalism in science was triggered by major developments in biology. As a discipline, biology had evolved in the early decades of the nineteenth century from medical physiology, natural history, botany, and zoology. From these disciplines, biology gathered together a body of knowledge that focused on the functional processes of the living organism. This focus resulted in the development of such fundamental concepts as cell theory and the germ theory of disease. Energized by the contributions of Darwin on evolution and natural selection, science became more visible to the general public, heightening attention to its application to social ends and its value for social and economic good. Changes occurred in attitudes about what constituted accepted and necessary areas of research in many disciplines.

Botany, like other sciences, underwent exceptional transformation during this period. In its early years, botany had concentrated on taxonomy. The emphasis on characterizing and classifying plants had left relatively little time for understanding functional processes. In the late 1800s, however, the emphasis in botany shifted. Reflecting the general shift toward physiology and experimentation in biology, the "new botany" placed an emphasis on the analysis of structures and systems of plants in order to understand their function.[1]

The new botany originated in Europe, particularly in Germany, and was transferred to the United States in several ways. First, in mid-century a small number of European botanists schooled in the new ways immigrated to America. Second, a small number of American botanists established contacts with European colleagues and kept up with the current literature from overseas. Finally, a handful of American botanists traveled to Europe to study under the masters.

The change of emphasis in botany from taxonomy to physiology was facilitated by substantive advances in research methods that were occurring in all biological sciences in the mid-1800s. Most notable among these advances was a revolution in laboratory technology and techniques. Microscopes, for example, which allowed a view into the microworld of plants and animals, evolved rapidly, and as their quality improved, so did their quantity. In the beginning of the century, a microscope was a relatively precious and unique lab accoutrement controlled by specialists. By the end of the century, it was a common tool that all botany students were required to master. The biological sciences were also transformed by the development of new laboratory procedures such as improvements in staining and, perhaps most important, the creation of pure-culture techniques.[2]

The power to peer deeply into living organisms produced a corollary desire to study lower plants, particularly fungi. Inspired by the expansive spirit of the new botany, scientists launched a new era of mycology. Although the practice of collecting was invigorated, in mycology, as with the higher plants, the focus was on physiology and morphology, not simply on discovering and classifying new species.

The new interest in fungi was obviously significant for the study of plant disease. In the 1850s German morphologist Anton De Bary finally proved beyond reasonable doubt the causal nature of fungi as plant pathogens. De Bary's research, along with the work of German agricultural expert Julius Kühn, provided the impetus for advanced studies on plant disease. Their groundbreaking results pointed toward the critical need for more fundamental examinations of fungal life cycles. The number of studies of pathogenic fungi grew notably in the 1860s and 1870s in Europe and, increasingly, in the United States, where mycology was slipping from the hands of gentleman horticulturists and amateur naturalists and becoming the property of professional scientists such as William Farlow of Harvard and Charles Bessey of Iowa Agricultural College.

The frontiers in the study of plant disease were also enlarged by the discovery of a new kind of plant pathogen—bacteria. The central American figures in the discovery of plant pathogenic bacteria were Thomas J. Burrill of Illinois Industrial University and Joseph C. Arthur of the New York State Agricultural Experiment Station. Working on fire blight of pome fruit in the 1880s, their breakthrough expanded the portfolio of plant disease studies and added to the opportunities and responsibilities of the growing corps of botanists and mycologists who were devoting at least part of their time to plant disease research. Moreover, their discovery strengthened the confidence of American scientists working in this new research area.

In the United States, the study of plant disease was linked to the powerful and utilitarian force of the economy. In the first half of the nineteenth century, American advocates of the movement for agricultural improvement had looked to the application of scientific principles as a way to enhance agricultural yields. Unfortunately, the botany of the time proved to be unequal to the challenge posed by most of the agricultural problems associated with plant disease. In the second half of the century, however, conditions were changing, and a successful and productive alliance between science and agriculture seemed possible. Agricultural improvers were interested in the practical value of the new biological knowledge and techniques. Botanists were beginning to see that the tenets of the new botany fit well with the study of agriculture, offering excellent opportunities for research on subjects that were intellectually stimulating, challenging, and rewarding.

Scientists, agriculturists, and politicians who wanted science and agriculture to mesh promoted the creation of institutions for teaching and research. Although agricultural problems were largely local, advocates of agricultural improvement hoped to solve problems such as plant disease through a system of national coordination and cooperation. Because they claimed large-scale economic benefits, they urged that the institutional bases they sought be funded by public money. Landmark 1862 federal legislation resulted in the creation of two of the three most important institutions created by the agricultural improvement movement—the United States Department of Agriculture and the land-grant college system. In the early 1870s, to complement these institutions, the government also funded state agricultural experiment stations.

At first, these three agricultural institutions were mired in controversy over how best to accomplish their goal of improving agriculture. Part of the debate concerned the basic premise of whether science should or could have a useful role in improving farming. Thus, when the wealth of new scientific knowledge, including the new botany, began to arrive in the United States, it found a welcoming audience in the agricultural institutions. Scientists embraced the new scientific advances as much-needed ammunition to win their war with vocational advocates. By the early 1880s, there was no longer a substantial debate as to whether science should play an integral role in agricultural education and research; rather, the discussion turned to exactly what role science was to play.[3]

As agricultural science and its institutions developed, so too did institutions supporting botanists and mycologists with strong interests in plant disease. From the 1860s, agricultural scientists were part of the developing national trend toward scientific specialization and professionalization.[4] In American universities and colleges, there were relatively few professors of subjects such as natural history, botany, and mycology who were advocates of the principles of the new botany. But men such as Charles E. Bessey of Iowa Agricultural College and later the University of Nebraska understood both the scientific and the practical significance of plant disease studies and altered their courses to include instruction on plant disease. Bessey and professors like him, who were groundbreaking researchers in their own right, eagerly proselytized for the young discipline, wrote influential papers and books that sought to present plant pathology to a scientific and popular American audience, and mentored the next generation of botany and mycology students who would focus on plant disease. In doing so, they reflected the fractionalization and recombination of botany and mycology into the nascent specialty of plant pathology.

A part of scientific professionalization in this era was a movement toward professional organization. Instead of confining themselves to casual participation in amateur societies of naturalists and agriculturists, the new professional botanists joined general-science associations that were national in scope. This bolstered the perception of their profession among other scientists as well as with the interested public. However, even in general associations they gathered along disciplinary lines. As their ranks expanded, they began to create unofficial networks. These linkages were a necessary component of professionalization, providing lasting lines of communication among administrators, teachers, students, and colleagues. The developing networks defined and addressed discipline-wide concerns and also provided research and career support for individuals.

By the 1880s, in hopes of affecting public policy on agricultural science at a state and national level, botanists and mycologists were using multiple forums to express their opinions. As members of state horticultural and agricultural societies and as professors at colleges and universities, they gained visibility and reputations for expertise. Within the established and prestigious American Association for the Advancement of Science (AAAS), a body that policy makers found useful for checking the general tenor of national scientific opinion, they constituted a well-organized minority. They also exploited a less established, but sometimes more dynamic, form of expression by participating in a series of national conventions dedicated to forging consensus on matters of agricultural science and education. Thus, botanists and mycologists united, rallied support, and brought sufficient pressure on the federal government to create a permanent unit of the USDA to coordinate attention to plant disease on a national scale. The USDA's Section of Mycology was a natural result of the tenets of new botany in America, as interpreted by agriculturally minded botanists and mycologists.[5] Although it was devoted to basic research, this research was undertaken with the belief that uncovering the secrets of the biological processes of plants and their pathogens would, by necessity, have an ultimate social and economic benefit.

Fig. 2.1.   George Engelmann, from the frontispiece to
The Botanical Works of the Late George Engelmann, St.
Louis, 1887, courtesy of the Missouri Botanical Garden,
St. Louis.

Fig. 2.2.   Asa Gray, courtesy of the Archives,
Library of the Gray Herbarium. Harvard
University.

Fig. 2.3.   William G. Farlow, courtesy of the
Archives, Library of the Gray Herbarium.
Harvard University.

Fig. 2.4. Bussey Institution, courtesy of the Harvard University Archives.

Fig. 2.5.   Charles E. Bessey, courtesy of the Iowa State
University Library, University Archives.

Fig. 2.6.　North Hall, Iowa State University, 1886, courtesy of the Iowa State University Library, University Archives.

Fig. 2.7.　Botany Classroom, Iowa State University, courtesy of the Iowa State University Library, University Archives.

Fig. 2.8. Thomas J. Burrill in laboratory with students, 1882 (L–R: Burrill, Fredericka Detmers, Annetta Ayers), courtesy of University Archives, University of Illinois.

Fig. 2.9. Bacteriological laboratory in University Hall basement, University of Illinois, 1880s, courtesy of University Archives, University of Illinois.

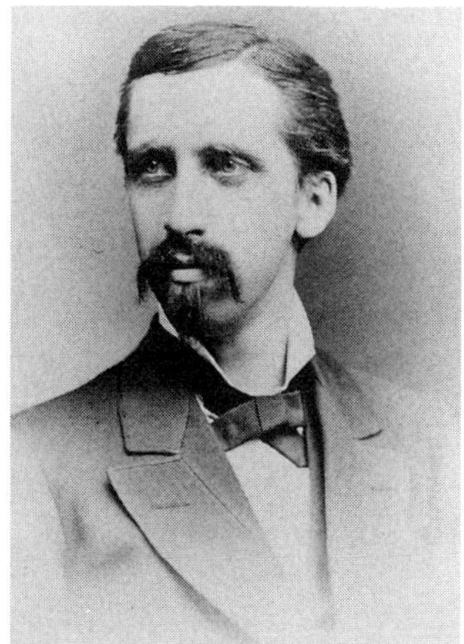

Fig. 2.10. Joseph C. Arthur, courtesy of the National Archives.

June 9. Pear blight
Found a flowering shoot of apple with about four inches of last years growth blighted with the customary abundance of bacteria. Also on flower spur of pear, but with only a very few bacteria. Several flower spurs of pear which are black show no bacteria under the microscope, and it is doubtful what killed them. Could find no hawthorne blighted. No shoots without flowers have yet been found blighted.
June 13 A foot or two of a Buffum is dead, probably continued from last year.

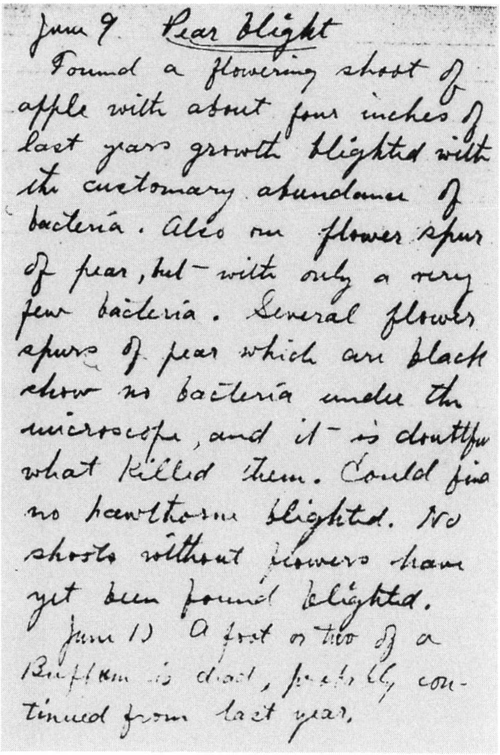

Fig. 2.11. Excerpt from Joseph C. Arthur's laboratory notebooks on fire blight research, 1884-1886, courtesy of the New York State Agricultural Experiment Station, Geneva.

Fig. 2.12.   Thomas Taylor, courtesy of the
USDA Systematic Botany and Mycological
Laboratory, Beltsville, Maryland.

Fig. 2.13.   Norman J. Colman,
courtesy of the National Archives.

Fig. 2.14.   Frank Lamson-Scribner, courtesy of the
National Archives.

CHAPTER 4

A Changing Botany: New Educational Opportunities

During the latter part of the nineteenth century, the study of plant disease in the United States experienced unprecedented growth. Born in the wake of monumental advances in biology and the shifting discipline of botany, the new science of plant pathology straddled the worlds of theoretical and applied science, offering American farmers desperately needed practical solutions to pressing disease problems. Nevertheless, in the first two and a half decades after the sweeping agricultural legislation of 1862, serious attention to plant disease seemed anything but assured.

Although Congress had created important institutions related to agriculture, it had not addressed the difficult and confusing task of deciding exactly what those institutions should do. Land-grant colleges, agricultural departments of existing colleges, state agricultural experiment stations, and the USDA provided the institutional underpinnings for the scientific study of plant disease in the United States, but these underpinnings were disorganized and lacked clear direction. Nevertheless, the stage was set to demonstrate the value of both the teaching and practice of what would become the science of phytopathology in America.

The most important gathering place for adherents of the so-called new botany and its corollary, the young science of plant pathology, was the American higher-education system. The first scientists who could properly be called "plant pathologists" (although none referred to themselves as such) labored over improved microscopes in laboratories and lectured in classrooms of American colleges and universities—particularly at the new land-grant colleges, which were devoted primarily to advancing agricultural and technical knowledge. During the two decades after the Civil War, plant pathology found the American university system to be an institutional home that nurtured and supported it in an encouraging, structured way.

THE CHANGING FACE OF BOTANY

Efforts to understand and control outbreaks of plant disease in the United States fell within the general movement for agricultural improvement, a movement that struggled with the relationship between agriculture and science.[1] While a small group of American agriculturists at mid-century debated in earnest whether microscopic fungi were the cause or the result of disease, botanists in Europe were making major scientific contributions to the knowledge of disease in plants. These European botanists ushered in the modern era of plant pathology and attracted the attention of American botanists ready to capitalize on the new opportunities in American colleges.

The event most responsible for bringing plant disease to the attention of botanists was the epidemic of potato late blight in Europe and America during the 1840s. The Reverend Miles Joseph Berkeley of England, a leading authority on fungi who had described hundreds of previously unrecognized species, applied his broad mycological knowledge to groundbreaking work on late blight. Furnished with information from other fungal theorists like Charles Morren in Belgium and James Teschemacher in America, Berkeley partially worked out the disease cycle of potato late blight and asserted boldly that it was caused by the fungus *Botrytis infestans*.[2]

Because Berkeley was a celebrated mycologist and an editor of the journal of the Royal Horticultural Society, his disease theory received widespread attention. His memoir on potato late blight in 1846 and his 173 papers published in the *Gardener's Chronicle* from 1854 to 1857 were major contributions to the developing fungal theory.[3] He even gave this area of study a name, "vegetable pathology." Berkeley did not, however, provide the experimental evidence that would close the case for the fungal theory.

Meanwhile, events that would have a tremendous influence on the study of plant disease were occurring in biology. Among the most significant of these was the development of the cell theory by M. J. Schleiden and Theodore Schwann in 1838 and 1839. One of the primary elements in the foundation of modern biology, cell theory led to an increasing interest in experimental studies of cellular morphology and chemistry, primarily from a medical point of view. The investigations of medical men, including Rudolf Virchow and Anton De Bary, drew on this work, extended it, and contributed significantly to the developing germ theory of disease.[4]

In 1853 the German mycologist Heinrich Anton De Bary finally resolved the debate as to whether fungi were the cause or the result of plant disease. In his landmark publication on rusts and smuts, *Untersuchungen über die Brandpilze,* De Bary provided conclusive experimental evidence that the so-called brand fungi were independent organisms and that these pathogenic microbes caused specific diseases.[5] De Bary's isolation and investigation of parasitic fungi supplied the first explanation for the cause of plant disease that concurred with ideas from experimental medicine. De Bary's work defined plant disease as a pathological aberration of plant physiology, reduced to the effects of another organism. Significantly, this was done nearly a quarter of a century before Robert Koch demonstrated the causal role of microorganisms in animal disease through his work with anthrax in cattle.

De Bary was at the center of a new movement in botany during the middle decades of the nineteenth century—an effort to move away from traditional, descriptive work and toward investigations of plant morphology and physiology. Aided by improved microscopes, German experimentalists like De Bary believed that it was necessary to study a plant not just as a dried and mounted herbarium specimen, but as a living plant in all of its external and internal relationships.[6] This fundamental shift in emphasis was essential for the development of plant pathology, a science centered on the relationship between the pathogen and the host plant.

During the middle years of the nineteenth century, De Bary took the lead in scientific mycology with his investigations of the parasitism of fungi, involving their reproduction, morphology, and physiology. In 1861 he described the complete life cycle of the late blight fungus, thus resolving the debate over the cause of the potato epidemic.[7] Two years later he returned to the wheat rust fungus. By experimentation, he established the life cycle of wheat rust and demonstrated the role of the barberry in the heteroecious life cycle of the fungus, that is, the involvement of two or more species of plants in the life cycle of one pathogen.[8]

But De Bary's most important contribution to the development of plant pathology may have been his role as teacher and mentor. As significant as his research was in establishing many of the general concepts of plant pathology, his influence on the men he trained was of equal or greater importance. His botanical laboratory at Freiburg, Germany, established in 1855, was one of only a half-dozen such institutions in the world at the time. Later, at the University of Strasbourg and then at the University of Halle, he attracted a steady stream of students from all over the world, including the United States. De Bary's meticulous methods of observation and experimentation and his untiring enthusiasm for teaching in his laboratory gave his students a new understanding of fungi and their relationships to plants.

Although De Bary's focus was on the cause of plant disease, his demonstration of fungal pathogenicity also gave credence to phytopathology's practical applications to agriculture. Works like German agriculturist Julius Kühn's 1858 textbook *Die Krankheiten der Kulturgewächse, ihre Ursachen und ihre Verhütung* summarized the research and discoveries of De Bary and others in the etiology of plant disease, particularly with respect to plant pathogenic fungi but not exclusive of other causal factors such as climate and soil conditions.[9] Together with Berkeley's celebrated work in the *Gardener's Chronicle* (1854–1857), Kühn's textbook introduced economic plant pathology to an eager public and established the foundations for modern applied plant pathology.

American advancement in plant pathology depended on the work of its botanists, who would have to lead the way by learning about European advances and by showing a greater interest in plant disease. Agriculturists and horticulturists in mid-nineteenth-century America lacked the necessary training to go beyond the empirical observations and experiments they were already attempting. America's botanical tradition was strong, but it was not oriented toward the type of experimental work on plant disease that was occurring in Europe. With the exception of the diversions of a few mycologists, medical botanists, and amateur naturalists such as Oliver Prescott and James Teschemacher, American botanists had concentrated overwhelmingly on the taxonomy of vascular plants, especially those collected during the myriad West-exploring forays at mid-century.[10] Moreover, the focus of American botanists was on wild rather than cultivated plants.

There were, however, a few scientists in a position to transfer to America knowledge of recent European developments in botany. George Engelmann, for example, arrived in St. Louis in 1835, shortly after obtaining an M.D. degree from the University of Würzburg, where he had written a dissertation on the nature and causes of morphological deviations in plants. While practicing medicine, Engelmann increasingly turned his attention toward botany. Associating with the foremost botanists in America, Asa Gray and John Torrey, he became the botanical czar of the Midwest. He had a hand in the description and classification of most of the botanical collections produced from the western expeditions of exploration, and he conducted his own fieldwork, from which he described many previously unidentified plants and became an expert on Cacti, Coniferae, and other groups.[11] Although Engelmann was essentially a vascular plant taxonomist, his depth of knowledge extended to other areas. By the 1860s he had begun to use his botanical expertise to investigate problems of economic plants.

Grape culture experienced an exceptional rise in popularity in the years just before and after the American Civil War. Grape vineyards sprang up all over the Midwest and into California, and with this new viticultural enthusiasm came increasing reports of disease.[12] Engelmann took an active interest in grape diseases caused by fungi, such as brown and black rots and mildews, which were devastating vines in parts of the Midwest. On September 16, 1861, as president and one of the founders of the Academy of Sciences of St. Louis, Engelmann delivered a paper carefully describing two species of fungi attacking native American grapevines. First, he discussed "brown rot," claiming it was caused by a species of *Botrytis*, "perhaps the same as Berkeley's *B. viticola*," or "near *B. acinorum*." Next, he discussed "black rot," which he believed to be caused by a different fungus, which botanists had not yet described. According to Engelmann, this fungus "evidently belonged near Ehrenberg's genus *Nœmaspora*, and ought to bear the name *ampelicida*."[13] Engelmann's description of these two species of fungi, with detailed descriptions of mycelia and spore structures, was superior to most other American investigations of pathogenic fungi in 1861. His botanical interest in plant disease led him to write several other important descriptive papers on grape diseases during the next two decades.[14]

But, even with his background in morphology, Engelmann was still a scientific amateur. For Engelmann, botany was a pleasant diversion from his medical practice, and this diversion extended only to observation and description. As impressive as his contributions to systematic botany were, his attention to plant disease fell short of the experimental work taking place in Europe. It was his attention to economic botany that made Engelmann a significant figure, particularly in the Midwest. After all, he worked in Missouri, on the edge of the Great Plains, far from the eastern United States, where most of the previous work on plant disease had been done.

Asa Gray, also a vascular plant taxonomist, was in a better position than Engelmann to transfer new ideas from European botany to America. As professor of natural history at Harvard since 1842, Gray was the foremost systematic botanist in America, and he made Harvard the mecca for most of the important botanical work in the United States. There, he assembled all the current books and established an unrivaled working herbarium. Before long, he had gathered around him a group of bright, young, eager students.[15]

Gray was influenced by developments in European science throughout his career. In 1832, as a botanical assistant to John Torrey, himself a noted botanist who had been inspired to study the vegetable kingdom by Samuel Mitchill and Amos Eaton, Gray had been introduced to the European-style, natural system of classification. Based upon reproduction features, this natural system gave Gray an early appreciation for the importance of plant morphology. He based his first book, *Elements of Botany*, on the natural system. Later in the same decade, Gray traveled to Europe to buy books and scientific equipment for his new assignment as chair of botany at the recently founded University of Michigan. While in Europe, he met prominent members of the scientific community, including Charles Darwin, who had just returned from his voyage on the H.M.S. *Beagle*. Impressed by Darwin's work, Gray later became a champion in America for Darwin's views on evolution.[16]

Above all, Asa Gray was a brilliant systematist. Yet he was just as influential as a teacher and organizer of his discipline. Gray had a grand vision for the future of botany in America. Although he believed that much work remained to be done in taxonomy, and he continued to follow this path, he encouraged his students to look to new directions in plant morphology, anatomy, and physiology. Clearly, "Gray was grooming students for the great task of bringing Europe's 'Scientific Botany' to America—a movement which stressed the study *of plants* rather than *about plants*."[17] Under the guidance of Asa Gray, who gathered such men as William G. Farlow, Charles E. Bessey, William J. Beal, John C. Coulter, and Liberty Hyde Bailey around him, America seemed assured of significant change in the plant sciences.

PLACING SCIENTIFIC BOTANY IN AGRICULTURAL EDUCATION
Launched with a mission of "teaching" agriculture, the new land-grant colleges created by the Morrill Act of 1862 were faced with a nearly nonexistent corpus

of knowledge from which to teach. The scientific disciplines were growing in quality and reputation, particularly in Europe, but with the exception of chemistry, most of them, such as entomology, geology, biology, and particularly botany, had not been adapted for application to the practical problems of agriculture.[18]

Under these circumstances, scientific instruction was not the usual practice in the fledgling land-grant colleges.[19] The dominant educational goal of the land grants was the creation of a "modern" farmer, a better and more efficient manager of his resources. There was little agreement on how to achieve this vague goal, but its supporters argued that if land-grant colleges "graduate youths who think they know something of vegetable physiology, agricultural chemistry, and the theories of Liebig, they will merely produce a considerable number of badly educated men."[20]

Even putting the question of their mission aside, the early land-grant colleges were having trouble attracting students. The number of students enrolling in land-grant colleges was very small, usually ten or fewer in the early years.[21] Students were expected to work several hours each day on their college's experimental farm, if there was one, or at some other physical labor. Needless to say, students were never enamored by this part of the curriculum. Young men who had grown up working on the farm did not relish the idea of going to college to learn "book farming," where not only were they expected to work on a farm but, in addition, had to pay for the privilege. J. C. Arthur felt that the two and one half hours of brush clearing required every day at Iowa State was abusive. He managed to redirect himself into the more "civilized" employ of the chemistry professor and later secured a transfer to the herbarium.[22]

Furthermore, farmers and educators suffered from a lack of rapport. An extreme, but not uncharacteristic, example of the extent of the schism was Henry McCandless, professor of agriculture at Cornell from 1871 to 1873, who habitually wore kid gloves and refused to touch any farm equipment.[23] Adding to the colleges' difficulties was the fact that qualified students were difficult to find because of the poor academic preparation provided by many high schools.[24]

The inability of the land grants to translate the practical message of "agricultural improvement" into a successful curriculum gave an edge to educators and scientists who favored a scientific mission. In 1869 Isaac Roberts, an Iowa State College professor of agriculture, moved into a scientific course plan by default. After he had used up his storehouse of practical farm wisdom and could find no texts from which to teach "farming," he decided he had no option but to turn to science as the subject matter for his lectures and began taking his students out on plant-identification expeditions.[25]

But supporters of the movement for scientific education in plant sciences did not have to depend solely on desperate professors to effect the change they desired. They advocated their cause through the scientific and popular press, and they organized their efforts nationwide with such activities as the 1871 Convention of the Friends of Agricultural Education in Chicago, where the concept and role of scientific education were debated.[26] The early supporters of scientific agricultural education looked to European universities as their models, particularly those in Germany that emphasized laboratory work and graduate research.[27] The views of the scientific education movement increasingly dominated the agendas of land-grant colleges.[28]

Several American universities had professors who taught in the plant sciences during the 1860s, but only two men, Asa Gray at Harvard and Daniel Cady Eaton at Yale, earned their livings exclusively in the field of botany. Botany was generally included along with other plant sciences, such as horticulture, as part of a natural history curriculum taught by professors who had to be masters of everything from anthropology to zoology.[29] Another botanist, John Torrey, taught both botany and chemistry at Columbia but made most of his money by assaying precious metals for the U.S. Assay Office in New York.[30] When Charles E. Bessey was a student at Michigan Agricultural College in the late 1860s, he informed the college president, Theophilus C. Abbott, that he intended to enter botany as a profession and was told, "Well, Bessey, I am glad of it, but you'll never get rich."[31] J. C. Arthur claimed that when he was Bessey's student at Iowa State Agricultural College in the early 1870s there were no pure botany professors in America.[32]

But things began to change during the 1870s. Universities and land-grant colleges experienced a rise in interest in scientific botany, and particularly in plant disease studies. Two of the most influential botanists responsible for that growth were William Gilson Farlow of Harvard University and Charles Edwin Bessey of Iowa State College and later the University of Nebraska. Both men were researchers and teachers. They contributed to plant sciences and taught the next generation of American scientists, as Asa Gray had taught them. They certainly were not the only individuals who forged the new science of plant pathology in the United States, but their careers intersected with every important figure in American plant science of the time, and many more around the world. Through teaching, organization, research, and outreach activities, Farlow and Bessey were central figures in the early development of phytopathology as a discipline in the United States.

From an intellectual pulpit at Harvard

Born in Boston, Massachusetts, on December 17, 1844, William Gilson Farlow was quickly recognized in school as a prodigy in several fields. Though an accomplished pianist, Farlow found science very much to his liking. He attended Harvard as an undergraduate and was a member of the Harvard Natural History Society where, it was said, his reputation with his fellow students was such that "his name was always mentioned by them with 'awed respect.'"[33]

After Harvard, Farlow wanted to continue in the study of botany, but his Harvard mentor, the venerable Asa Gray, advised him to secure some other sort of livelihood because botany was an unsure career. So Farlow decided on a medical degree, a common path for scientifically inclined gentlemen of the day. He was so accomplished in his studies at Harvard's medical school that when he came up for final examinations in May 1870, the only question asked of him was "Where do you intend to practice, Mr. Farlow?"[34] But Farlow never intended to practice medicine. Rather, he returned to botany, accepting an appointment as Asa Gray's assistant in July 1870.[35]

Gray once described Farlow in a letter as "very fungously inclined."[36] After two years as an assistant to Gray, Farlow had a focus—lower cryptogams. Next he had to deal with the problem of trying to learn more about this subject in America, where lower plants were not studied rigorously. Something had to be done. As Farlow later observed, it was "ridiculous that one who had only just finished his medical studies and knew nothing about cryptogams beyond what he had read in leisure moments or had picked up in the field should attempt to teach the subject."[37] Therefore, in 1872, following the path of many American scientists before him, Farlow went to Europe for advanced studies that were not to be found in the United States. For the next two years, he did intense work on cryptogams—lichens with J. Müller of Geneva; algae with E. Bornet and

G. Thuret at Antibes. He spent the bulk of his time, however, in Strasbourg with Anton De Bary.[38]

The German university system impressed Farlow. He noted the differences between the American and European systems. Professors in Germany, he noted, generally delivered lectures to small groups of students and encouraged and expected scientific research. "The different proportion of instructors to students," Farlow wrote, "affords the clue as to how the Germans are able to afford to do so much high scientific work. . . . It is for this object that they are appointed professor while in America the professors are only able to do any purely scientific work in intervals between long courses of elementary instruction which is done in Germany in the gymnasium."[39]

If Farlow found De Bary a man "often interested in abstract propositions which no one but a German cares anything about," he nonetheless discovered in him a great scientist and teacher.[40] During Farlow's time in Strasbourg, De Bary was working on his now-famous book, *Comparative Anatomy of the Vegetative Organs of Phanerogams and Ferns*, and left much of the instruction to his advanced students.[41] Under the tutelage at De Bary's laboratory, Farlow received an education on cryptogams available nowhere in America.

An important part of Farlow's education was gaining a familiarity with fungal pathogens. During his two-year stay in Strasbourg, Farlow did systematic, anatomical, and physiological studies on a wide range of pathogens, including the fungi that caused rusts and ergot of cereals, and the fungi that caused potato diseases. His work also involved cultivating his own rusts and inoculating barberry leaves with rust spores.[42] This serious attention to science impressed the young Farlow. He not only found the German botanical laboratory much better furnished than the laboratories he had seen at Harvard but also learned the value of disciplined teaching. De Bary and his assistants were strict taskmasters, rigorously training Farlow in, among other things, the careful microscope techniques that he would later emphasize with his own students.[43] In De Bary's lab, Farlow once wrote, "the only thing worth living for is to study the development" of lower cryptogams.[44]

As Farlow neared the end of his stay in Europe, he began writing to Gray to express his concern that he might not be able to use his education upon his return to America. Gray reassured him that a job was waiting at Harvard. Writing to Farlow at Strasbourg, Gray suggested that his opportunities were strengthened because "your winning card, as it is what is *most wanted* in U.S. is Cryptogamy—especially low Cryptogamy." Gray went on to say that "*fungi* will be the most telling card. There you will have an exhaustless and a *popular* field —in which, well prepared—you can make a mark here."[45] Asa Gray was prescient indeed.

Farlow returned to America in 1874 and was "looked upon very much as one would be who had returned from a journey to Thibet [*sic*] or Central Africa."[46] American scientists and doctors often went to Europe for graduate instruction, but their destination was generally Great Britain, particularly Edinburgh. A number of American chemists had been markedly influenced by the labs of Justus Liebig in Giessen, Germany.[47] But Farlow was among the first students of botany to take the path to Germany that would soon be well worn by Americans.

Upon his return to America, Farlow assumed the post of assistant professor of botany at the Bussey Institution, a branch of Harvard created to be a school of horticulture and agriculture. The institution had been planned in 1835 by the philanthropist Benjamin Bussey, who willed his estate at Roxbury to Harvard for the purpose of creating a school of practical agriculture and horticulture. Complications with his bequest, however, delayed its actual founding until 1870.

The Bussey Institution was established to teach students whose interests were in the applied aspects of botanical studies. Its goal was to train scientists to solve practical agricultural problems by applying scientific principles. The institution predated the first true agricultural experiment station in America. Although it was similar in concept to the land-grant colleges, it was funded by private money and responsible only to the wishes of Harvard officials, so its mission was less influenced by the special interests that advocated agricultural education follow practical, nonscientific lines.[48]

During his tenure at the Bussey Institution from 1874 to 1879, Farlow spent a portion of his time at the Cambridge campus of Harvard, where he taught cryptogamic botany two days a week in a rudimentary lab in Lawrence Hall. Farlow seems to have held the first official academic position devoted to cryptogamic studies in the United States.[49] During the 1874 to 1875 term, he offered a course entitled "Fungi, especially those injurious to vegetation."[50]

In addition to teaching at the Bussey Institution, Farlow transmitted his knowledge of cryptogams to an American scientific audience through a number of noteworthy publications. Some of his papers appeared in the first volume of the *Bulletin of the Bussey Institution*, published by Harvard in 1876, and consisted of scholarly and original pieces on potato rot, sooty mold of citrus and olive trees in California, downy mildew of grape, and black knot of plum and cherry.[51] The next year, Farlow published the results of his examinations on onion smut in the annual report of the Massachusetts State Board of Agriculture.[52] As a follow-up to the work he had done with De Bary, he began to study rusts, the Uredineae, the results of which he published in 1878 in the *Proceedings of the American Academy of Arts and Sciences*.[53] In 1879 Farlow wrote on the parasitic rust fungi of forest trees in the *Botanical Gazette*.[54]

These studies all showed the striking influence of Farlow's European training. His attention to detail and comprehension of the taxonomy and life histories of the pathogenic fungi represented a benchmark in descriptive mycology in the United States. His early papers on potato rot and black knot, for example, were meticulous in their explanation of the germination, growth, and propagation, and so forth, of the parasitic fungi associated with these diseases. Furthermore, to present the most current scientific information available of what "a microscopic study of the *Peronospora* has given us," Farlow not only used his own observations of the potato rot fungus, but he cited from the recent work of De Bary.[55] His paper on black knot was even more original. Although Charles H. Peck, the state botanist of New York and a mycologist himself, had done some important descriptive work on the black knot fungus in 1872, Farlow was the first to give a thorough life history of *Apiosporina morbosa*.[56]

Farlow's precise descriptions of the different hyphal characteristics and modes of reproduction of plant pathogenic fungi showed clearly that his primary concern was classification. He was, after all, trained in descriptive mycology. Still, Farlow realized the economic value of his scientific inquiries. As much as, if not more than, anyone else in America, Farlow knew that controlling plant disease depended on a proper diagnosis of causation and an accurate understanding of the growth and reproduction of the pathogenic microorganisms. He stressed this message in his publications while at the Bussey Institution.

In his 1876 paper on potato rot, Farlow wrote that one of his primary goals was to make agriculturists aware of what science knew and did not know about "the habits of the *Peronospora infestans*." Details about the biology of the fungus, that it required moisture for germination and attacked the potato plants in mid-summer, had offered some clues as to how to arrest the spread of the disease, such as planting tubers in properly drained soils, planting early instead of late, and planting vigorous growers and ripeners. But Farlow cautioned that until the discovery of the sexual, or overwintering, stage of the fungus, actual prevention of potato rot appeared unlikely. This life-history puzzle was a "question which science has still to answer, and the direction in which investigations will have to be made in the future," he explained. Thus, Farlow shrewdly reinforced the growing idea that the farmer needed the scientist.[57]

Farlow's insight into the practical side of his mycological endeavors elicited the hearty approval of some of his colleagues. Fellow mycologist Charles H. Peck, for example, wrote Farlow that "we need more such papers as the one on potato rot for it touches a point that most people can appreciate." Peck urged Farlow to "by all means follow up this line of investigation."[58]

For better or worse, however, the economic aspects of plant disease research failed to dominate Farlow's attention. In fact, he was quite relieved to depart the troubled Bussey Institution in 1879 and move over to Harvard proper to begin his long career as professor of cryptogamic botany. In Farlow's mind, this change was a positive step. His work at the Bussey Institution had often been too practically oriented to satisfy his broad scientific interests. He said he had felt "hampered and interfered with" at Bussey and was happy to move full-time to the Cambridge campus, where he could devote himself to mycology.[59]

But Farlow had made quite a mark on the American botanical scene during his time at Bussey. In 1879 Samuel W. Johnson of the Connecticut Agricultural Experiment Station called Farlow "the only American who has thoroughly studied microscopic fungi."[60] As one of the early leading practitioners of mycology, Farlow had made the Bussey Institution and Harvard an important center for cryptogamic studies. Moreover, Farlow instilled respect in his students for the type of accurate and meticulous study that he had learned in De Bary's laboratory.[61] In turn, his students had their own impact on the development of a science in plant pathology in America.

Much of the fundamental work in mycology that Farlow started at the Bussey Institution was carried forward by his students in plant pathology. Byron D. Halsted, Farlow's first Ph.D. student, continued his mentor's work on black knot. He would later go on to be an important contributor to the study of plant disease at Iowa State University and at the New Jersey Agricultural Experiment

Station at Rutgers.[62] Halsted would later say that he was "proud of having been Dr. Farlow's first pupil in plant disease."[63] Farlow's onion smut research was taken up by Roland Thaxter, Farlow's close friend and successor at Harvard, who would later do definitive work on the cause and prevention of this disease during a brief but illustrious career at the Connecticut Agricultural Experiment Station. Farlow's students also continued important studies on the heteroecism of rusts and continued to identify the alternate hosts of the rust fungi.[64]

After his years at the Bussey Institution, Farlow was not as directly involved in economic plant disease problems. He was not to be a moving force in the continued emergence of plant pathology either in the areas of experimentation or publishing. He did, however, train a number of students, such as Thaxter and Halsted, who made very important contributions to the development of the science. Furthermore, his role as a pioneer in American science, in general, cannot be underestimated. During his long career he served as president of the National Academy of Sciences, the American Association for the Advancement of Science, and the Botanical Society of America. William Farlow used his notable intellect and position at Harvard to bring the study of cryptogams and their role in plant disease into the spotlight of American academe. He also influenced the general development of science at a pivotal time in the United States.[65]

PIONEERING BOTANICAL EDUCATION
AT LAND-GRANT INSTITUTIONS

While Farlow was laboring to establish plant pathology as a science in the eastern United States, Charles Bessey was exploring some of the same paths in the Midwest. Although he never had an academic platform to match the one that Harvard provided Farlow, Bessey was a leader in popularizing the so-called new botany. He was a key figure in establishing plant pathology as a reputable field of scientific endeavor. His teaching skills, administrative brilliance, and prolific writing propelled him and the institutions he served to the forefront of the botanical sciences and the study of plant disease.

Charles Edwin Bessey was born on May 21, 1845, in Wayne County, Ohio. His father, Adnah Bessey, was a teacher, and Bessey studied under him as a youth. Acquiring his own certificate to teach when he was only seventeen years old, he entered Michigan Agricultural College in July 1866, and, like many teachers of the day, he paid his way through college by teaching.[66]

Michigan Agricultural College boasted several notable botanists around the time Bessey took his training. Chief among them was Albert N. Prentiss, who later went on to teach at Cornell.[67] Bessey also developed a relationship with the college president, Theophilus Abbott.[68] Prentiss and Abbott were instrumental in turning Bessey to a career in botany. He graduated with a bachelor of science in November 1869.[69]

The struggling land-grant colleges were having trouble finding qualified professors. Through his connections with Prentiss and Abbot, Bessey secured an assistantship in horticulture at Michigan Agricultural College immediately after graduation and was placed in charge of the college greenhouse. In December 1869, however, when he had been in this position barely a month, Bessey received an offer from Iowa State College of Agriculture at Ames to teach botany and horticulture. He accepted and became a professor of natural history.[70]

Bessey's early years at Iowa State were instrumental in developing his career in plant science. His first botany class, with an uncommonly large number of forty-three sophomores, began within a month of his arrival in Ames. Like every other botany instructor in the country, Bessey taught from Gray's *Lessons in Botany*, but it was soon clear that Bessey was not just another botany teacher.[71] As a professor of natural history, Bessey was responsible for zoology at Iowa State, and his connection to animal life stimulated his rethinking of botanical science in light of the many advances occurring at that time in biology.[72] Furthermore, Bessey understood the economic potential of botany, particularly in the area of plant disease. By the 1871–72 school year, he devoted several weeks of his Cryptogamic Botany course for juniors to the study of "rusts, smuts, moulds, and other parasitic forms . . . and means for preventing their ravages."[73]

In 1872 Bessey's work at Ames earned him a master of science degree from Michigan Agricultural College, and he was promoted to the rank of full professor at Iowa State College. Then in August, he was elected to the American Association for the Advancement of Science at the annual meeting in Dubuque, Iowa. The Dubuque convocation offered Bessey the opportunity to meet many noted botanists of the day, of whom the most prominent was, of course, Asa Gray. Bessey's ensuing relationship with Gray was to alter his scientific career.[74]

In December 1872 Bessey went to Cambridge, Massachusetts, to study with Gray and his colleague, George L. Goodale, a vegetable physiologist. At Harvard, Bessey learned more about exciting new scientific ideas coming from Europe while reinforcing his already apparent predilection for experimental botanical science. He immersed himself in anatomy, morphology, and physiology —cell structure, tissue and tissue systems, and chemical processes.[75] His work with Gray helped propel Bessey beyond "life among species and phyla."[76] He studied at Harvard again in the winter of 1875–76, when he met and worked with William Farlow, newly returned from Europe and already doing research on fungal pathogens.[77]

The influence of his experience at Harvard showed in several areas of Bessey's career over the next few years. It encouraged his interest in plant disease and the practical relevance of economic botany. He published his own first attempt at pathological work with "On Injurious Fungi," printed in the Iowa State report in 1875, and followed this with a second report in 1877. Although these two articles were largely taxonomic and descriptive, Bessey did attempt to provide practical advice on how the pathogen affected the host and which plants were susceptible to the fungi. Most important, these were the first of many "popular articles in which he described in depth troublesome fungi for Iowa farmers."[78]

Bessey had understood the value of laboratory science prior even to meeting Gray. In 1871 he purchased a Tolles Student Microscope, which he considered adequate for "all except the very nicest work."[79] Microscopes marked a "mature" scientific program, regardless of whether there was any real use for them. Indeed, Michigan Agricultural College had had a microscope, but kept it locked in a case in a corner of the botany classroom. Bessey received permission from Prentiss to examine it, and he later claimed that "this was all the practice I had with the instrument while in the college."[80] Bessey later brought the first microscopes to the University of Minnesota in 1881 and to the University of Nebraska in 1884–85.[81]

By integrating laboratory work into his science courses, Bessey's practice was very different from the usual demagogic style, which consisted of lecturing from textbooks and having students recite the information back. Still, Bessey's experimental penchant did not alter instruction immediately at Iowa State. J. C. Arthur claimed that the availability of a microscope in 1871 did little to separate lab work from traditional recitation.[82] Although Bessey's students did do microscopic work on samples prepared by Bessey himself, opportunity for experimentation was obviously limited by resources—after all, the college possessed only one microscope.

But Bessey continued to move in new directions. By 1873 he was already exhibiting his lifelong trait of "taking good-natured pleasure in baiting the scientific conservatives," or what Bessey called "stirring up the animals."[83] Unlike many of his American contemporaries, he could read German botanical writings, and he saw the "one-sidedness of the current botanical instruction."[84] He wrote to William J. Beal that any "college which proposes to keep up with the current must provide botanical and zoological laboratories."[85] Acting on this concept, after his first return from Harvard in 1873, he startled his colleagues by hanging a sign on the door of his makeshift lab that read "Botanical Laboratory."[86] In this room, he had his microscope, a table, jars of material preserved in alcohol, reagents, scalpels, needles, razors, and other light tools that would allow students to prepare mounts.[87] Not surprisingly, in a college that was completely "housed in one building, professors, students, laboratories, library and everything," the creation of a botanical lab was criticized by some as "a mere bit of boasting or buncombe."[88] At first, working in Bessey's lab was a privilege allowed to only a few students, but by 1874 it was a requirement for his botany course, and students were expected to prepare their own microscopic samples.[89] Undaunted by the criticism, Bessey continued to develop his lab, and by 1876 he had seven microscopes.

Founded in 1873, Bessey's Botanical Laboratory was perhaps the first true botanical laboratory for undergraduate students in the United States.[90] Harvard's Botanic Gardens and Cornell's botanical building, dating from 1872, may have predated Bessey's lab by a few months, but, as J. C. Arthur noted, the laboratory at Iowa State was certainly among the earliest to be established.[91] Regardless of the question of firsts, Bessey had brought laboratory instruction to the Midwest. He was at the center of a remarkable current of events that would move botanical instruction from the lecture hall to the laboratory by the 1890s.[92]

Bessey's prestigious scientific connections at Harvard also helped launch his influential writing career. In 1878, on the recommendation of George Goodale at Harvard, Henry Holt and Company approached Bessey about writing a general botany text. Bessey received Asa Gray's blessing for the project, even though he felt compelled to add that "it is intended to be one of my rivals in that field."

As the core text of *Botany for High Schools and Colleges* (1880), Bessey adapted a series of lectures on botany that he had delivered in California in 1875. And he borrowed heavily, if not outright translated, the anatomy and physiology sections of Julius von Sach's influential 1868 publication *Lehrbuch der Botanik*.[93] It was the flagship text of the so-called new botany, and a book that, Bessey said, "marked an epoch in botany."[94]

Bessey's *Botany for High Schools and Colleges* was also to become an important book, for it vastly expanded the attention given in previous texts to cryptogamic botany, physiology, and morphology. Gray, whose classic texts had dealt with those subjects only sparingly, reviewed Bessey's book favorably, but could not resist putting Bessey in his academic place. "It speaks well for the progress of science in the United States," Gray wrote, "when a professor in a college in so new a State as Iowa . . . can produce so creditable a book as this."[95] J. C. Arthur was predictably more effusive in his praise, calling the book "the first rays of the dawn of a new era for American botany."[96]

Bessey's next book, *Essentials of Botany* (1884), which catalogued the changing botany more completely, received even more praise. John M. Coulter's *Botanical Gazette* reported that "no text book ever gave better promise of meeting a long felt want than this."[97] Joseph T. Rothrock of the University of Pennsylvania, an advocate of laboratory education, also praised the book but added a note of caution, warning botanists not to veer too far into nonsystematic studies, "of being content to study cells and cell growth and aggregation, without being able to name the plant on which the observations are made.[98] *Essentials of Botany* went through seven editions and was one of the most widely used botanical texts in America.[99] Both of Bessey's botanical textbooks went on to become standards for years, first supplementing, then replacing, Gray's earlier texts.

By the mid-1880s, from the unlikely seat of a frontier agricultural college, Bessey had earned a stellar reputation in America's elite scientific and educational circles. His presence had helped move Iowa State College into the ranks of respected academic institutions, one of his primary ambitions. To reward him for his publications and the outstanding work he had accomplished for botany and the state, the University of Iowa had granted him a Ph.D in 1879.[100] Iowa State was eager to hold on to its rapidly rising star, but Bessey was already contemplating leaving Ames. In 1876, and again in 1881, he considered job offers in California. In 1881 he was actively recruited by the University of Minnesota.[101]

Events reached an impasse in 1883, with the summary dismissal of Bessey's friend, Iowa State's president A. S. Welch, and an increasingly uncomfortable political atmosphere at the college.[102] The following year, Bessey chose to leave Iowa State College for the posts of professor of botany and dean of the college of agriculture at the University of Nebraska. Bessey had negotiated with the University of Nebraska before. He had earlier refused a position of professor of botany; he wanted a more active role in administrative affairs. When the deanship was added to the package, Bessey was off to Lincoln.

Bessey's years at the University of Nebraska further displayed his merits as a teacher and writer, and also demonstrated his organizational and administrative talents. When he joined the university, the College of Agriculture was so neglected that it was commonly known as Botany Bay.[103] In a relatively short time, Bessey turned the college into one of the best institutions of scientific education in the country. At Nebraska, he would also serve as dean of the College of Literature, Science, and the Arts from 1888 to 1891 and as acting chancellor on three different occasions.[104] If these responsibilities were not enough, he also became the Nebraska state botanist, a position that further elevated his political clout by placing him in contact with the governor's office.[105]

During these productive years, plant disease work was a constant rather than a defining feature of Bessey's career. He began working on plant disease after his experience at Harvard with Gray, Farlow, and Goodale in the early and mid-1870s. Throughout his career, he contributed a number of valuable papers on a variety of different disease problems, both at Iowa State College and the University of Nebraska. His papers dealt with many of the most significant economic plant diseases caused by fungi, such as rusts and smuts, mildews, and blights.[106]

Bessey's plant disease publications were primarily informational. His audience was often the state horticultural or pomological society. Always the teacher, Bessey strove to educate the farmer and the orchardist in plain words. Stressing the ultimate purpose of plant disease control, his publications dealt not with complicated science, but rather with the basics of fungal life. His stated description of one paper was typical: "an inquiry as to their general nature and structure [fungi], in order that we may the more intelligently discuss proposed remedies and methods of prevention."[107] Bessey's control recommendations reflected the best that science had to offer at the time: cultivation practices, fungicides, and selective breeding. In the case of rusts of wheat, he prophesied that disease-resistant cultivars would one day be found.[108]

Bessey was an articulate spokesman for pathology and the economic potential of plant disease research. He did as much as anyone to sell the idea that pathology warranted a home in agricultural botany. His focus, like that of most of his contemporaries, was on diseases caused by fungi. His work never moved seriously in the directions of diseases caused by bacteria, which would arrive on the scientific forefront in the early 1880s, or diseases caused by viruses, which would take their place as significant plant pathogens at a later date. He nonetheless promoted attention to these areas throughout his career.

Like Farlow, Bessey did not devote all of his energy to plant disease research, but his high ideals in education and his professional standing in science elevated the standing of the study of plant disease in American colleges and scientific organizations. His research on plant disease reflected his opinion of the role of science in society, particularly in education. Scientists, he felt, should set the experimental agenda at colleges based on generalized concepts of nature. Agricultural science, Bessey thought, was defined too narrowly, most often by the region in which it was taught. It needed a broader base. He believed that science should have a practical application, but need not be slavish to immediate political or economic forces. Science would thrive only if it had a practical benefit, but it should not have to define that benefit beforehand.[109] Bessey wanted to attract young men to science who were interested in what he called the "culture-value" of the profession, not the "money-getting value."[110]

Furthermore, Bessey was a teacher and a mentor to many students who went on to do great things. His pedagogical method was new in America: He believed that scientific instruction ought to be based on empirical study with mandatory laboratory work, not on recitation from textbooks. Then too, Bessey was interested in the quality and rigor of his curriculum. He had no interest in lowering standards to entice more students into land-grant colleges.[111] Just before his death, Bessey estimated that he had taught over 4,000 students in his long, distinguished career.[112] One of his most notable students, Albert F. Woods, went on to lead the USDA's Division of Vegetable Physiology and Pathology.

All through his career Bessey cemented a network of connections within major scientific organizations. His active participation in the professional life of his discipline elevated not only his own prestige but that of botany and plant disease studies in general. From the time of his induction into the American Association for the Advancement of Science, Bessey rarely missed a meeting and was elected a fellow in 1880. He served as vice president of the AAAS and chairman of Section G (Botany) in 1893, 1894, 1902, and 1907; and he was elected president of the AAAS from 1910 to 1911. Bessey was a charter member of the Botanical Society of America and served as its president in 1895.[113] He was also a member and served as president of many other organizations, including the Microscopical Society and the Society for the Promotion of Agricultural Science. He served on numerous committees, including one chaired by President Eliot of Harvard in 1892 that sought to redefine high school instruction in natural history and botany.[114] Bessey was chosen botanical editor of the *American Naturalist* in 1880 and held the post until 1897, when he assumed a similar job for the journal *Science*.[115] From these posts he continually harangued the botanical community to increase the professionalism of their discipline.[116]

Charles E. Bessey is sometimes referred to as the father of plant pathology in America. This title might overstate Bessey's role, but certainly he was a towering presence in botanical sciences in the late 1800s. Had Bessey been at Harvard or Yale, he would be remembered as one of the most important scientists and educators in American history.[117] Even from his posts on the outskirts of the American prairie, Bessey managed to serve in high office in national organizations numerous times, and in 1906 he was considered seriously for the position of secretary of the Smithsonian Institution.[118] In his long, prosperous career working at agricultural colleges and through professional organizations, Charles Bessey, like William Farlow, had a tremendous impact by advancing plant science as a career and as a discipline. The careers of these two men represented the passing of an age of the amateur naturalist and agriculturist and the advent of a new era dominated by the academic specialist.

CHAPTER 5

Fire Blight and New Realms of Research

In the late 1870s, while Charles Bessey and William Farlow, among others, were bringing the scientific study of plant disease to the colleges and universities of the United States, scientists were beginning to explore entirely new areas of research on plant disease. One of these new areas originated when a study of fire blight of pome fruits led to the discovery that plant disease could be caused by bacteria. The identification of bacteria as plant pathogens was initially disruptive for scientists who had just begun to understand the nature of disease-causing fungi. But on the other hand, the new discovery reinforced the idea that plant disease problems could be understood, and perhaps one day solved, through scientific research.

Up to and beyond the time of the breakthrough with fire blight, scientific work on plant disease centered on fungi. This was entirely reasonable: Most plant maladies with major economic consequences during the nineteenth century, such as rusts and smuts of grains, mildews of grapes, and late blight of potatoes, were caused by fungi. Furthermore, fungi were more visible than bacteria. Fungi on plants often could be seen with the naked eye and the intricacies of their fruiting bodies could be observed with the aid of a magnifying lens or simple microscope. Bacteria, on the other hand, were smaller than could be seen with most nineteenth-century microscopes. And even when bacteria were observed, they were sometimes mistaken for microscopic fungi.

Some plant disease workers suspected that the origin of fire blight was a fungus. They founded this suspicion not on research, but rather on their general belief in fungal theory. On the other hand, there were still some who cited traditional causal theories such as insects and frozen sap. This lack of consensus generated a great deal of interest in the disease among American scientists.

Fire blight was a worthy subject for American researchers. It was causing widespread devastation, and, furthermore, it was a uniquely North American disease. The resulting level of research activity meant that it was no accident that a scientist in the United States discovered a bacterium as a plant pathogen. Europe was not burdened by a plant disease of bacterial origin of the same destructive magnitude as fire blight. The farmers and gardeners of Europe were troubled by many of the same diseases as those in the United States—rusts, smuts, mildews, and late blight—but those diseases were of fungal origin.

The groundbreaking research on fire blight of pome fruits highlighted the scientific potential of the new agricultural institutions in the United States, particularly the land-grant colleges and state agricultural experiment stations. The pioneering experiments of Thomas J. Burrill, a professor at Illinois Industrial University, and Joseph Charles Arthur, a botanist at the New York State Experiment Station at Geneva, combined to open an entirely new discipline in plant pathology, the study of bacterial plant disease.

LANDMARK STUDIES

Like William Farlow at Harvard and Charles Bessey at Iowa Agricultural College and the University of Nebraska, Thomas J. Burrill was an advocate and practitioner of scientific education, particularly botanical education, during his many years at the University of Illinois, both as a professor and an administrator. But Burrill's more significant contribution was extending the scope of the science of plant pathology in the United States into an entirely new realm. In the 1870s, at a time when the role of bacteria as pathogens of animals was just beginning to be understood, Burrill's landmark studies on fire blight of pears and apples indicated that bacteria could be pathogens of plants. In many ways, he helped create the conditions that would allow Americans to achieve world prominence in phytopathology by the early years of the twentieth century.

Thomas Jonathan Burrill was born in Pittsfield, Massachusetts, in April 1839, but his family moved to Illinois when he was a young boy. In 1858, Burrill began high school, earning the money he needed to continue his schooling by teaching.[1] After graduation, he entered Illinois State Normal University. At the time, Illinois State Normal was the principle institution of higher learning in Illinois, and Burrill's professors in the natural sciences were very qualified. They included Benjamin D. Walsh, state entomologist; George W. Vasey, president of the Illinois State Natural History Society and later a USDA botanist; Jonathan Baldwin Turner, one of the leaders in the national land-grant college movement; and Joseph A. Sewall, botanist and curator of the Museum of the State Natural History Society, who had been a student of Asa Gray and Louis Agassiz at Harvard. Sewell, more than anyone else, stimulated Burrill's love of natural history.

While a student at Illinois State Normal, Burrill cultivated his interest in botany and horticulture, particularly fruit trees, by working during school breaks at the Phoenix Nursery in Bloomington. After graduating with a degree in natural history in 1865, Burrill became superintendent of public schools in Urbana, Illinois, still maintaining his interest in botany.[2] In 1867, he accompanied the first expedition of exploration led by Major John W. Powell, a fellow Illinoisan, to the Rocky Mountains of Colorado. Then in April 1868, Burrill secured a post at the new Illinois Industrial University (renamed the University of Illinois in 1886), the state's land-grant institution. Hired to teach algebra, within seven months he was professor of natural history and botany, a job more suited to his training and interests. Six months later he was professor of botany and horticulture.[3]

At Illinois Industrial University, Burrill was one of the first botanists in America to introduce information about plant disease into his courses. In a presentation before the university, he gave the following reason for his course content: "A new field of labor is also opening before the student of botany, that of the vegetable diseases. Perhaps nothing pertaining to plants is so little understood; but the importance of the study and the increased facilities of late years for microscopic observations, will undoubtedly call more attention to the subject."[4] And since a variety of plant diseases were "believed to be the result of fungus plants," he found a place for cryptogamic botany in his classes.[5]

Burrill taught as many as eight courses a year, including botany, biology, microscopy, and physiology. In 1869 he launched his students into the following curriculum: one term of structural and physiological botany, then two terms of systematic botany, including information on the microscope, cryptogamic botany, fungi, and vegetable diseases.[6] The use of the microscope suggests that Burrill's laboratory instruction was similar to Bessey's in Iowa. For a text, Burrill used M. C. Cooke's mycology-oriented *Rust, Smut, Mildew, & Mould.*[7] He eventually also included Berkeley's *Introduction to Cryptogamic Botany.*[8] Notably, both texts were European in origin. Burrill had recently begun to study German and was influenced by the course of investigations in Germany.

By 1874 Burrill was teaching plant pathology formally. His courses on general horticulture, pomology and forestry, floriculture, botany, vegetable physiology, and microscopy all included sections on injurious fungi.[9] He quickly made a reputation for himself as one of the leaders of a very small group of botanists in America who had "expertise" in plant disease. In 1872, the story (though possibly apocryphal) goes, he was traveling on a train with Asa Gray and a group of ten or twelve botanists, heading for Dubuque, Iowa. They were going to the same AAAS meeting where Charles Bessey was to meet Asa Gray for the first time. When the train stopped for a wreck to be cleared on the track ahead, Gray entered a nearby cornfield to gather some smutted corn. Returning to the train, Gray asked the gathered botanists, "Who can tell me what this is?" Aware that one of his colleagues knew all too well, he added, "Burrill, keep your mouth shut."[10]

Based on a survey he had done of plant diseases caused by parasitic fungi, including rusts, smuts, and powdery mildews, the first of Burrill's publications on plant disease, an 1874 monograph on powdery mildews for *North American Pyrenomycetes*, was edited by J. B. Ellis and B. M. Everhart. That same year, he published "Aggressive Parasitism of Fungi."[11]

Burrill's research record on fungal pathogens was solid, but it was his research on the bacterial origins of fire blight that led him to become one of the most celebrated botanical figures of the nineteenth century. This work extended the frontiers of plant disease studies at a critical time in the development of the young science. His groundbreaking microbiological research, coupled with his influence on an entire generation of young botanists, certainly make him a candidate for the title "father of American plant pathology."

Fire blight had annoyed and confounded fruit growers in the United States since the latter part of the eighteenth century. By the 1840s, the disease had spread with fruit culture from the eastern United States into the Ohio Valley. Public interest was high, and fire blight received more coverage in the agricultural press than any other horticultural problem during the period.[12] One of Burrill's professors, Jonathan B. Turner of Illinois State Normal, took an early research interest in fire blight. Turner's investigations "found nothing, save that the . . . *apparent* origin of the disease . . . was the work of fungus, or extremely minute animalculae, invisible with a common microscope."[13] Turner's suspicion that fungus was the cause of the disease was common enough for his day, but his mention of "extremely minute animalculae" would prove to be prophetic.

Although Burrill began to study fire blight as early as 1873, he did not present his first major paper on the disease until September 1876. In this paper, presented to the board of trustees of Illinois Industrial University, Burrill wrote that his investigations had uncovered "oscillating corpuscles" in the diseased tissue when viewed under a microscope.[14] He did not, however, identify these microscopic corpuscles as bacteria.

In a second paper on fire blight, presented to the Illinois State Horticultural Society in December 1877, Burrill again reported that the "cambium of the blighted branch . . . is filled with very minute moving particles, very similar to those known as Spermatia in fungi."[15] But Burrill still did not know exactly what these particles were. While he thought that the "theory of the fungus origin of the fire-blight of pear . . . is well founded," he still had no proof that any specific fungus caused the death of pear trees.[16]

Presented before the Illinois State Horticultural Society in December 1878, Burrill's third communication on fire blight was a very significant event in the history of plant disease studies. In it, Burrill tentatively identified the "oscillating corpuscles" as "moving atoms known in a general way as *bacteria.*" Burrill had introduced these particles with a knife into the bark of a healthy pear tree and observed that "in many cases" fire blight followed. This allowed him to suggest tentatively that bacteria might be the cause of fire blight. He wrote, "Does it not seem plausible that they [the particles] cause the subsequently-apparent change? It does to me, but this is the extent of my own faith; we should not say the conclusion is reached and the cause of the difficulty definitely ascertained. So far as I know the idea is an entirely new one—that *bacteria* cause disease in plants—though abundantly proved in the case of animals."[17]

In 1880 there was a severe outbreak of fire blight. In the summer of that year Burrill began an important series of experiments to explore further whether a bacterium actually caused fire blight or whether bacteria were just a by-product of the disease. He inoculated 69 trees—36 pear, 29 apple, and 4 quince.[18] Burrill's inoculum was either the exudate or sections of bark from infected shoots. He collected the exudate in the morning from new lesions only, diluted it with water, and examined it microscopically to make sure it was free from fungi. To determine whether transmission was external or internal, he applied the "virus," as he referred to the exudate, only externally, to the leaves of the plants he was inoculating. (At that time, the term *virus* was used to refer to a poisonous chemical or substance and not to a specific type of pathogen.) His results showed that external application of the blight exudate never resulted in disease, which indicated clearly that the disease could be produced only by introduction of the "virus" internally into the plant.[19] The infection experiments were successful enough for Burrill to state that the "introduction of the virus introduced the cause of the disease, and the potency of the virus was quite positively due to the living bacteria."[20]

Burrill also showed definitively that the blight of pear, apple, and quince were all interrelated.[21] In his work, he made tentative connections of the fire blight bacteria with Pasteur's *vibrion butyrique*. He referred to Pasteur's concept that this microscopic organism might be a common element in many diseases, allowing that if this were so, "it would not necessarily . . . invalidate its agency in producing this disease of the pear; but it might render less hopeful the discovery of remedial treatment."[22]

Burrill presented the results of his inoculation studies in three different papers between August and December 1880. First, he presented "Anthrax of Fruit Trees; or The So-Called Fire Blight of Pear, and Twig Blight of Apple, Trees" to the AAAS in Boston.[23] Also in August, he reported his findings to Regent Theodore Draper of Illinois Industrial College.[24] In December he presented a paper essentially the same as the AAAS paper to the Illinois State Horticultural Society.[25] Charles Bessey, as the botanical editor of the *American Naturalist*, reported on Burrill's work with interest, suggesting that it had not reached a wider public because the AAAS's proceedings took nearly a year to reach publication.[26]

Although Burrill described the bacterium in 1881, it was not named for two years.[27] Burrill claimed that "other interests . . . so engaged my attention and time, that the work was not so extensively prosecuted as I heartily wished it had been, after becoming gradually convinced of the possible complete demonstration of the perplexing problem."[28] Initially, in 1883, the bacterium was named *Micrococcus amylovorus*, but it was later renamed *Erwinia amylovora*, after Erwin F. Smith, a pioneer researcher on bacterial diseases of plants, who would contribute much written support to Burrill's reputation as founder of American bacteriology.

An aspect of fire blight that continued to be debated was the point of infection. Burrill had conjectured that the microorganism entered the tree through wounds or punctures.[29] He did not consider flower infection as a possibility until it was suggested to him by George P. Peffer, a grower from Wisconsin, who wrote on blight and apple tree blossoms for the Wisconsin State Horticultural Society.[30] Even then, the point of infection remained debatable. Peffer himself was dubious of bacteria as the cause and preserved the notion that, whatever the cause, fire blight first required an injured or wounded tree to take hold.[31]

Anytime a scientist is credited with a significant first, it sparks a debate over whether the scientist deserves that credit. Scientists rarely, if ever, extract theories from thin air, and Burrill was no exception. His work benefited from a long history of interest in causal theories for fire blight, and even from a few previous examples of research on bacterial diseases of plants outside of the United States.

Before Burrill's work, there were many theories about the cause of fire blight, variously blaming frozen sap, insects, lightning, and fungi. The frozen sap theory held prominence in the 1840s and continued to have supporters into the 1870s.[32] Insects, often the *Xyleborus* beetle, received the blame in the first half of the nineteenth century and into the 1870s.[33] By the 1860s and 1870s, however, the more often blamed culprit was fungus. In 1863 J. H. Salisbury and C. B. Salisbury of Ohio blamed *Sphaerotheca pyrus*, a fungus that causes a powdery mildew, for the blight.[34] They claimed that the same fungus was responsible not just for the blight in pear, but also in apple and quince. At the 1867 meeting of the American Pomological Society in St. Louis, Thomas Meehan of Pennsylvania also named a fungus as the causal agent.[35]

The fungal theory gained additional credence by several experiments that demonstrated the transmissibility of the blight from infected to healthy trees. Reuben Ragan did inoculation tests with pear blight in 1845, and four years later Herman Wendell of New York reported that he had "discovered blight to be either contagious or epidemic."[36] In 1868 E. S. Hull, the Illinois state horticulturist, transmitted the infection between trees. He then stated that "fire-blight . . . is induced by an extremely minute fungus, seen only by the aid of a powerful microscope."[37] It is likely that Hull was actually observing bacteria of the species *Erwinia amylovora* but could not identify what he was seeing. Burrill was aware of Hull's experiments when he began his work on fire blight.[38] He originally supported the fungal theory himself, later blaming his misidentification of the bacteria as spermatia of fungal bodies on "the want at the time of a proper glass."[39]

From the work of others, Burrill knew that fire blight was transmissible. He also knew that the blight of pear, apple, and quince probably had the same cause and that as succulence increased, so did susceptibility.[40] Furthermore, he knew from the work of researchers such as E. S. Hull that the diseased tissue was host to minute bodies. It was at this point, however, that the accepted wisdom went astray. Since most observers at the time could not clearly differentiate bacteria from cryptogamic fungi, many scientists conjectured that the cause of the disease was a fungus.

Burrill was able to absorb the accepted research and move beyond it because of extraordinary developments occurring in biology. During the 1850s, in France, Louis Pasteur was working across the range of chemistry, medicine, and biology to demonstrate the role of microorganisms in fermentation. His work finally persuaded most biologists that the theory of spontaneous generation was false. In 1876, in Germany, Robert Koch and Ferdinand Julius Cohn published experiments showing for the first time that anthrax in cattle was caused by a rod-shaped bacterium.[41] In the late 1870s H. J. Detmers of the USDA conducted a search for a bacterial cause of hog cholera. Burrill witnessed many of Detmers's experiments himself.[42]

Furthermore, experimental techniques were improving rapidly. In 1876 Koch developed a poured-plate method for isolation and purification of bacteria. The next year, Fannie Hesse, the wife of Walther Hesse, one of Koch's coworkers, suggested agar as the solidifying component of culture medium instead of gelatin.[43]

Nevertheless, when Burrill began to study fire blight in the late 1870s, bacteriological work was still in its infancy and was nearly nonexistent in the United States, particularly in terms of plant disease. But there was work occurring overseas, and Burrill was aware of some of it. He knew, for example, that Eilhard Mitscherlich, a German chemist, had reported to the Imperial Academy of Science in Berlin as early as 1850 that he had observed "active liquid" degrading potato cell walls, which he blamed on *vibrios*, one of the six named genera of bacteria.[44] It is probable, however, that Mitscherlich's results were so vague that he was not considered to have accomplished certain requisite connections between disease and bacteria needed to deserve credit for the discovery. In his historical review of bacteriology in *Bacteria in Relation to Plant Disease*, E. F. Smith did not even mention Mitscherlich.[45] In his own results, Burrill wrote that the idea that bacteria could cause disease in plants was new "so far as I know."[46]

While Burrill was conducting his research, however, other investigators overseas were at least noting the presence of bacteria in conjunction with plant disease. In a study of sugarcane in Bahia, Brazil, in 1869, for example, F. M. Dränert commented that diseased cane produced "yellow material," which "appears as a micrococcus."[47] In 1866 M. S. Woronin in Russia observed bacteria in root nodules of lupine, and in 1879 E. Prillieux in France found bacteria in the cavities of seeds in "rose-red" disease of wheat. In the early 1880s, O. Comes noted bacteria associated with several plant diseases. With the possible exception of Dränert's, most of this work, however, was inferior to Burrill's because no inoculations were done.[48]

Although Burrill carried out extensive inoculation tests in his fire blight experiments, he did not use pure cultures. H. H. Thornberry claims that pure-culture techniques had not been invented when Burrill did the bulk of his fire blight work.[49] K. F. Baker maintains, however, that William Farlow had used pure-culture techniques at Harvard prior to Burrill's work on fire blight and, furthermore, that Burrill was aware of the methodology.[50] Then, too, Burrill had little to say about cures or preventives for fire blight. Indeed, there was little he could recommend beyond some traditional methods long used by growers for a wide variety of diseases. He wrote, "No claim is here made to the discovery of prophylactic or remedial methods which shall change fruit-growing from failure to success. . . . While something is to be advanced in the way of improvement in the means usually adopted to save our pear trees, the information already given may be of more service in saving us from crude hypotheses and useless labor, fighting in the dark an unknown foe."[51]

Burrill did, however, dismiss the popular notion that mutilating the trees, such as with root pruning, as Hull had advised in the 1860s, could combat fire blight—other than to remove the disease by killing the tree. Furthermore, he warned that the careless pruner's knife could serve as a means of creating a wound for possible infection or transferring the infection from tree to tree. He recommended washes of lye, linseed oil, lime and sulfur solutions, and even carbolic acid—not because he had experimental proof that they could prevent or cure fire blight, but because they "do no injury, and may be beneficial."[52]

Burrill's experimental work on the cause of fire blight was the pinnacle of his career in scientific research. He did little follow-up work, interpreting as approval the absence of dissent about his theories. "There seems to have been no attempt," Burrill wrote in 1883, "to disprove the conclusions as published in . . .

1880, nor has there been any evidence . . . that anything besides bacteria does this deadly work in the tissues of our pear-trees."[53]

But his research on fire blight was not the end of Burrill's bacteriological work. He attributed a variety of plant diseases to bacterial origins, including diseases of Lombardy poplar, butternut, and maple, and he even suggested erroneously that bacteria were responsible for the "poison" of poison ivy.[54] In 1885 he investigated bacterial origins of "corn blight."[55] In 1886 and 1887 he worked on broomcorn and sorghum diseases. For this work, he did pure-culture experiments.[56] He was followed in this work by his assistant Merton B. Waite.[57] Although he once surmised that a bacterium was the cause of peach yellows, a pernicious disease that had defied explanation for years, in 1883 he recanted his earlier statements, insisting that "it is impossible for me to have an opinion as to whether bacteria have, or do not have, anything to do with this disease called yellows."[58]

Burrill did not limit himself to bacterial studies. He continued to do research on fungal diseases, conducting and supervising research on the etiology and control of apple bitter rot as well as work on apple scab, twig blight, black rot of grapes, potato rot, potato scab, and blackberry and raspberry diseases.[59] Additionally, he studied diseases of cherry and lettuce, produced two papers on fungicides, and in 1884 he wrote the notable and authoritative book *Parasitic Fungi of Illinois—Uredineae.*[60]

Burrill spent his entire career at the University of Illinois. By the 1890s, he was engaged increasingly with administrative duties, which left less time for research and writing. At various times, he held the posts of vice president of the university, dean of the college of natural sciences, dean of the graduate college, and interim regent of the university. He also served as acting regent four times, and as the president of the board of directors of the Illinois Agricultural Experiment Station, as well as the station botanist.[61]

Burrill's teaching influenced a generation of students throughout American academe and science, particularly in the field of microbiology. A bacteriology laboratory established at the University of Illinois in 1892 "offered the best facilities extant at the time in any American University (exclusive of a few medical colleges)."[62] Burrill's students and experiment station assistants included many who would have significant influence on botany and plant pathology, including George P. Clinton of the Connecticut Agricultural Experiment Station, Benjamin M. Duggar of the University of Wisconsin, Arthur B. Seymour of Harvard, and Merton B. Waite of the USDA.[63]

Over a period of two to three decades, Burrill also influenced numerous young plant pathologists and botanists who were not his students or immediate colleagues. He was a voluminous correspondent and was inundated by letters from many young scientists who asked for advice and wisdom. One of these was William Trelease of the University of Wisconsin, who later became head of the Shaw Botanical Garden in St. Louis, an institution that had been founded through the efforts of George Engelmann and Asa Gray. Trelease corresponded extensively with Burrill about fungi, benefiting from the elder's expertise and willingness to share. Eventually he filled Burrill's place at the University of Illinois.[64]

But Burrill had the greatest influence on the field of bacteriology. He wrote two important general works on the subject: *Bacteria and Their Effects* (1882) and *Bacteria: An Account of Their Nature and Effects, Together With a Systematic Description of the Species* (1882).[65] Erwin F. Smith, the great advocate of American phytobacteriology of the late nineteenth and early twentieth centuries, said after Burrill's death, "Others, elsewhere, in these same early days made similar announcements, but were less fortunate or less painstaking, since no one in later days has been able to confirm their findings, whereas Burrill's discoveries have been confirmed a hundred times, and relate to one of our most serious orchard diseases, known for a hundred years."[66]

Thomas J. Burrill did not single-handedly discover bacterial pathogenicity in plants. He was, like all scientists, part of a process, and the process would continue after him. Conclusive evidence that bacteria caused diseases of plants would have to be provided by other scientists, chief among them Joseph Charles Arthur at the state experiment station in Geneva, New York.

PROOF THAT BACTERIA CAUSE PLANT DISEASE

State agricultural experiment stations, which were to host important advances in plant pathology in the last quarter of the nineteenth century, first appeared in the 1870s. The movement to found them was centered in Connecticut with a group of Yale scientists, beginning with John Pitkin Norton in mid-century and followed afterward by Samuel W. Johnson and Wilbur O. Atwater.

Agricultural experiment stations were a European idea. After many fruitless years of trying to promote them in America, Johnson and Atwater finally achieved success in 1875, when they secured some state funds to create the Connecticut Agricultural Experiment Station.[67] The mechanisms of funding established a precedent of government support for research in agricultural science, which in this case was largely agricultural chemistry.

According to Johnson, a successful experiment station required three founding principles: a state charter and subsidy; a sole mission of scientific research; and a promise of self-government, that is, control of the research agenda by the station staff.[68] Advocates of experiment stations fought over these principles and only partially achieved them in the other experiment stations that were created around the country in the decade after the establishment of the Connecticut station.

Like the land-grant colleges to which they were often attached, experiment stations foundered. They frequently suffered from small budgets, poorly defined missions, and a combination of apathy and misunderstanding by their agricultural constituencies, which included both farmers and legislators. Some stations without state funding had difficulty sustaining operations. New York's first station, founded at Cornell in 1879, refused state funds out of fear of state control, and had to shut down operations in 1885.[69]

A chronic bone of contention was control of the experimental agenda.[70] Station scientists felt strongly that stations should be places of unencumbered research. The general public, on the other hand, often did not understand the merits of pure scientific research and hesitated to give scientists such control. If they wanted experiment stations at all, they wanted them to perform practical functions for improving agriculture, such as analyzing soils, assaying fertilizers, and identifying weeds.[71]

In some ways, the station advocates had brought this predicament on themselves when they stressed the potential economic benefits of the stations in order to sell them. Samuel Johnson, for example, had promoted the experiment station idea to the Connecticut legislature by emphasizing the role that a station could play in battling rampant fertilizer fraud.[72] The result of the sales pitch, however, was that many station directors found their research schedules overburdened by requests from local farmers for what the staff scientists considered mundane and nonproductive analytical work.

Nevertheless, some state agricultural experiment stations performed excellent scientific research, of which none was more important than the fire blight work of Joseph Charles Arthur of the New York State Agricultural Experiment Station at Geneva, New York. In 1884, Arthur, a young botanist fresh from college, met Edward L. Sturtevant, director of the Geneva station, at the annual meeting of the AAAS and was offered the position as staff botanist.[73] During the next three years at the New York station, Arthur made an international name for himself in the nascent field of plant bacteriology. He was credited by his contemporaries with doing the first institutional, systematic study of plant disease in the United States.[74]

Joseph Charles Arthur was born in Lowville, New York, on January 11, 1850.[75] Arthur developed an interest in plants during his childhood on the Iowa prairies, but he received little encouragement from his parents. Favoring a more lucrative future for their son, they argued that the pursuit of "botany could not . . . bring in enough to keep a cat alive."[76] But Arthur was not to be dissuaded. He entered Iowa Agricultural College in 1869 and was part of the first graduating class in 1872.[77] This was the beginning of a long line of firsts for Arthur. In 1878 he presented Iowa Agricultural College's first master's thesis (not on plant disease, but "On the Structure of *Echinocystis lobata*").[78] In 1886 he received Cornell's first doctorate awarded in a scientific field, which was also the first doctorate awarded in the United States for research on plant disease, for his work on fire blight. He was the first botanist at the state experiment station in Geneva, New York, and, in fact, the first botanist at any state experiment station. Then, later, he was the first botanist at Purdue University. Significantly, Arthur was also the first scientist to be known as a "phyto-pathologist."[79]

At Iowa Agricultural College, Arthur was the protégé of Charles Bessey. When he entered the college in 1869, Bessey had not yet joined the faculty, and botany was not a part of the curriculum. After Bessey arrived in December 1869 and began instruction in botany, particularly as it related to vegetable physiology and plant disease, he and Arthur entered into a professional relationship that would last throughout their lives.[80] Arthur continued to correspond with Bessey throughout his career, just as Bessey corresponded with his mentor, A. N. Prentiss.

Arthur's connection with Bessey served him well. In 1878, after his graduation from Iowa Agricultural College with a master's degree, Arthur had difficulty securing a suitable appointment. On his way to a post with a New Jersey greenhouse, he stopped in Baltimore to look around the campus of the newly opened

Johns Hopkins University, a school openly modeled on the German university system, and was unexpectedly ushered in to see President Daniel Gilman. Gilman knew Arthur's name and reputation through Charles Bessey, who had delivered a series of lectures in California at the invitation of Gilman, then president of the University of California. Gilman offered Arthur a fellowship at Johns Hopkins, enticing him with the lure that William Farlow was coming from Harvard to deliver a series of lectures. Arthur accepted promptly and set about retrieving his luggage from New Jersey.[81] Arthur was so impressed by Farlow's lectures at Johns Hopkins that, after his fellowship ended, he went to Harvard for a summer course. Next, he became an instructor of botany at the University of Wisconsin from 1879 to 1881, and then briefly at the University of Minnesota.[82]

When Arthur took the position at the Geneva station in 1884, he arrived to find an agricultural institution with limited scientific resources. He had access to Sturtevant's microscope, a baroque glass "covered with all sorts of brass instruments, scapels [*sic*], needles, etc. with ivory handles," but otherwise the station was ill-equipped.[83] Arthur's worktable was a board placed over the radiator by a window. He had a few test tubes, cotton, and simple apparatus, but the facilities were crude by the standards he had seen at Harvard, Johns Hopkins, and Iowa Agricultural College. Arthur complained that Frank Lamson-Scribner, while an assistant botanist with the USDA, came to Geneva "with the intention of spending a few days there but he did not find much and so did not stay. There was nothing to stay for."[84]

What Arthur did have at Geneva was enormous control over research decisions. "I find I am free to plan and execute any investigation I see fit," Arthur confided to Charles Bessey in 1884. This control gave him an advantage over many of his colleagues around the country. And although he was not sure exactly what he wanted to do, he had "concluded to look into the life and habits of some of the plants that cause diseases of cultivated crops." Arthur asked Bessey to suggest fields of profitable research to take advantage of the fact that he had "the station at my command."[85]

In the spring of 1884 Arthur decided to focus on fire blight, setting himself the task of showing "that Prof. Burrill's work was substantial and to proclaim in favor of his hypothesis of the parasitic nature of bacteria in plants."[86] Deciding that the first order of business was to follow up Burrill's disease-transmission experiments, he wrote, "The experiments of Professor Burrill showed that the disease alluded to was invariably accompanied by a specific form of bacteria. . . . Although from these and subsequent investigations the theory has been quite generally accepted that the bacteria are the cause of the disease, no rigid proof of it has yet been brought forward."[87] Arthur's subsequent experiments demonstrated quickly and irrefutably that infection occurred from an infusion. He was able to create the infection both by placing diseased wood in water to create a bacterial solution and by transferring the bacteria-containing exudate straight from a blighted twig to a healthy shoot.[88]

The next year, in the spring of 1885, Arthur embarked on a series of "pure culture" experiments that would prove definitively the role of bacteria in the cause and transmission of fire blight. Since Robert Koch's famous paper in 1881 on methods for isolation and characterization of pathogens, pure-culture technique was quickly becoming an accepted practice in disease studies.[89] If Arthur hoped to contribute definitive proof, he would have to follow this technique. At the suggestion of Cornell's A. N. Prentiss, Arthur began to separate the bacteria from "the juices which accompanied them."[90] He created what he referred to as "artificial" or "fractional" cultures through two series of six serial transfers with sterilized infusion of cornmeal and "porous earthenware vessels." In July he did inoculation tests with both his cultured bacteria and the bacterium-free filtrate on unripe Bartlett pears. The pears injected with bacteria became blighted; those injected with the filtrate did not. As Arthur reported, "The evidence is thoroughly satisfactory and conclusive. The bacteria accompanying the disease of trees known as pear blight when fully isolated will produce the disease, while the juices in which they live will not. They are therefore the direct cause of the disease."[91] At the 1885 AAAS meeting in Ann Arbor, Michigan, Arthur presented his results as absolute proof of the bacterial origin of fire blight and a confirmation of Burrill's earlier work.[92]

Next, Arthur extended his experiments by growing bacteria in and on a variety of media. He found they grew quite vigorously on cornmeal, hay, starch, and various vegetable substances. But his attempt to create the blight by transferring other bacteria from rotting plants failed. When he mixed *Micrococcus amylovorus* with other bacteria and injected them into healthy trees, the resulting exudates from blighted tissue contained only the one blight-causing bacterium. Arthur's experiments further confirmed that one type of micrococcus was responsible for fire blight. He also confirmed that the blight organisms in pear, apple, and quince were of the same species.[93]

Arthur described *Micrococcus amylovorus* in more detail than Burrill, and he studied its life history. The bacterium was globular, oval, 1 by 1.5 microns or 0.00004 by 0.00006 inches. It was smaller than most bacteria, such as those in common putrefaction, but was not as small as *Streptococcus diphtheriticus*, the cause of diphtheria.[94]

Two of the major mysteries of fire blight were how it progressed inside a tree, and how it was transferred between trees. Arthur had some success in tracing its progress inside a tree. He saw that the bacteria attacked the starch of cells first, then the cellulose of cell walls, and finally liquefied the whole tissue. This was best observed in tender, succulent tissue. He theorized that the "progress . . . is doubt-less largely due to simple displacement as multiplication takes place, although aided by the limited activity of the organism and the movement of the sap."[95]

But Arthur was unable to determine with certainty how the bacteria moved between trees in nature. He smeared the "virus" over the outside of branches, leaves, and fruit. He tied diseased branches with healthy ones. He arranged an "apparatus . . . to draw air across diseased branches upon healthy ones." He even watered potted pear trees with water that was white with living, blight-causing bacteria. None of these procedures succeeded in transmitting the pathogen and causing blight.[96]

Nevertheless, with more time and observation, Arthur began to form theories about transmission. He named three vulnerable points for blight bacteria to enter trees. After observing English hawthorn, Arthur suggested that the "germs" might enter trees in the spring through the flowers and tender surfaces of young buds. This theory recalled the discussions between Burrill and George Peffer. "The short period that the flowers are open," Arthur wrote, "is time enough to seal the doom of many limbs, and even whole trees."[97] According to Arthur, the second vulnerable entry point was the growing tips of branches or a developing bud, and the third was a crack or injury to the bark.[98]

Almost as an afterthought, Arthur allowed that insects could "now and then" be the agent of blight transferral.[99] Nearly another decade would pass before Merton B. Waite would follow up on Arthur's work and confirm the central role of insects in the transmission of the bacteria that cause fire blight.

In 1886 Arthur wrote an article for the USDA in which he discussed "preventatives and remedies" for fire blight, but he could offer little more than Burrill had five years before. He noted quickly that "no method is known or has yet suggested itself of rendering the tree insusceptible to the disease."[100] Spraying had so far showed little promise. Arthur noted that the internal surface of the flowers "are so well protected by the stamens and other organs that the antiseptic used does no service."[101] Washes, another old method of delivering germicides, were also problematic. First of all, washes could not be applied to flowers or growing shoots, even though "excellent results may reasonably be expected when made to the trunks and larger branches."[102] Some growers had reported success with a traditional solution of sulfur mixed with lime. Further study needed to be made, but Arthur suggested using linseed oil for an elastic coating to which sulfur and 1 percent carbolic acid should be added.[103]

In the end, however, Arthur had little help to offer growers for prevention or cure of fire blight. Pending a major breakthrough in chemical control, the best method of decreasing the damage of fire blight was careful cultural practice. Infected branches and limbs should be pruned immediately and burned. The knife used in pruning should be disinfected frequently with carbolic acid to prevent it from becoming an agent of pathogen transmission itself.

While noting the lack of rigor in some of Burrill's work, Arthur confirmed Burrill's role as a pioneer in the developing field of plant bacteriology. Bacterial pathogenicity was not yet accepted universally, but in 1886 Arthur submitted his research on fire blight to Cornell University as a thesis and received his doctorate.[104] Further experimental work to confirm and clarify the nature of bacterial diseases of plants would come later, notably from Merton B. Waite and Erwin F. Smith, both with the USDA. By the turn of the century American practitioners were recognized as world leaders in plant bacteriology.

REMAINING CHALLENGES

The work of Thomas Burrill and Joseph Arthur on fire blight represents an important period in the formative years of American plant pathology. In the early 1880s, institutions such as land-grant colleges and state agriculture experiment stations reflected the increasing interest in plant disease and the advancing scientific knowledge concerning pathogenic microorganisms and the diseases they caused. The arguments over the role of scientific research in a university setting had largely been settled in favor of those who saw research as part of the mission of higher education. However, the usefulness of the experiment stations was still a subject of debate in the early 1880s. These institutions continued to struggle with the dilemma of satisfying immediate, parochial needs while still engaging in the long-term research that would be required to provide many of the needed answers.

Arthur was not the first scientist to study plant disease at a state experiment station or an institution like it. Harvard's Bussey Institution and Yale's Sheffield Scientific School were similar establishments. In fact, F. H. Storer, dean of the Bussey Institution and a professor of agricultural chemistry at Harvard, said that Bussey was "the nearest thing we in Massachusetts have to the experiment station."[105] The Bussey Institution and the Sheffield Scientific School were adjuncts to their universities—places where scientific research could be done but with some necessary application to the universities' course work.

Burrill's and Arthur's cutting-edge plant disease research would eventually make the United States the world leader in plant pathology. Prior to their work, original experimentation on plant disease was usually done by Europeans, particularly in German universities. But in American institutions modeled on the European plan, that is, research- and graduate-oriented universities and experiment stations, American plant scientists found exciting research opportunities to study unique disease problems such as fire blight. They began to solidify their role as leaders in new areas of botanical and biological study.

In the early days, the efforts of American scientists were constrained somewhat by inconsistent funding and support. To solve these problems, they would have to turn to funding sources other than private citizens, society grants, college administrators, and state legislators. They would have to turn to the federal government.

❧ CHAPTER 6 ❧

Early Years of the
United States Department of Agriculture

In the two decades after Congress passed laws to establish the land-grant colleges and the United States Department of Agriculture, most significant work on plant disease occurred in colleges, universities, and occasionally state agricultural experiment stations. From its beginning in 1862, the USDA had been interested in plant disease, but investigating plant disease was not central to its purpose. Furthermore, the department had neither the expertise nor the equipment to study plant disease with any degree of scientific precision.

Like the land-grant schools, the USDA had begun its existence without a clearly defined mission. The law that created the USDA had directed the chief of the department, then called a commissioner, to "acquire and preserve . . . all information concerning agriculture" through "practical and scientific experiments." But the ways in which this was to be done were unclear. The department spent its first two decades mirroring the agricultural work of the Patent Office: collecting information, publishing bulletins of agricultural interest, creating statistical tables on agricultural products, and distributing seeds. Scientific experimentation was not a priority.

The shortcomings of the USDA were painfully obvious to agricultural interests. As the editors of *The Country Gentleman* commented in 1863, they could not "entertain any very brilliant anticipations from the establishment of governmental laboratories."[1]

POLITICS AND SCIENCE

Despite the USDA's shortcomings, promoters of federal aid to agriculture continued to harbor hopes for the young department. The key element, many thought, was the appointment of a commissioner who would encourage the development of the department along scientific lines. Unfortunately, the early commissioners had more concern for politics than for science.[2] Nevertheless, at lower levels within the department, science gradually crept into regular activities. The law establishing the department had instructed the commissioner to "employ . . . chemists, botanists, entomologists, and other persons skilled in the natural sciences pertaining to agriculture." Although the law went on to specify that the commissioner should hire these scientists "for such time as their services shall be needed," these positions were essentially permanent.

Despite his name—Isaac Newton—the first commissioner of agriculture was not a scientist, but rather an agriculturist from Pennsylvania and a personal acquaintance of President Lincoln's. Newton's appointment went unheralded by much of the agricultural press and agricultural societies, but he did bring a number of scientists into the department, thus setting the stage for the development of discipline-oriented divisions.[3] He appointed horticulturist William Saunders as superintendent of the propagating gardens; Charles M. Wetherill, a former student of Justus Liebig, as chemist; and Townend Glover as entomologist, a position similar to the one Glover had held in the old Agricultural Bureau of the Patent Office.[4]

Newton established a formal list of seven departmental priorities. Since his top priority was the collection of agricultural statistics, his most useful appointment was that of statistician Lewis Bollman.[5] The "promotion of botany" was in sixth place on his priorities list, just above "establish a library and museum."[6] Botanical sciences were not important considerations beyond the establishment of propagating gardens and test plots of various crop plants.

Many of the early scientists at the department discovered that it was very difficult to work in a politically charged environment. For example, distinguished chemist Charles M. Wetherill was discharged within a few years of his appointment on the charge that he was devoting too much time to chemical work for other departments of the government.[7] Charles C. Parry, appointed to the newly created position of chief botanist in 1869 by the second commissioner, Horace Capron, was forced to retire by Commissioner Frederick Watts, who took over from Capron in 1871. Watts charged that Parry had failed to add sufficiently to the "practical" mission of the USDA.[8] Parry complained about his dismissal to powerful scientific friends, including Asa Gray, John Torrey, and Joseph Henry, and urged them to launch a letter-writing campaign decrying the poor state of science in the USDA. Letters in scientific journals raised awareness of the plight of USDA scientists among the nation's scientists and educators, but they had little effect on the politics of the department. Parry was not restored to his post.[9]

Chief entomologist Charles Valentine Riley entered and left the department under Watts's successor, William Gates LeDuc. Prior to his appointment to the USDA in 1878, Riley had been very active in the scientific community, both as an editor in the agricultural press and as the state entomologist of Missouri. In 1877 he had helped spur Congress to create the United States Entomological Commission, with a primary mission of studying the Rocky Mountain locust, a pest that had ravaged field crops in the West for the previous four years.[10] In 1879, under pressure from Commissioner LeDuc, Riley resigned from the USDA, but he continued with the United States Entomological Commission, which had become part of the Department of the Interior.[11]

Riley's forced departure from the USDA was one of the catalysts that brought the condition of USDA science to the forefront of the debate over the direction of agricultural science. Riley became a vigorous and public critic of the status of science in the USDA. In 1879, at the first meeting of the American Agricultural Association, he charged that scientific work at the department was "inferior to . . . many private associations and State institutions."[12] In 1881 Riley regained his job at the USDA under LeDuc's successor, George Bailey Loring, who was much more agreeable to the entreaties of scientists.[13] During this second term of employment, he brought the Entomological Commission from the Department of the Interior to the USDA.[14]

The role of science at the USDA became an important topic of discussion among American agricultural scientists. These scientists, working primarily at universities or land-grant colleges, were quite vocal in their desire to see the department transformed into a scientific bureau.[15] And they were becoming better organized and more sophisticated in their methods of bringing pressure to bear in the political arena.

Throughout the 1870s and into the 1880s, a cohesive community of agricultural scientists and educators gradually formed. An important forum for these scientists was the agricultural convention, which allowed those of like mind to discuss and debate their agendas. Such conventions usually failed to coalesce into lasting organizations. In fact, many an organization vanished after its first annual meeting, though its members would reassemble several years later at the first annual meeting of a different organization.

One such agricultural convention was the first meeting of the new American Agricultural Association in 1879. Established as a national organization, the American Agricultural Association considered itself the heir to the United States Agricultural Society that had been so instrumental in the 1850s movement to create the USDA. The association was committed to bringing more agricultural science into the USDA, while maintaining the independence of the land-grant colleges and experiment stations. Its membership was dominated by land-grant college and state experiment station personnel. The first meeting of the association was one long diatribe against the weakness of the federal government's efforts on behalf of agriculture. It was at this meeting that Charles V. Riley lambasted the state of science at the USDA. But even though its members were drawn from the best of agricultural science, the association could not maintain enough cohesiveness to convene a second national meeting.[16]

Specific organizations might have faded away, but the momentum to organize continued. Deliberately, scientists and educators began to wrest control of the nation's agricultural policy debate from members of state agricultural or horticultural societies and agricultural journalists. George Bailey Loring, the new commissioner of agriculture in 1881, was more popular with the agricultural science community than his predecessors had been. He sensed a growing desire from quarters such as the land-grant colleges and popular agricultural groups such as the Grange for the department to take a more active role in agricultural science. In 1882 and 1883 he convened meetings of agricultural scientists, educators, and interest groups. These meetings were well attended and furthered the role of the USDA as a central actor in agricultural-science policy, if only as a peacemaker between warring factions.[17]

Plant disease and the early usda

During Loring's tenure as commissioner, the movement for a greater scientific focus for the USDA included a small but audible contingent calling for increased study of plant disease. In the late 1870s and early 1880s, many botanical scientists and educators located at land-grant colleges and experiment stations favored more federal support for economic botany. Although they were certainly not willing to surrender their own autonomy, these botany professionals wanted the government to play a part in coordinating the nation's scientific agricultural research. As envisioned by the promoters of the study of plant disease, this federal role would include not only increased funding of land-grant colleges and experiment stations, but also of Department of Agriculture laboratories established in Washington and elsewhere around the country.

The USDA had not ignored plant disease completely in its early years. During the 1860s the department's annual reports included papers on various crop diseases. In the first USDA report, for example, horticulturist William Saunders outlined his duties and included among them, "to investigate more thoroughly the various maladies and diseases of plants."[18]

The earliest USDA reports, which preceded the work of Farlow, Bessey, Burrill, and Arthur at universities, land-grant colleges, and experiment stations, were equal to the best material published in the agricultural press. In fact, much of what the USDA published during these years was drawn directly from the agricultural press or from European sources. Most of this information was in small sections within larger articles and was essentially descriptive. Some articles also included recommendations for control measures, generally repeating the practical advice that had filled the agricultural journals since the early 1800s.

The early publications of the USDA demonstrated the growing acceptance of fungi as causal agents of plant diseases. An article on wheat cultivation in the 1862 report included a brief descriptive section on the fungal origins of smut.[19] That same year, Lewis Bollman, USDA statistician, wrote a report on wheat that discussed rust- and smut-causing fungi, borrowing heavily from John H. Klippart's 1858 "Essay on the Origin, Growth, Diseases, Varieties, &c., of the Wheat Plant."[20] Two years later, in the 1864 report, Bollman wrote about "mould" in an article on "The Hop Plant."[21] In both of his articles, Bollman gave brief descriptions of these diseases caused by fungi and recommendations for control measures typical of the day. For wheat diseases he suggested treating seed with bluestone (copper sulfate) washes. And because hop mold had a propensity to grow in warm, damp weather, he offered advice on proper sites for planting and on destroying "every spotted leaf."[22]

Some of the early USDA articles on plant disease went into detail about the nature of crop-destroying fungi. In 1865, for example, William Saunders compiled a sizable amount of material on grape culture that included information on "mildew," both in terms of cause and control. He described the appearances of mildew caused by both the fungus known at the time as *Peronospora viticola* and by an as-yet undescribed fungus from the genus *Erysiphe*. He distinguished among the grape cultivars that the mildews attacked and outlined a plan to use sulfur to treat both of these fungus-caused diseases of grape.[23]

The report for 1865 also included a long article, "The Grape Disease in Europe; Its Origin, History, Phenomena and Cure," written by Henri Erni, the USDA chemist.[24] Erni described *Oidium tuckeri*, a fungus that was then ravaging European grape stock, and prescribed two treatments. The first treatment was sulfur, popularly regarded as a cure-all, which Erni claimed "withers and dries up" the powdery mildew fungus. He could not explain why sulfur had this effect, but he claimed that it had been observed microscopically. The second treatment was hydrosulfide of lime, which had been used successfully against *Oidium*, or so Erni claimed, by a Dr. Turrel of France.[25]

Ironically, the same report that published these descriptions and control measures for fungal grape diseases also published articles based on older, traditional ideas about the nature of plant disease. William C. Lodge of Delaware wrote a piece on fruit tree diseases that said nothing of fungi. For black knot of plums and cherries, he blamed "a disease of the sap imparted from either the soil or atmosphere." And he concluded that he had "discovered a sovereign remedy for nearly all diseases of our fruit trees" which "is nothing more than common salt."[26] Clearly, if the USDA was to play a more active role in agricultural science, it would have to educate its chief constituents, farmers and politicians—and even some of their contributors—about recent scientific concepts like fungal pathogenicity as proved by Anton De Bary.

A USDA MICROSCOPIST AND FUNGUS RESEARCH

In 1871 USDA Commissioner Frederick Watts hired Thomas Taylor as the department's microscopist. Taylor's charge was to make use of the department's one and only microscope, but he would also introduce into the USDA agenda the experimental study of fungal plant disease. His access to a microscope allowed Taylor to publish plates showing microscopic bodies and life stages of various fungi.

Thomas Taylor, a native of Scotland, was born in 1820. He exhibited an interest in science and technology early in life and did research on electric lights and batteries, a popular topic in Great Britain in the early nineteenth century. While still in his twenties, he invented a pneumatic battery that was used for demolition and mining. When he came to the United States in 1851, he experimented with the transmission of electricity through water without wires. During the Civil War, Taylor's eclectic talents gained him employment designing and testing ordnance for the War Department.[27]

When he went to the USDA in 1871, Taylor quickly began work on fungal diseases of plants. His first project was his "Report on Fungoid Diseases of Plants." In this report, Taylor described several plant diseases, including those of grape, pear, peach, and lilac.[28] However, he spent most of his time and effort on the grape mildew caused by *Botrytis vitis viticola*. He carried out simple experiments to test the contagiousness of the disease by placing mildewed leaves in various proximities to healthy leaves.[29]

Taylor also searched for the presence of fungi in connection with several other diseases, often making the simplistic assumption that fungal presence proved causation. After laboring to find a fungus in a solution made from the bark of a peach tree with yellows, Taylor concluded incorrectly that "the fact that the threaded mycelium of fungi has been found in profusion on the liber of the peach-tree is sufficient to explain the cause of . . . disease."[30] He also worked on a pear disease, not identifying it specifically, but only as "an internal decomposition or rot." Again, to Taylor, the "presence of entozoa and fungi in the pears . . . explains the cause of their rapid destruction."[31] In many of his studies Taylor was misguided, but not to any greater extent than many other investigators at the time, who were also seeking fungal causes for plant disease.

Taylor's most important contribution came from his observation of grape mildew, *Oidium*, on European vinestock in USDA greenhouses. In the summer of 1871, he was the first to view the "perfected fruit" of the *Oidium*.[32] Assisted by William Saunders, he tested optimum growing conditions for the powdery mildew fungus on the host plant and also evaluated some control measures using sulfur.[33] He was credited for this work by the prominent British botanist M. J. Berkeley.[34]

For several years after 1871, Taylor wrote articles on plant disease for the department's annual reports and for agricultural journals. As a microscopist, he wanted to observe fungi and document their physiology and life-history. He first studied potato late blight.[35] He next turned to black knot of plum and cherry, drawing the admiring attention of Charles Peck, the noted mycologist from New York who was considered a black knot expert.[36]

But given the practical mission of the USDA, Taylor also had to make some suggestions and experiment with control measures, as he had with *Oidium*. He made the traditional recommendations of sulfate of copper for treating *Peronospora infestans*, the causal agent of late blight in potato. He also claimed to have destroyed onion rust with applications of the fumes of nitrous acid, but added that "the extent to which this treatment may be safely applied to the smaller grains will be the subject of future experiment." Typically, however, Taylor could do little more than decry the lack of research on a disease or promise that the USDA would soon begin experiments on treatment.[37]

Taylor became a useful tool of the USDA. As growers became aware of the department's potential for assisting with disease problems, they often turned into supporters who would petition Congress for increased funding. When, for example, the USDA received "numerous letters" from Florida orange growers calling for assistance with a "new form of disease" where "the branches become covered . . . with a rust-like substance," Taylor wrote a survey piece in which he announced his detection of a mildew of the genus *Antennariae* in the blighted material.[38] But Taylor concluded his article by saying, "It is to be regretted that the Department has no fund at its disposal for original investigations in the field. Funds should be at the disposal of the Commissioner of Agriculture to enable him to send scientific persons to make thorough examinations when they are deemed necessary."[39]

The commissioners quickly recognized Taylor's value, and they began to promise their demanding constituents that diseases would "occupy the attention of the Microscopist."[40] In response to "numerous letters from cranberry-growers," Taylor studied the rot and scald of cranberries. Dispatched to the cranberry bogs of New Jersey and Massachusetts, Taylor produced a report on cranberry diseases, amounting largely to a discussion of cultural practices involving soil nutrition, field location, and irrigation.[41] Similarly, New Jersey viticulturists who "suffered severely from the rot of the grape" pressured the USDA to study their malady. Usually, Taylor stayed in his Washington laboratory, and growers around the country sent him samples of a variety of diseased fruits and trees for microscopic examination.[42]

Taylor was applauded for his technical ability to create slides of fungi for the microscope that were "delicate & fragile in its tissues."[43] He also earned attention for his drawings from microscopic observations of various fungi.[44] While serving as USDA microscopist, Taylor earned a medical degree from the University of Georgetown; his dissertation was entitled "Cryptogamic Parasites and Infectious Germs Considered in Their Relation to Man and the Lower Animals."[45] He also corresponded with noted botanists, including Berkeley and Cooke of Great Britain, primarily for the exchange of samples.[46]

Unfortunately, the strength of Taylor's science was sometimes debatable. He maintained a professional relationship with many of the mycologists and botanists of his day, usually sending out samples of fungi for identification, but he was not a trained mycologist. As the study of plant disease attracted more university-trained scientists, Taylor's work could not measure up to the improving standards. In fact, by the late 1870s and early 1880s, he was abandoning plant disease research for areas such as information and growing techniques for edible mushrooms.[47] He also used his microscopic skills to study the adulteration of foodstuffs, an early effort by the government to ensure food purity.[48] His plant disease research, however, was quickly becoming another example of the poor science at the USDA. Although Charles Peck applauded Taylor's attention to black knot of plums and cherries and claimed that he "shows conclusively the connection between the *Cladosporium* and the sphaeria; a connection which I have long suspected," Peck added, "I fear he has done it unwittingly."[49]

By the late 1880s, Taylor experienced great disfavor among the "professional" plant pathologists who came to the USDA. Beverly T. Galloway, who was chief of the USDA's plant disease research in the late nineteenth and early twentieth centuries, wrote that Taylor's microscopic observations "showed many things not even dreamed of in the realm of plant pathology or mycology," and added sarcastically that "the Doctor shows in one of the old annual reports an amazing number of 'fungoid' forms some of which are clearly zoological and others botanical such as cell crystals and plant hairs of many kinds."[50] Erwin Frink Smith, one of the most celebrated USDA phytopathologists of the late nineteenth and early twentieth centuries, went so far as to write that Taylor "has not contributed to Science one single fact of any importance."[51]

But Thomas Taylor, whatever his level of scientific accomplishment, was the most visible researcher on plant disease during the first twenty years of the USDA. His articles may have been gleaned largely from the publications of others, but that was the traditional clearinghouse purpose of departmental reports, just as it had been previously for the Patent Office. He worked in a federally funded position, which implies that his work was viewed by the government as beneficial to the nation, and part of his work included the use of a microscope to study plant disease.

It was not until the late 1870s, several years after Taylor began his work, that microscopes became a common teaching tool in America's colleges. Therefore, for his first five years, Taylor was doing something relatively unusual. But as men like Bessey, Burrill, Beal, and Goodale made all their botanical students into microscopists through laboratory instruction, Taylor's uniqueness faded. Compared to the work of trained scientists, his work was scientifically unambitious. As the century wore on, science passed him by. A self-trained generalist in an era of increasing specialization, Taylor was dismissed in 1895 at age 75, after twenty-four years of service, when Secretary of Agriculture J. Sterling Morton abolished the Division of Microscopy as an anachronism.[52]

Taylor's most significant, if unintentional, contribution may have been his demonstration late in his career that the gap was widening between the work on plant disease at the USDA and the science being done by the likes of Farlow, Burrill, and Arthur in other institutions across the United States. On the other hand, the USDA had advantages that agricultural colleges and experiment stations did not—the chance for centralized coordination of research and experimentation and, even more important, the chance at secure funding. Private endowments from families like Cornell, Purdue, and Bussey were limited, and maintaining such ties required a measure of political acumen that scientists found distracting and sometimes demeaning.[53] In 1882 and 1883, when Commissioner Loring summoned land-grant college and state experiment station personnel to debate the path that agricultural science should take, leadership had fallen to those who wanted the federal government to be more active.

The daily hardship of keeping the agricultural colleges and experiment stations operating in the face of overwhelming apathy by most farmers and legislators had slowed the efforts of the educators and scientists to create organizations to influence government science. Gradually, however, a powerful scientific community was forming in America. This community would soon bring about a new direction for research in the USDA.

CREATION OF THE SECTION OF MYCOLOGY

In the 1880s, calls for the federal government to support the study of plant disease came from some prominent scientific voices. William Beal, professor of botany at the Michigan Agricultural College, remarked in 1883 that "in the study of effectual remedies against fungi . . . there is still much demand for more knowledge."[54] He added that the federal government had appropriated only "a small sum, considering its importance" to battle the effects of the ravages of plant disease.[55] Charles Bessey wrote in 1883 that "if the investigation of injurious insects be considered a government duty, will not the same reasoning show it to be equally its duty to provide for the similar investigation of the numerous parasitic fungi which injure and often entirely destroy farm and garden crops?"[56]

It was a livestock disease, however, that finally forced the USDA to play a more active role in scientific research.[57] In the early 1880s, cattle pleuropneumonia had disrupted the wealthy and influential beef industry of the United States. By 1884 both science and business interests exerted pressure on a reluctant Congress to appropriate money to the USDA to create the Bureau of Animal Industry. The law required that the bureau chief be a competent veterinary surgeon, not just a political appointee.[58] The requirement that a science professional be placed in a position of influence and power was an important breakthrough.

The precedent of the Bureau of Animal Industry, a purely scientific and disease-oriented bureau in the USDA, in combination with the growing clamor for more federal money to aid plant disease studies pushed events forward. In 1884 and 1885, a call for federal action on plant disease came from a significant source—the American Association for the Advancement of Science (AAAS), one of the premier professional organizations for American scientists. A Botanical Club had formed inside the AAAS in 1883, and figures like Charles Bessey and Thomas Burrill were beginning to move into positions of prominence in the organization. In 1884, at the annual meeting in Philadelphia, the AAAS formed a Committee on the Encouragement of Researches on the Health and Diseases of Plants. Consisting of six noted botanists and horticulturists, most of whom were already outspoken proponents of federal support for plant disease research, the committee included Burrill and Bessey, who were vice president and secretary of the Biology Section of the AAAS, respectively; William Farlow, Charles Peck, William Beal, whose opinions had already been published; and

J. T. Rothrock of Pennsylvania, who was on his way to becoming one of the nation's early leaders in forestry. The committee's chairman was J. C. Arthur, botanist at the New York State Agricultural Experiment Station at Geneva, who was only a year away from presenting the results of his experiments on bacteria and fire blight to the AAAS.[59] These were scientists at the pinnacle of their field—men of expertise and growing influence who were eager to use the status and organizational power of the AAAS to push their science in new directions and into new institutions.

The Committee on the Encouragement of Researches on the Health and Diseases of Plants was assigned to write and address a petition to the Commissioner of Agriculture to plead the case for consistent, government-sponsored research on plant disease. After nine months of correspondence, in April 1885 the committee produced its petition.[60] All members except Farlow signed the official letter. Even Farlow, whose reasons for not signing the letter are not clear, certainly supported the general aims of the petition, and, in fact, Farlow continued to offer his views privately to the commissioner in support of federal plant disease investigations.[61]

Only three paragraphs in length, the petition clearly expressed the expectations of its authors. Addressed to "The Honorable Commissioner of Agriculture," it declared that the undersigned committee "respectfully urge that in planning the work of the department for the coming years, you make provision for the thorough scientific study of the diseases of plants, especially those due to the parasitic fungi, believing that by so doing you will greatly increase the efficiency and value of the Department."[62] The committee went on to urge the department to find "an officer" who would begin research on plant disease and "whose training has been such as to enable him to call to his aid all the knowledge and appliances of the best modern scientific methods."[63]

The petition urged the USDA to establish a bureau similar in stature and operation to the Bureau of Animal Industry, but devoted to the study of plant disease. It recommended that a qualified scientist be in charge who would respond to the scientific imperatives of disease problems rather than the institutional pressures of political survival. It argued that the department would profit by such a bureau, as it would bring valuable knowledge to farmers and expand agriculture in the United States. Unfortunately, there existed neither a horticultural or field crop producers' association that could match the economically powerful cattlemen who benefited from the Bureau of Animal Industry. Nor was there a single plant disease crisis in the 1880s that equaled the disruptive consequences of pleuropneumonia.

However, there was a growing cry to develop controls for a number of serious diseases, particularly those endemic to fruit, which formed the basis of an influential industry. Support also came in 1885 with the appointment of Norman J. Colman as the new commissioner of agriculture. Colman was very sympathetic to the interests expressed in the committee's petition and had a long record of support for agricultural science and education in his native Missouri. The publisher and editor of *Colman's Rural World*, he was also a stock breeder, a pomologist, the president of several agricultural organizations, and an experienced political administrator, having served four terms as a Missouri state legislator and one term as lieutenant governor.[64] His experience made him well suited and highly motivated to increase the USDA's scientific mandate. His appointment was applauded widely by agricultural scientists.[65]

Colman replied promptly and positively to the AAAS committee's petition. By May 1885 he had appointed Frank Lamson-Scribner, a botanist from Maine who was teaching at Girard College in Philadelphia, as assistant botanist with an unofficial portfolio to study harmful fungi. On May 29, 1885, Scribner wrote to Charles Bessey at the University of Nebraska and explained that his "line of work" at the USDA would involve "the investigation of parasitic fungi—those species that affect injuriously our field & garden crops."[66] This mandate made Scribner the first scientist employed by the USDA with the specific and exclusive assignment of studying plant disease.

It is likely that Colman and the AAAS committee coordinated their actions. Because hiring an assistant botanist did not require new appropriations from Congress, Colman was able to establish the ad hoc position to study plant disease without a political fight, but the committee's petition gave him the evidence that the American scientific community demanded and supported the appointment. Colman hoped that Scribner's plant disease research would demonstrate both the need and the value of the work at a federal level. This would give him ammunition for future budget battles as he proceeded with his plans to expand the USDA.

Scribner was not an experienced mycologist. While at the University of Maine in the early 1870s, he had studied botany from Gray's *Manual of Botany*, with some exposure to mycology from M. C. Cooke's work.[67] Still, this limited instruction hardly made him the most qualified figure to undertake an indexing or exhaustive study of injurious fungi. Nevertheless his appointment elicited little open criticism from other botanists and horticulturists. Perhaps no other qualified scientist was interested in an uncertain position with an organization that had such a poor scientific reputation. Scribner would reminisce late in his life that Commissioner Colman polled the botanical community to find the right man for the assignment and his name was the overwhelming favorite.[68] However, it is likely that George Vasey, the USDA botanist, was instrumental in Scribner's selection, because they shared an interest in agrostology and Scribner had worked for him in the past.

Aside from arranging the department's collection of fungi and preparing a circular for distribution, the exact nature of Scribner's new duties was not specified. In his letter to Charles Bessey describing his position, Scribner had asked Bessey to "kindly draw up a plan of operations that will, in your opinion, make the work most effective," adding that the help of mycologists was "essential to the accomplishments of valuable results" in the USDA's new undertaking.[69] Scribner's request for help shows the poor state of institutional preparation at the dawn of federal plant disease research. But seeking help from Bessey, a political ally and a leading scientist in the field, also indicated Scribner's good judgment. The exchange was a precursor to the close association that the USDA was beginning to forge with plant pathologists at universities and experiment stations around the nation.

After securing some advice, Scribner began his work by preparing a long review article on fungus diseases of plants in the USDA report for 1885. In the review article, Scribner outlined the basics of fungal life-history and discussed diseases such as smuts, blights, rusts, and mildews. He also wrote on control measures, but this information, in the years just prior to the popularization of Bordeaux mixture as an effective fungicide, mostly involved altered cultural practices. He did discuss briefly the same specific chemical treatment recommendations for grape diseases that Saunders, Erni, and Taylor had mentioned previously, including the application of lime solution and copper sulfate.[70] In the same year's USDA report, William Saunders wrote on fungal diseases of fruit and potato.[71] Both Scribner's and Saunder's reports signaled the readers that the subject of plant disease was now to be a significant part of the USDA mission. This was a first step to educating the general public about fungal pathogenicity and the role of science in plant disease control.

The members of the AAAS committee corresponded frequently with Colman and Scribner, helping to shape the future of the USDA work on plant disease. William Farlow wrote to the commissioner to recommend that Scribner become not just another USDA pamphleteer, but rather a truly scientific researcher. Colman answered, "I fully recognize the correctness of your view that the requirement of a completed knowledge of many of those plant diseases is a slow and tedious process and an investigator should not be judged wholly by the amount of work he may present to the public as much of the time spent in investigation may give only negative results."[72]

This was an interesting admission by the chief of a department that had been founded on the promise of practical service. Colman clearly supported agricultural scientists who would do more than identify weeds and analyze fertilizer samples. However, he was sufficiently visionary to recognize the political liabilities inherent in embracing the nascent field of agricultural science. The department had existed for twenty years by fostering expectations of immediate, concrete benefits to America's farmers. Basic scientific research did not yet fit clearly into that mission. Colman had to balance his powerful political constituencies with a growing scientific one.

Together with AAAS committee members and other botanical and horticultural groups, Colman embarked on a yearlong strategy to convince Congress that Scribner's ad hoc position was an economic necessity and must be funded as a permanent component of the USDA.[73] The national scientific voice that had filtered through the convention movement was finally united in a single cause. And this scientific community had a commissioner who would fight for their agenda from inside the government.

Colman wanted the 1886 USDA budget to include a new Section of Mycology inside the Division of Botany. To bolster his case, he solicited the support of botanists around the country. The Botanical Club of the AAAS advised him that "the members of the club hereby pledge themselves to use their influence in inducing their representatives in Congress to make a liberal appropriation . . . in accordance with your estimates."[74] George Vasey, the chief of the Division of Botany, asked Charles Bessey for "any influence that can be brought to bear . . . from men who are not directly connected with the Government, but who are known for their scientific work," and sent Bessey a list of congressmen to whom he should direct his respected opinions. [75]

Letters poured into Washington from botanists and horticulturists around the country. Bessey arranged for the Nebraska State Horticultural Society to address a resolution to that state's congressional delegation in support of an increased appropriation to the USDA for "the study of fungi injurious to vegetation."[76] Thomas Burrill wrote the chairman of the House Agriculture Committee, William Hatch of Missouri, that "the losses by fungous parasites equal, aye surpass, those by insects." Burrill then asked, "Shall not Congress recognize this fact?"[77]

In 1886, Congress voted for the appropriations that officially created the Section of Mycology, and Frank Lamson-Scribner became its chief. The successful campaign was significant. It demonstrated that scientists, with organization, could influence government officials to create supportive institutions for increasingly specialized disciplines such as plant disease research.

The establishment of the Section of Mycology set the stage for the Department of Agriculture's entry into scientific research into plant disease. Like the land-grant colleges, the department had gone through a period of debate and struggle, searching for a definition to its mission. But now the USDA had wandered from the path of practical and political services alone onto the path of scientific research.

Part Three

THE RISE OF PLANT DISEASE RESEARCH

*I have long looked upon this period as epoch-making in
agricultural progress, especially in the science of plant pathology, for
the practical results achieved proved beyond question the value of
scientific work applied to agricultural practices.*
—Beverly T. Galloway

*The study of plant diseases has had a larger development in
this country than anywhere abroad owing to the fostering care of
the National Government.*
—Erwin F. Smith

By the turn of the twentieth century, plant pathology was a very different science than it had been in 1885, when Frank Lamson-Scribner was hired by the United States Department of Agriculture as the first federal employee with the specific mission of studying plant disease. In part, the transformation of plant pathology was the result of changing economic and social conditions in the United States and the increasing activism of the federal government in American agriculture and science.

The United States' rapid growth in the late nineteenth century had fundamental consequences for agriculture. As America transformed into an industrial nation, subsistence agriculture became less important. Commerce became not just an opportunity, but a necessity. Specialization and intensification became more pronounced as regions of the country focused on crops that could be grown for market in quantity. A boom in transportation, particularly railroads, helped to foster this trend, and this transportation boom was aided by the movement toward farm mechanization, powered by both the horse and the machine.[1]

Agriculture was perceived as one of the chief necessities of national prosperity, and the federal government took an active interest in its well-being. The importance and size of the USDA increased, and USDA scientists struggled to apply a growing body of knowledge to practical agricultural problems. In the 1880s science and agriculture were uneasy partners. It was unclear exactly what the former had to offer the latter, particularly within the USDA, where practical service to farmers had always been the central objective.

For twenty years, many scientists, agriculturists, and politicians had advocated a central place for science in education, particularly at land-grant colleges, and they now began to push for more scientific content in the USDA mission. They believed that agricultural problems, including diseases of plants and animals, could be solved only by the application of scientific principles from biology and chemistry. The establishment of problem-oriented bureaus in the USDA and of federally funded state agricultural experiment stations was evidence that their philosophy was gaining support.[2]

Nevertheless, the value of science to agriculture had yet to be firmly established. Frank Lamson-Scribner faced this dilemma in 1886, when he became head of the newly created USDA Section of Mycology. Although his job was to study the diseases of plants caused by fungi, merely studying them would not suffice. His funding sources also expected cures and controls for these diseases. The expectation of the practical application of scientific research was one of the central differences between the plant pathology that Scribner embarked on and that which had been practiced before him. William Farlow, for example, rarely felt pressure at Harvard University to produce information that would lead directly to effective control measures for plant disease. In order to answer his charge to apply his science to agriculture in useful and practical ways, Scribner searched for innovative ways to combine laboratory work with field-tested disease-control strategies. It was Frank Lamson-Scribner, as much as anyone in the United States, who fused science and agriculture to create the applied science of plant pathology.[3]

The federal government became the chief advocate and sponsor of this applied science. In controlling plant disease, federally funded plant pathology scored notable successes in the last fifteen years of the nineteenth century. The most lasting legacy was the widespread use of chemical fungicides. It was fortunate for Scribner's applied pathology that Bordeaux mixture, a French copper sulfate and lime fungicide, appeared in 1886 as a promising control for grape diseases. Spearheading its introduction to the United States and bolstering its effective use through rigorous scientific research, Scribner was largely responsible for an agricultural innovation that changed American agriculture. Spray treatments were already in use for insect pests and for a few fungus diseases, but the usefulness of copper sulfates combined with improved sprayer technology launched a new era of chemical-aided agriculture.

Scribner's early successes with applied plant pathology boosted the government's efforts to expand the field. Under his successor, Beverly T. Galloway, federal plant pathology grew rapidly, incorporating the study of many new diseases and experiments with numerous control measures. To answer the demands from growers for help with disease problems, the USDA created research laboratories in different parts of the country. In addition, the federal government funded a system of state agricultural experiment stations, which were technically independent or were attached to land-grant colleges.

Within a decade, the experiment stations were brought into a growing network of agricultural scientific institutions with the USDA at the center. The intervention of the federal government in experiment stations was favored by station scientists, who wanted to do research that was unfettered by the demands of local constituencies for mundane services such as identification of weeds or the certification of fertilizers. Applied plant pathology found a home at these agricultural experiment stations.

The scientific advances made by plant pathologists during this era were both applied and fundamental. Some plant pathologists discovered new aspects of the basic biology of fungal and bacterial pathogens. At the same time, new vistas of disease research were opening up, particularly with regard to understanding host plants. The mid-1890s brought a fresh emphasis on plant physiology that would lead to a clearer perception of the life processes of the healthy plant and its interaction with the pathogen. Another area of plant pathology involving the host plant that grew in importance just before the turn of the century was the practice of breeding for resistance. Although this idea was not new —disease-free cultivars had long been a focus in agriculture during disease outbreaks—it now came under scientific scrutiny. USDA scientists became extremely interested in searching for new plants and cultivars and testing them for disease resistance.

The movement into new areas of research brought new difficulties for plant pathologists. The more new, unknown challenges that researchers took on, the greater the chances that they would run into disease problems that could not be understood or solved with current knowledge. This was a particularly vexing problem for scientists in the USDA, who were under constant pressure to produce practical solutions for farmers. Long-term, time-consuming scientific programs always have the potential of producing nothing but wrong answers. Such blind alleys could create friction in an applied science that was based on service. Fortunately, however, successes in combating plant disease were visible enough to demonstrate the value of continuing scientific research.

By the turn of the century, the science of plant pathology was developing into a recognized discipline, separate from its scientific progenitors like botany, mycology, and bacteriology. Plant pathology was establishing its own body of knowledge; its own group of practitioners, who were certified in the science through education and training; and a growing network of professional associations. American plant pathologists now had the confidence and knowledge to challenge the roots of their science in Europe. From 1898 to 1901, Erwin F. Smith, of the USDA, engaged the German bacteriologist Alfred Fischer in a highly publicized academic debate over whether bacteria could cause plant disease. By the time it ended, Smith had settled the matter in favor of his science and had brought worldwide recognition to American plant pathology.

Fig. 3.1. Frank Lamson-Scribner spraying grape vines with Bordeaux mixture on test plots at the USDA in Washington, ca. 1885, courtesy of USDA Systematic Botany and Mycological Laboratory, Beltsville, Maryland.

Fig. 3.2.   Beverly T. Galloway,
courtesy of the National Archives.

Fig. 3.3.   Erwin F. Smith making sections of wilted watermelon,
courtesy of Special Collections, National Agricultural Library,
Beltsville, Maryland.

Fig. 3.4.   Merton B. Waite working on pear blight,
attic of the Old Main Building, USDA, ca. 1893,
courtesy of the National Archives.

Fig. 3.5.   Effie A. [Southworth]
Spaulding, courtesy of University
Archives, University of Southern
California.

Fig. 3.6. Sub-Tropical Laboratory, Eustis, Florida, ca. 1890s, courtesy of Special Collections, National Agricultural Library, Beltsville, Maryland.

Fig. 3.7. Office of USDA Division of Vegetable Physiology and Pathology, ca. 1890s, courtesy of the National Archives.

Fig. 3.8.   Herbert J. Webber in the Sub-
Tropical Laboratory at Eustis, Florida, at work
with the microtome, 1895, courtesy of Special
Collections, National Agricultural Library,
Beltsville, Maryland.

Fig. 3.9.   Walter T. Swingle, courtesy of the National
Archives.

Fig. 3.10.   Newton B. Pierce, courtesy of the
National Archives.

Fig. 3.11. Albert F. Woods studying the Bermuda Lily disease in the Laboratory of the Division of Vegetable Physiology and Pathology, courtesy of the National Archives.

Fig. 3.12. USDA Plant Pathologists, ca. 1890s, (L-R, front: David G. Fairchild, Palemon H. Dorsett, Beverly T. Galloway, Erwin F. Smith; back: Walter T. Swingle, Merton B. Waite, Mark A. Carleton, Albert F. Woods), courtesy of the USDA Systematic Botany and Mycological Laboratory, Beltsville, Maryland.

Fig. 3.13.   Roland Thaxter, courtesy of the Archives,
Farlow Reference Library of Cryptogamic Botany,
Harvard University.

Fig. 3.14.   Connecticut Agricultural Experiment Station, 1889, courtesy of the Connecticut Agricultural Experiment Station, New Haven.

Fig. 3.15.   George P. Clinton spraying for potato blight, 1903, courtesy of the Connecticut Agricultural Experiment Station, New Haven.

Fig. 3.16.   Byron D. Halsted, courtesy of
Special Collections and University Archives,
Rutgers University Libraries.

Fig. 3.17.   Lewis R. Jones, courtesy of
the National Archives.

Fig. 3.18.   Potato diseases and spraying, 1899, from Vermont Agricultural Experiment Station Bulletin No. 72.

Fig. 3.19. Henry L. Bolley, courtesy of
the Department of Plant Pathology,
University of Minnesota.

Fig. 3.20. George F. Atkinson, courtesy of the Iowa State
University Library, Department of Special Collections.

Practical Service to Farmers

In 1886, when Frank Lamson-Scribner was named head of the new Section of Mycology of the United States Department of Agriculture, he became the first scientist employed by the U.S. federal government with the specific charge of furthering the understanding and control of plant disease. The purpose of the new section was to investigate "the diseases of fruits and fruit trees, grains and other useful plants, caused by fungi."[1] To meet this challenge, Scribner would have to move disease control from the hands of agricultural empiricists into the laboratories of scientists who studied the life histories of fungi and sought to understand the complex relationship between pathogens and disease.

Scribner believed that science was the cornerstone of plant disease control. In his first USDA report, "Fungous Diseases of Plants," in 1885, one of his central messages was the requirement for more scientific work on the nature of fungal life cycles and the associated plant disease cycles. But just knowing "that a disease . . . is due to a species of fungus," Scribner insisted, was hardly enough information "to advise remedies or preventives." He emphasized that dependable and efficacious treatments would follow only after careful studies of "the life history of the parasite, its method of nutrition, growth and propagation, and the varied forms or conditions under which it perfects its spores or fruit; the manner of distribution, exactly how it comes upon or enters the affected plant, and its means of continuing its existence from year to year." With America's "crops . . . damaged to the extent of many millions of dollars annually by the attacks of fungi," Scribner added, science was more vital than ever to accelerating and extending plant disease investigations in the United States.[2]

Scribner's task was formidable. Placed in an institution established on the promise of practical service to American farmers, he would be evaluated not by his fundamental scientific discoveries, but by his ability to provide prompt solutions to plant disease problems. He had to find a way to merge science with practice. He needed a practical success story to prove the value of scientific research, which he viewed as essential. Wisely, Scribner chose to improve his chances for demonstrating the correctness of his beliefs by concentrating his energies on the diseases of one crop, the grape.

GRAPE DISEASES AND THE SECTION OF MYCOLOGY

During the second half of the nineteenth century, grape diseases were among the most visible and injurious plant diseases in the United States. Their destructiveness corresponded with the rise of a commercial grape industry, which had been expanding steadily since the 1830s. Although grapes were grown in many states, Ohio, Missouri, New York, and California had assumed the lead in production for market. The bulk of the harvested crop went into the production of wine.[3] In 1880 USDA officials estimated that the American wine industry would produce nearly 30 million gallons of wine at a value of $20 million.[4]

In the central and eastern states, grape growers increasingly found themselves confronting the destructive powers of black rot (*Guignardia bidwellii*) and downy mildew (*Plasmopara viticola*). In Ohio, the nation's leading wine-producing state in 1859, fungal pathogens played a major role in rendering vineyards "largely moribund" by the 1870s.[5] When Missouri took the lead in wine production in the 1870s, its growers also encountered black rot and downy mildew. One Missouri viticulturist told the American Pomological Society in 1879 that during four successive years black rot had "destroyed almost the entire crops of most vineyards in our region . . . planted at an expense of over a million of dollars."[6] USDA plant pathologist Albert F. Woods would later write that in the central and eastern United States, fungal diseases "practically destroyed the grape industry."[7] Although California vineyards were spared the onslaught of these two diseases because of different environmental conditions, California vines, too, would suffer from their own calamitous grape diseases.

By the 1880s most informed observers were attributing black rot and downy mildew to pathogenic fungi. European studies, along with those of American mycologists such as William Farlow at Harvard University and George Engelmann in Missouri, had furnished scientific information on grape diseases. Although much of this work fell into the category of descriptive mycology rather than economic botany, increasing knowledge of the nature of the fungi that caused the grape diseases encouraged the promotion of several useful disease-control strategies. These included the removal and destruction of diseased grape leaves and berries, and trellising and sheltering vines. Information on disease control was disseminated largely through agricultural and horticultural societies and the agricultural press. Unfortunately, such disease-control recommendations were impractical for large vineyards, which were challenged by unprecedented disease outbreaks during the second half of the nineteenth century.

Although entomologists had scored victories against the Colorado potato beetle and apple cankerworm in the 1870s and 1880s with chemical spray compounds such as Paris Green and London Purple, plant disease workers still had very few reliable chemical weapons in their arsenal. Since the late 1840s, many viticulturists had used handheld bellows to dust their vines with sulfur. This was a successful treatment for the less harmful powdery mildew (*Uncinula necator*), which occurred on the leaves, stems, and berries of grapevines.[8] But sulfur was one of the rare success stories and its success was only weakly connected to an understanding of the biology of the pathogen or the overall disease cycle.

Furthermore, black rot and downy mildew did not respond to treatment with sulfur. The reason, which mycologists and some perceptive agriculturists were discovering, was that not all pathogenic fungi, and, therefore, not all diseases, were alike. There were significant variations in how fungi survived, in their life cycles, and in how they did their damage. In the case of the two fungi associated with the grape mildews, scientists were accumulating evidence that the causal agents of powdery and downy mildews prospered in different climatic conditions, were morphologically distinct, and differed in their life cycles on the host plant.[9] While the fungus that caused powdery mildew grew mostly on the outside of the leaf, the fungus that caused downy mildew established itself mostly on the underside of leaves and grew extensively inside the tissues of the host plant. Because it was difficult to reach the mycelia of the downy mildew fungus with applications of sulfur, Scribner noted in 1885, "little can be done to check" its "ravages."[10]

It was a breakthrough in the control of downy mildew in Europe that gave Scribner and the Americans an important boost at a critical time. Downy mildew had been ravaging French vineyards since its recent introduction from North America. Pierre Marie Alexis Millardet, a professor of botany at Bordeaux and a former student of Anton De Bary, had been working on the life cycle of the causal fungus, *Plasmopara viticola*, since he and J. É. Planchon had discovered it in France in 1878. In 1882, the possibly apocryphal story of the discovery goes, Millardet noticed that some grapevines in the Médoc that had been sprinkled with a mixture of copper sulfate and lime to prevent pilferage were free from disease, whereas untreated vines in the same vineyard were diseased. He began to investigate the effect of combinations of copper salts and lime on controlling the disease.

Regardless of the truth of this tale, certainly Millardet already knew, as did many of his contemporaries, that copper sulfate had proved effective against the germination of fungal spores and subsequent infections, particularly with covered smut or bunt of wheat. In the United States, in California, for example, in the 1870s and 1880s, informed agriculturists made recommendations to use "bluestone" in addition to sulfur in the treatment of grapevine mildew.[11] Adding lime reduced the danger of injury to the grape foliage. From 1882 to 1884, the results of Millardet's work with copper sulfate were nothing short of stunning.[12]

News of the French success with the chemical remedy for downy mildew arrived in America just as Scribner was launching his plant disease work at the USDA. USDA entomologist C. V. Riley, who had gone to France to help viticulturists combat the insect phylloxera, made one of the earliest American references to Millardet's "Bordeaux mixture" in his paper "The Mildews of the Grape-Vine," presented to the American Pomological Society in September 1885. According to Riley, the mixture adhered well to the leaves and proved effective after only a single application.[13] Grasping the potential of copper sulfate and lime against grape disease, Scribner tentatively endorsed its use against downy mildew in his 1885 annual report. Not only was there now a prospect of eradicating downy mildew specifically, but mycologists also realized that copper sulfate and lime might hold promise in arresting the diseases caused by other pathogenic fungi, including the nearly forty species of *Peronospora* in the United States.[14]

In 1886, with the French experience as a guidepost, Scribner initiated original investigations for using Bordeaux mixture in the United States. Lacking sufficient in-house laboratory facilities and experimental plots, he called on the assistance of American growers to support the USDA's experimental efforts. In May, he sent two circulars to growers, hoping to gain their cooperation in testing the different chemicals being promoted for control of grape diseases and also to obtain statistical data on the range and extent of diseases of grape and other plants in the United States. The first circular included five different suggestions for chemical treatments. Bordeaux mixture, or, as it was popularly referred to, "the copper mixture of Gironde," was "highly recommended."[15] The second circular elicited "a more definite knowledge as to the distribution of and the losses occasioned by" grape diseases.[16] Although they were only marginally effective, the circulars represented a landmark effort by the USDA to gather and coordinate information on major plant disease problems.

A limited response to the first circular left Scribner with insufficient data to draw conclusions on the relative efficacy of the five proposed remedies. Although Scribner said that he had every "reason to believe that many made a trial of one or more of the remedies proposed," he admitted "that few responded to the request that the results of these trials be reported to the Department." Scribner blamed the poor response on two factors. First, American grape growers had little experience with systematic experiments. Second, many Americans were reluctant to use anything besides sulfur, which they believed to be a general "cure-all" for plant disease.

On the other hand, Millardet's Bordeaux mixture did obtain some positive results. One grower from Middle Bass, Ohio, reported that "sulphate of copper and lime were applied as directed in remedy No. 3 [Bordeaux mixture] with results that convince one that, with proper application, this remedy will prove more beneficial than anything yet known here for Mildew and Black-Rot."[17] Moreover, the response to the request of the second circular provided additional evidence of the principal diseases of the vine that were "widespread and devastating" to American viticulturists. USDA officials estimated from these returns that "the entire loss from mildews and black-rot cannot, on an average, be much less than 40 per cent. annually."[18]

Scribner used the response to the circulars as the basis for several important publications on grape diseases. The most significant of these was USDA Bulletin No. 2., "Report of the Fungous Diseases of the Grape Vine." This bulletin, a milestone in American plant disease studies, was the first scientific, book-length treatment dealing with the major known diseases affecting a single host. It signified that the federal government was taking the lead in publishing scientific material on plant disease. Besides contributing a wealth of general information on the diseases of the grape caused by fungi, the 1886 publication illustrated Scribner's commitment to plant disease control as "the primary and ultimate object of the work of this section."

Much of the information contained in UDSA Bulletin No. 2 was based on the results of European research, but Scribner refused to take the French cure for granted. The successful control of grape diseases in America, he believed, would require more systematic and thorough testing of fungicides under the distinctive conditions provided by soil, climate, plants, and microorganisms that occurred in the United States. The Section of Mycology had yet to acquire the equipment, personnel, and facilities to do scientific inquiry at this level, and the approach of having growers perform experiments had been but a partial success. In the meantime, Scribner was skeptical of placing too much stock in the reported American successes of copper sulfate and lime remedies, and he continued to recommend sheltering vines, bagging grape clusters, and removing and destroying plant debris as the most reliable control strategies.

Commissioner Colman acknowledged the limitations of the Section of Mycology and once again came to the aid of plant disease interests. In 1887, he won approval from Congress to enlarge the Section of Mycology into the Section of Vegetable Pathology, with a "small appropriation" and an assistant, Erwin Frink Smith, from Michigan.[19] Scribner became chief of the new section.

With an enhanced capability and mission, Scribner pushed forward to learn more about grape diseases and adapt copper-based fungicides for the United States. He pondered questions about the most suitable time in the growing season for spraying, the possibility that diluted mixtures could be just as effective and cheaper than full-strength mixtures, and the benefits that could be derived from improved spraying equipment. To acquire more reliable evidence on the problems before him, in 1887 he established several special "stations." Choosing vineyards in Vineland, New Jersey; Charlottesville, Virginia; Neosho, Missouri; and Denison, Texas—regions hard hit by both black rot and downy mildew— he designated the vineyard owners as USDA "special agents," with the duty of testing copper salts as possible disease remedies. To guarantee consistent results, the department supplied these agents with both chemicals and equipment for the field tests.[20]

While struggling to initiate practical fungicide experiments, Scribner turned to life-history research on the causal agent of black rot, making discoveries about the nature of the black rot fungus that would have profound implications for applied plant pathology. For many years, botanists had associated black rot on the fruit (*Phoma uvicola*) and the leaf spot disease of grape (*Phyllosticta labruscae, Phyllosticta viticola*, etc.) with separate fungi. Scribner himself had made this association in early publications.[21] However, his understanding of black rot began to change during the summer of 1887, when Pierre Viala of the National School of Agriculture at Montpellier arrived in the United States on a special mission from the French Ministry of Agriculture.

Viala was an authority on viticulture and vine diseases. His main purpose in coming to the United States was to search for American grape stocks that were resistant to phylloxera and would grow well in France. Commissioner Colman directed Scribner to accompany Viala on field excursions and furnish whatever facilities the department could provide. Between June and September, Scribner and Viala visited vineyards in New Jersey, Maryland, Virginia, North Carolina, New York, Ohio, Texas, and California.[22]

During their travels, Scribner and Viala compiled a comprehensive file of American grape diseases, including black rot and leaf spot. Viala had speculated in earlier mycological studies that black rot of the fruit and the leaf spot were caused by the same fungus, and his excursions to the vineyards of the United States convinced him that this theory had merit. At first, Scribner was reluctant to accept Viala's hypothesis, but his travels with Viala and additional research in Washington convinced him that Viala was right. His studies of fungal life-history clarified the situation.

The fungus overwintered primarily as perithecia on diseased grapes found on the soil surface. Through careful experiments Scribner found ascospores forcibly projected up to four centimeters distant from perithecia. Ascospores germinated and gave rise to infective mycelium within a few hours. When placed on healthy leaves still on the vine, they produced leaf spots characteristic of black rot within eight to twelve days.

In the field, leaves were attacked as they emerged in the spring. Afterward, spores from pycnidia on infected leaves were dispersed to the developing grapes. Thus, Scribner and Viala demonstrated clearly that the initial point of infection for the black rot fungus was the grape leaf, for leaves often showed symptoms several weeks before the fruit.[23] It was not surprising, therefore, that growers and "special agents" in 1887 reported poor disease-control results. They were applying copper sulfate too late. After the fungus was well established on the leaves, it was too late to prevent the spread of the black rot fungus to the grape berries.

During the 1888 growing season, Scribner focused his experiments at the Vineland, New Jersey, research station under the direction of Colonel Alexander W. Pearson, a skilled viticulturist and superintendent of the Vineland Wine Company. Scribner instructed Pearson to apply different copper sulfate compounds to grapevine foliage, shoots, and berries early and repeatedly during the growing season.[24] The emphasis was on the application of these chemicals to the foliage before the first appearance of the black rot on the fruit.

In marked contrast to the year before, the 1888 experiments at Vineland were a stunning success. The treatments proved highly effective against black rot, and Bordeaux mixture clearly outperformed Eau Céleste and the other copper compounds. Scribner was able to proclaim success at last. In August, he wrote to Hermann Jaeger, special agent at Neosho, Missouri, declaring that "the treatment made at Vineland . . . with the Bordeaux mixture renders *it no longer doubtful* that by *proper application . . .* we can *subdue* or *even entirely prevent the black rot.*"[25]

The 1888 experiments were a major victory for the USDA, for American plant disease workers, and, indeed, for American viticulturists. Subsequent field tests at the other research stations in the United States confirmed the findings at Vineland, and the conclusions of the effectiveness of Bordeaux mixture against black rot were corroborated shortly thereafter by M. Prillieux, inspector-general of agricultural education in France and professor at the Institut Agronomique.

Scribner had shown the value of combining careful research in the laboratory with practical disease-control experiments in the field. Moreover, his collaboration with Viala demonstrated the value of international cooperation. By the end of 1888, the prospect of controlling two major grape diseases with fungicide applications was encouraging. Scribner had not only verified and extended important European work on the treatment of downy mildew, but also had contributed substantially to the understanding and prevention of black rot. Viala, in recognition of Scribner's achievement, wrote late in 1888 that "the discovery of an efficacious course of treatment for Black Rot will permit . . . the development of American viticulture on a new basis."[26]

In three short years, Scribner's work on black rot established a model for applied plant pathology in the USDA. He had mobilized resources beyond his scientific expertise to attack specific agricultural problems, much as federal entomologists and animal scientists had with locusts and cattle pleuropneumonia. He had initiated the practice of using growers to supply statistics on crop health and to perform experiments, and he had used circulars, bulletins, annual reports, and the agricultural press to disseminate the results of USDA plant disease research. His invention of a system capable of large-scale, chemical control of plant disease was a victory for federal science and elevated the status of plant pathology in the USDA. His work would play a significant part in launching applied plant pathology in the United States, and would also contribute to the fundamental alteration of agricultural practice through the advent of chemical fungicides for plant disease control.

In 1888, with USDA-sponsored plant disease research off to a successful start, Scribner decided to leave the USDA for a higher salaried position as professor of botany and horticulture at the University of Tennessee and botanist and horticulturist at the Tennessee Agricultural Experiment Station. But before he left, he recruited a number of qualified scientists to continue and expand the work of the Section of Vegetable Pathology. Scribner's legacy of a scientific approach to solving practical disease problems and his insight in bringing together scientists who would become leaders in USDA plant pathology would continue to serve American agriculture for many years to come.

Transition and expansion

After Scribner's departure in the middle of 1888, his former assistant, Beverly Galloway, took over as chief of the Section of Vegetable Pathology. At the time, the section was merely the framework of the organization that in 1891 would become the Division of Vegetable Pathology, and in 1894 the Division of Vegetable Physiology and Pathology. In late 1887, when Galloway had arrived as Scribner's assistant, the section occupied only one room on the third floor of the old Administration Building in Washington, D.C., and the staff was quite small. There were three assistant mycologists—Galloway, Erwin Frink Smith, and Effie Southworth—plus three or four clerks and a typist.[27] But under Galloway's dynamic leadership, U.S. plant pathology reached new heights. He understood that the section's value lay in practical service to its main constituency, the growers, but he refused to neglect the basic scientific research that would lead to significant breakthroughs in years to come. A master of scientific and political administration, Galloway would become an influential voice in government, science, and education, and also one of the central figures in the rise of American plant pathology in the late nineteenth and early twentieth centuries.

The trajectory of Beverly Thomas Galloway's career demonstrated the evolution of the professional plant sciences and plant pathology in the last two decades of the nineteenth century. Born to farming parents on October 16, 1863, in Millersburg, Missouri, Galloway at first embarked on a career as a pharmacist. But in 1880 he gave up his pharmacy practice to enter the University of Missouri's Agricultural College at Columbia.[28] There he studied under Professor Samuel Mills Tracy in botany and horticulture. He received his bachelor of science degree in 1884.[29]

Expanding career opportunities in the plant sciences encouraged industrious students such as Galloway to consider further education. In 1885 he began graduate work in botany at the University of Missouri, where he continued to study under Professor Tracy, took courses in chemistry, biology, and foreign languages, and assumed responsibility for greenhouse management.[30] Before long, however, he expressed an interest in plant pathology. He presented papers about plant disease at the Missouri State Horticultural Society and coauthored several mycological articles with Tracy and others, such as the noted New Jersey mycologist Job Bicknell Ellis.[31] He displayed a regard for the public-service aspect of disease research, publishing "practical instructions" on disease prevention in local newspapers.[32]

Gaining access to the network of scientists and professionals shaping the new discipline, Galloway corresponded from the University of Missouri with the leaders in American plant disease research, such as Farlow at Harvard and Burrill at Illinois.[33] Commissioner of Agriculture Norman Colman, a fellow Missourian, learned of Galloway's talents when Galloway published his first paper on a plant disease, "*Puccinia graminis*—Rust of wheat, oats, etc." in *Colman's Rural World* in June 1885. When the USDA expanded the Section of Vegetable Pathology in 1887, Galloway was a logical choice to become Scribner's new assistant. He served in that position only a short time, before taking over as chief of the section from the departing Scribner in 1888.

As chief, Galloway continued the grape disease research begun under Scribner. Black rot and downy mildew were responding quite well to fungicidal treatments and provided models for further experimentation with chemical control.[34] Underscoring the importance of grape diseases, Galloway "personally conducted" fundamental research on black rot. He attempted to clarify the method of the spread of the black rot fungus from leaf to berry, as proven earlier by Scribner.[35] In addition, he demonstrated that Bordeaux mixture could protect up to 99 percent of the grape crop from black rot.[36] Viticulturists across the United States, learning of the disease-fighting capability of copper sprays, moved quickly to adopt their use.

By 1888 the USDA was poised to begin expanding its vistas of plant disease research beyond grape diseases to the understanding and chemical control of other pathogenic fungi. Galloway shepherded this expansion. USDA Commissioner Colman wrote in 1888 that "for the past two years the money and energy of the section have been especially devoted to experimenting on the fungi affecting the grape, but requests are coming from farmers and gardeners all over the Union that we investigate the causes of, and try the effects of fungicides upon, other diseases."[37] The diseases discussed in the 1888 commissioner's report, in addition to those of the grape, included downy mildew of potato, black rot of tomato, brown rot and powdery mildew of cherry, leaf blight and cracking of pear, apple rust, celery leaf blight, leaf spot of rose and maple, and leaf rust of cottonwood.[38]

Taking advantage of the ground that Scribner had prepared, Galloway began to develop his own system of federal applied plant pathology because, as he wrote, "we fully appreciate the fact that practical results are what the farmer, gardener, and fruit-grower are after."[39] He solicited input from growers to guide him in deciding which disease problems needed study. But Galloway wanted reciprocity from these agriculturists to aid in the growth of his section. He asked each one to "write to your Congressman urging him to . . . use his influence in securing the appropriation for the *investigation of plant diseases*."[40] The proof of Galloway's success in reaching out to the growers was that the section experienced significant increases in the amount of mail it received—from 500 letters in 1887 to 3,000 in 1890—and it also received increased appropriations.[41]

Galloway was not satisfied with the "grower" agent system that he had inherited from Scribner. Although he understood the necessity of using agents, he also saw the limitations of such a system. Plant disease research was changing rapidly and growing more complex each year, and Galloway realized that many of the growers who were co-opted for research were not always adept at performing careful experiments or even at carrying out the specific instructions for chemical preparations and applications. In keeping with what he saw as a need for scientists to begin doing more of the actual field research, Galloway asked the USDA to establish its own experiment station, arguing that most state agricultural experiment stations were not yet engaged in the study of diseases caused by fungi and, as a result, could not contribute significantly to pathological work. A federal experiment station, a supportive Commissioner Colman wrote in 1888, "would be of untold value," and the USDA "could confidently expect that the practical results would be threefold what they are at present."[42]

Although Galloway's desires for a station were left unfulfilled, he continued his expansion of pathological studies by encouraging more experimentation by grower agents and by improving the section's facilities. In 1888 the grower agents of the section reported from only five states; in 1889 that number had doubled. The investigations also expanded from an emphasis on the fungicidal treatment of downy mildew and black rot of grape to other grape diseases, and to diseases of apple, pear, quince, peach, melon, strawberry, blackberry, tomato, and potato.[43] Meanwhile, Galloway acquired a small greenhouse and devoted a great deal of attention to expanding the USDA fungal herbarium, which increased from 3,000 to 14,000 specimens in just a few years.[44] To keep up with the continued growth in number of these specimens, he brought Franklin Sumner Earle and Flora W. Patterson into the department to assist with the herbarium work.[45]

More important to the understanding and control of fungal diseases was the work of an enlarging core of able assistants, including Effie A. Southworth and David G. Fairchild. Southworth held the distinction of being the first woman researcher in the USDA.[46] The department had employed women prior to her arrival, but only to mount specimens.[47] Southworth had worked in federal plant pathology before Beverly Galloway. She received her bachelor of science degree at Michigan in 1885, after which she accepted a two-year appointment as fellow and instructor at Bryn Mawr College.[48] She was hired by the USDA in 1887 on the recommendation of Erwin Frink Smith, who was Scribner's special agent at the time.[49]

Southworth was an expert microscopist, and at the USDA she became heavily involved in the study of fungal pathogens. Over several years, as assistant pathologist, she worked closely with Galloway to prepare numerous mycological publications. Perhaps her most significant contribution was an 1891 *Journal of*

Mycology article on anthracnose of cotton, describing, for the first time, *Colletotrichum gossypii.*[50] In 1893 Southworth left the USDA to become an assistant in botany at Barnard College. Later she would move to the West Coast to join the botany faculty of the University of Southern California.

David G. Fairchild, another Galloway assistant, was the nephew of Byron D. Halsted, a noted New Jersey plant pathologist and a frequent contributor to USDA publications. Fairchild received a science degree from Kansas State College in 1889, working under W. A. Kellerman. After joining the USDA in 1888, Fairchild worked closely with Galloway, both in the laboratory and in the field, contributing a great deal to the expansion of the section's scope.

In 1890 Galloway and Fairchild made a series of fungicide tests near Washington, D.C., on black rot of grape, pear scab, and leaf blight of pear, cherry, and strawberry.[51] Their work on black rot and the application of copper fungicides saved an unprecedented 93 to 99 percent of the grape crop; untreated vines failed completely.[52] The next year, Fairchild also worked with Galloway on the chemical control of orchard diseases, including scab and leaf blight of pear.[53]

Tied closely to advances in the chemical control of fungal diseases were improvements in the technology of sprayers for applying fungicides. Scribner had come to understand the importance of better spraying equipment during his extensive work with grape diseases. He introduced the French knapsack sprayer to the United States and had six of them made for use at his different grape disease research stations.[54]

When Galloway was placed in charge of the section in 1888, he became a driving force behind new sprayer technology. American manufacturers at first showed an indifference to producing sprayer equipment.[55] To counter the lackluster response by American industry, Galloway personally oversaw the design and construction of a knapsack sprayer by two private companies. He took the lead in making chemical sprayers, particularly the knapsack version with improved pumps and nozzles, more accessible to growers.[56]

Eventually, the number of sprayers available commercially to growers increased. The American sprayer industry continued to develop through the 1890s with increased specialization of equipment for specific crops and the variations of sprayer sizes from the knapsack model to steam- and gasoline-powered machines for large acreages.[57]

Galloway promoted USDA applied plant pathology through a prolific use of publications. In addition to publishing articles in the USDA's annual report, he issued many bulletins and circulars on specific diseases and their treatments. The bulletins often contained striking photographs of orchards or fields ravaged by diseases. He might, for example, juxtapose a picture of a decimated orchard labeled "Pear Leaf Blight—Untreated" next to a picture of a healthy orchard with the caption "Pear Leaf Blight—Treated."[58] Galloway also promoted the section's value by giving information on the amount of crop yield saved from diseases by chemical spraying.[59] He did not hesitate to estimate the amount of dollar profit accruing to growers if they followed the section's suggested fungicidal treatments.[60] He took satisfaction in pointing out how many copies of each publication were printed and how quickly they were requested and distributed.[61] This information was powerful ammunition when he went before the secretary of agriculture or Congress to justify the utility of the section's work or to make staffing and budget requests.

Galloway's contribution to more strictly scientific publishing included launching the first true serial dedicated to phytopathology in the United States. In March 1889 the Section of Vegetable Pathology assumed responsibility for the publication of the *Journal of Mycology*, and the journal's mission shifted almost exclusively from reporting systematic mycology to publishing information on plant disease caused by fungi. Galloway valued the journal, not only as a venue to showcase the applied work of his section, but also as visible proof of the fundamental research being done in the laboratory. Distributed to "botanists of all countries," the new journal advertised to the scientific community that the section was interested in fundamental scientific questions.[62] Although Galloway and his assistants in the section wrote most of the articles, he also sought contributions from experts such as his former boss, Frank Lamson-Scribner at the Tennessee Agricultural Experiment Station, and Byron D. Halsted of the New Jersey Agricultural Experiment Station at Rutgers. Such cooperation was part of Galloway's commitment to coordinate the section's activities with the pathological work being carried on at the nation's agricultural experiment stations. It was another example of his positioning the section in a central role in the growing community of plant disease researchers in the United States.

ENLARGING THE SPHERE OF RESEARCH

Galloway continued to gather a nucleus of talented assistants in the Section of Vegetable Pathology. Southworth and Fairchild fulfilled the primary goal of the section, that is, the understanding and control of fungal diseases. But other assistants allowed Galloway to move research in new directions and thus expand the scope of federal plant pathology. Notable among these scientists were Merton B. Waite, Newton B. Pierce, and Erwin F. Smith. The work of these scientists reached out both geographically and scientifically to explore aspects of plant pathology beyond those associated with fungal pathogens.

Joining the section in 1888, Merton B. Waite was the first of the section's assistants to investigate bacterial diseases of plants. Perhaps his most notable success was unlocking the mystery of the transmission of fire blight of pear and, thereby, initiating the study of insect vectors associated with plant disease. Waite had studied at the University of Illinois under Burrill, so it was not surprising that he set out to do work on bacteriology when he joined the section.[63] Borrowing equipment from the Section of Animal Industry, in 1889 he set up the section's first bacteriological laboratory. Although he was aware from his work with Burrill that fire blight was caused by a bacterium, as a worker in the Section of Vegetable Pathology he did try spraying copper fungicides for fire blight, reasoning that the method had proved effective on many other diseases.[64] In 1889 and 1890, he traveled to Georgia, where pear blight was especially widespread and severe, and spent several months testing fungicidal sprays. These treatments had no apparent effect.[65]

Seeking to better understand the pear blight organism, known at the time as *Micrococcus amylovorus,* Waite divided his time between field and laboratory work.[66] He isolated the bacterium in the lab to use for his inoculations and, according to E. F. Smith, provided the best description of the organism to date.[67] From his initial observations, Waite theorized that the bacteria "multiply in the nectar of the flowers, and are carried by insects from one flower to another."[68] To verify his hypothesis, he devised a field experiment whereby insects were permitted to visit certain trees on which flowers were infected artificially with the bacterium. The insects were then allowed to visit the flowers of other nearby, healthy trees. Waite discovered that the pathogen was indeed spread to nearby pear trees, and these trees became diseased. Other trees whose flowers were protected by bagging showed no symptoms of fire blight. Waite's experiment was an important breakthrough in pathogen transmission studies and a major contribution to plant pathology.

Another bacterial researcher in the section was Newton B. Pierce. Pierce was among the first of a new generation of scientists trained specifically for a career as a federal plant pathologist. He studied plant science under Volney M. Spalding at the University of Michigan, bacteriology under F. G. Novy at Michigan, and entomology at Harvard University.[69] Spalding maintained close contact with his former student E. F. Smith at the USDA, and he tapped into this connection when some of his students approached the end of their graduate careers.[70] Spalding wrote to Smith in November 1888 about the possibility of federal research on a mysterious ailment of the vine in southern California. It "makes me ache to have some one ready to take it up," Spalding confided, "and if Galloway is not in too much of a hurry I have little doubt that in due time he can be provided with a trained man from our laboratory."[71] That trained man was Newton B. Pierce, who joined the USDA in May 1889.

Pierce launched his study of the California vine disease immediately. This highly destructive disease had first attracted attention in southern California around 1884, specifically in the grape districts around Santa Ana and Anaheim. Scribner and Viala had observed the disease in their travels in 1887, but they had offered no solutions.[72] In March 1889 a viticulturist from California paid a direct call on the secretary of agriculture to insist that "this disease, if not speedily checked, bids fair to wipe out of existence one of the leading industries of the country." The USDA pledged assistance, and shortly afterward, Galloway announced plans to send "an expert" to California for thorough investigations.[73]

Pierce arrived in Santa Ana, California, on May 23, 1889, and over the next ten months he devoted his full attention to southern California vineyards and the mysterious vine disease. His early doubts about the popular fungal theory as the cause of the vine's troubles and his tendency toward slow, meticulous research raised the ire of growers who cared "little for the name or history of the disease; all they want to know is how to cure it."[74] He found himself embroiled in a maze of causal theories, control strategies, and state politics.[75]

Pierce's investigations that year included mapping the disease's distribution, studying the "non-parasitic agencies" of climate and soil, considering parasitic insects and fungi, and trying to determine if the disease was contagious.[76] By June 1889 Pierce had found bacteria "present in the diseased vines," and he began to suspect that bacteria were the cause of the malady.[77] Confident that fungi and insects could be discounted, he initiated more bacteriological investigations.

But Pierce simply did not have the facilities in California to do the necessary level of research. Inoculation tests "with these germs" were begun in California but had to be "transferred to Washington, owing to the all-pervading nature of the disease in this region, and the difficulty of keeping plants from it." He also relocated grafting experiments to the department.[78] In March 1890 he prepared to return to Washington, D.C., without the proof he needed to declare that bacteria were responsible for the "relentless and complete" death of the vine in California. Similarly, he had found no preventive or remedy.[79] Pierce had found the California vine disease to be a stubborn enigma—highly contagious but without an identifiable causal agent.

Indeed, the scientists of the Section of Vegetable Pathology were beginning to discover that broadening their lines of investigation held enormous potential for trouble. Expansion of disease work meant encountering biological mysteries, many of which resisted explanation and treatment. The conflict between the section's mission of practical disease control and the time-consuming requirements of research in applied plant pathology deepened. Nowhere was this dichotomy more evident than in the struggle of E. F. Smith with peach yellows.

American farmers had suffered the effects of peach yellows for over a century. Plaguing the Northeast in the late 1700s to mid-1800s, yellows also became the bane of orchardists in the Midwest. By the late 1870s peach yellows had become so destructive in Michigan that the state enacted a law requiring the burning of all affected trees. Similar laws were passed in the 1880s in New York and Ontario, Canada, and there was support for such laws in Maryland and Delaware.[80] The laws did not eradicate yellows from affected areas, but in some regions they "assisted the growers to meet the destroyer in successful combat."[81] E. F. Smith, a future leader of American plant pathology, devoted an early and frustrating portion of his research career with the USDA attempting to unravel the mysteries of this disease.

Erwin Frink Smith was born on January 21, 1854, in Gilbert Mills, New York. In 1870 his family moved to Michigan, where he entered public school at the age of eighteen. His educational path was not particularly unusual for the time, and Smith, a bright and eager student, had the benefit of tutoring in literature, languages, and the sciences.[82] As he learned more and began to develop an interest and expertise in science, he became intrigued with diseases, both human and vegetable. For a time, he worked for the Michigan State Board of Health, where he was exposed to revolutionary work on the human germ theory of disease, which was coming from European scientists such as Louis Pasteur, Joseph Lister, and Robert Koch.[83] He attended classes at the Michigan Agricultural College before entering the University of Michigan, where he graduated with honors with a bachelor's degree in science in 1886. During his years at the university, he had gravitated toward the study of botany and plant disease under the professorship of Volney Spalding. He also developed a correspondence with notables in plant science, including Farlow and Burrill.[84] In 1885, before he had even obtained his bachelor's degree, he was elected to the American Association for the Advancement of Science at the same meeting where J. C. Arthur's Committee for the Encouragement of Researches on the Health and Diseases of Plants reported that it had helped convince the Department of Agriculture to begin formal work on plant disease.[85]

At the University of Michigan, Smith specialized in the study of *Peronosporaceae*, the family of fungi responsible for both the potato rot and grapevine mildew.[86] His first major paper, in 1885, was on potato late blight, drawing the life history information on *Phytophthora infestans* from De Bary's classic work. The article was published in several forms in the *Michigan Crop Report*, the *Annual Report of the Wisconsin Agricultural Experiment Station*, and the *Report of the Commissioner of Agriculture*.[87]

Smith's work on potato rot, in addition to the research he did on black rot of grape, brought him to the attention of Frank Lamson-Scribner. Smith had assumed that after graduation he would either continue at the University of Michigan or seek a position with one of the many experiment stations being initiated across the country. However, a letter from Scribner and an invitation to accept a temporary position with the Department of Agriculture changed Smith's life.

In the fall of 1886 E. F. Smith joined the USDA as Scribner's assistant in the Section of Mycology. His first duties included preparing part of the seminal Bulletin No. 2, "The Report of the Fungous Diseases of the Grape Vine" and creating a map showing the distribution of potato rot in the United States.[88] His knowledge of foreign languages allowed him to translate "Prevention of Mildews

—Results of Experiments with Various Fungicides in French and Italian Vineyards in 1885," which contained Millardet's account of the discovery of Bordeaux mixture.[89]

In July 1887 Smith received a new assignment as special agent to study peach yellows.[90] For the next seven years, Smith's work would be dominated by peach diseases, particularly yellows and peach rosette. Peach yellows had stymied even the likes of Burrill, who was convinced that it was bacterial in origin, but had not been able to support his theory with hard evidence. Yellows and rosette were to frustrate Smith as well. The discovery of the causal agents defied his most detailed observations and meticulous research.

In 1887, with the assistance of a college friend, Liberty Hyde Bailey, a former assistant at Gray's herbarium at Harvard and now a professor at the Michigan Agricultural College, Smith began his extensive fieldwork on peach yellows in Michigan. With Bailey's help, Smith procured the research expertise of other scientists, such as William Rane Lazenby of Ohio State University, and L. R. Taft of the University of Missouri.[91] He traveled through Michigan, Pennsylvania, New Jersey, Delaware, and Maryland, observing diseased trees in peach orchards, talking with growers, and collecting soil, wood, and leaf samples. He sent many samples back to Washington for chemical analysis, but Smith himself spent little time there. In fact, when Galloway joined the section, he found that Smith "spent nearly all his time in the field mostly in Delaware and Maryland. Usually during the winter he would come in for a month or two to check up his field activities."[92]

In his search for the causal agent of peach yellows, Smith's method was systematically to discard suspects. Over the years many different insects, weather patterns, and cultivation practices had received the blame, but by 1888 the most popular theory was that peach yellows resulted from poor nutrition and soil exhaustion. This concept had been advanced vigorously by Charles A. Goessman and David Pearce Penhallow of the Massachusetts Agricultural Experiment Station at Amherst.[93] Smith questioned this theory, as did several notable horticulturists, and Smith set out to test it empirically.[94]

Over a period of more than a year, Smith collected, summarized, and analyzed data on weather and cultural practices and did extensive testing on soil chemistry and fertilizers, including moving a train carload of soil from a diseased orchard in Delaware and placing it around yellows-free trees in Michigan.

No diseased trees resulted from this somewhat unusual maneuver.[95] Thus, in 1889 Smith tentatively ruled out soil exhaustion as the cause of peach yellows. In 1891 a series of fertilizer experiments finally allowed him to state that "the Goessman-Penhallow method of treatment was founded on an error."[96]

Smith had suspected from the beginning that the yellows was a contagious disease caused by a microorganism. Smith was aware of rudimentary inoculation tests using diseased buds conducted by William Prince as early as 1828 and by Noyes Darling in the 1830s.[97] In November 1887 he began to search for a pathogen by comparing healthy and diseased peach twigs under the microscope. Deciding that "larger fungi are out of the question," he turned his attention to bacteria. He anticipated growing the disease-causing bacterium in pure culture and doing inoculation tests along the same lines as Burrill's and Arthur's pioneering work on fire blight.[98]

But Smith never made this breakthrough. He found himself continually frustrated in the search for the mysterious cause of peach yellows, of peach rosette, and of another disease called "little peach," which he encountered in Michigan. And while he struggled with the biological aspects of peach yellows in the field, he found himself continually under fiscal and administrative pressure from the USDA. Smith's techniques and approach were far more involved and intricate than anything Scribner had experienced or was in a position to support indefinitely.

Smith responded to repeated calls from Washington for results with pleas for more time. Scribner generally relented to Smith's desires, but he warned that the department needed something for its investment and that funding would not last forever.[99] Smith's thorough scientific method did little to help Scribner in his quest for practical solutions. As Scribner wrote about Smith in July 1888: "I directed him early in December to prepare a report on the subject of yellows as it is known today. And to state briefly just what he had done or accomplished in the field and laboratory. . . . Mr. (Smith) signally failed to meet my expectations. Instead of making a report covering the ground desired, he devoted all the allotted time to the consideration of a single portion of it which, however well done in itself, was of no earthly use to us."[100]

When the money ran out in February 1888, Smith encouraged an impressive array of horticulturists and scientists to plead his case to the USDA. The influential Maryland orchardist, W. S. Maxwell, concerned over the cancellation of the yellows work, wrote that "this will never do. I am writing to gentlemen of influence who will get our Congressmen interested in the matter."[101] A few interested parties offered to find alternative funding, and many offered positions to Smith.[102] But Smith also had his doubters. Charles W. Garfield, a horticulturist friend of Smith's who had taught at Michigan Agricultural College, advised him to find a different career, writing to Smith that "I should not sacrifice myself on the altar of Science were I in your place. . . . I would drop yellows or blues and delve in a field where payment would be made for service rendered."[103] But Smith ignored Garfield's advice, deciding to wait for a resumption of government funding.

Scribner and Colman recognized that Smith was a tireless worker with an ability to master a subject. Although he might resist following orders that did not fall in line with his research goals, he was too good a scientist to lose. In addition, the outcry from horticultural interests demonstrated the value that these interests placed on Smith and the section's work. In the summer of 1888, the department again retained Smith's services to finish the work on peach yellows.[104] Published in March 1889, his preliminary report, Botanical Division Bulletin No. 9, titled "Peach Yellows," contained a thorough history and a discussion of the soils and weather conditions associated with the disease. Although it contained an extended examination of symptomology, Smith was still unable to determine the causal agent.[105] Nevertheless, Smith submitted his work to date on yellows to the University of Michigan, and it earned him a doctorate of science.[106]

Smith continued his research on yellows and rosette for several more years, publishing his final report in 1894. In this report, which was much shorter than the preliminary report of 1889, Smith admitted that "with our present knowledge the cure of peach yellows appears to be impossible."[107] Despite his extensive and painstaking research, Smith was no closer to unveiling the cause of the disease than when he began his research in 1887. "No fungus has been found associated with it constantly," Smith wrote, "and it is almost certainly not a bacterial disease, statements to the contrary resting upon evidence no careful mycologist or bacteriologist would for a moment be willing to accept."[108]

Smith's 1894 report had none of the self-confidence or excitement of his preliminary report of 1889. The process of eliminating causal agents begun in 1888 had succeeded in eliminating every cause known to science. In some ways, the

1894 report was an admission of defeat both for Smith and for the plant disease research and control system initiated by Scribner and institutionalized by Galloway. The department had invested a great deal of time and effort into investigating the problem of peach yellows, but had come up with neither a cause nor a solution.

Smith's peach yellows investigation was a rare setback for the early USDA. Yet in many ways the studies on peach yellows represented a degree of scientific maturity that was uncommon in American plant pathology at that time. Although Smith did his share of spraying Bordeaux mixture in attempting to solve the problem, he certainly did not qualify as what Harvard mycologist Roland Thaxter would pejoratively refer to as a "squirt gun botanist." In fact, by refusing to speculate about the cause of yellows, Smith had jeopardized his career.

The "failure" to solve the mystery of peach yellows emphasized the need for more basic research to support disease control. Beverly Galloway was beginning to formalize an institutional model to do the experimental work that nonprofessional, "grower" special agents had done in the late 1880s. In the future, he would acquire an even larger staff, with many of the best plant pathologists in the country, to work in Washington as well as to establish sophisticated laboratories around the country. Over the next decade, the system created by Frank Lamson-Scribner and expanded by Beverly Galloway would run well, propelling the study of plant disease to new heights. With increased ties, both formal and informal, to state agricultural experiment stations, Galloway would come to initiate or coordinate much of the plant disease research in the country.

CHAPTER 8

Plant Pathology Nationwide

In the 1880s and 1890s rapidly increasing market opportunities in the United States encouraged agricultural specialization and created economic incentives to provide solutions to plant disease problems. At the same time that Frank Lamson-Scribner and Beverly T. Galloway were giving plant pathology a firm scientific and financial base in the federal government, state agricultural experiment stations across the United States were establishing laboratories dedicated to understanding and combating plant disease. Research at these stations was generally aimed at seeking solutions to pressing agricultural problems, with general lines of work defined by state or regional needs.

J. C. Arthur's research on fire blight at the New York station at Geneva from 1884 to 1886 was among the earliest systematic work on a plant disease by an experiment station scientist. By the early 1890s, however, a significant portion of the research in plant pathology in the United States was being done at experiment stations from Connecticut to Alabama and from Texas to North Dakota. This expansion of research at the state and regional level was made possible by the federal government's commitment to both financial and organizational support. Agricultural experiment stations were important institutions in creating a professional network linking personal acquaintances, educational institutions, the agricultural press, scientific organizations, and the U.S. government into the new profession of plant pathology.

The hatch act

The first state agricultural experiment stations were created in the United States in 1875 by Connecticut and California.[1] In the next twelve years several other states established experiment stations. Most were funded by state appropriations or as part of land-grant colleges, but some were supported by private donations and subscriptions. Since experiment stations often served as the official state agency for certifying fertilizers and/or seeds, some survived by collecting a licensing fee or tax for such services.[2] But political and financial support was always tenuous, and the early experiment stations teetered constantly on the edge of financial extinction.

Experiment station personnel were involved in the long-running debate over whether agricultural scientists should focus on immediate, practical service or on basic research. Because chemistry was the dominant agricultural science at the time, staff scientists were most often chemists. The emphasis on practical service meant that they spent much of their time analyzing the composition of commercial fertilizers. Station scientists with botanical training often found their time occupied with identifying weeds and examining seeds for purity.

Although there were few opportunities for mycologists or botanists who wanted to study plant disease, there were some exceptions. In the late 1870s, at the experiment station at the Massachusetts Agricultural College, Charles A. Goessman and David P. Penhallow studied peach yellows.[3] But they directed their research toward soil nutrition and chemistry rather than pathology. In the mid-1880s, in work more directly related to biology and pathology, J. C. Arthur and Emmett Stull Goff at the station in Geneva, New York, investigated a variety of disease problems, with a general focus on diseases of tree fruit and garden crops of upstate New York. Between 1880 and 1883, researchers at the California station investigated diseases affecting grapes in local vineyards. This work, however, was financed by a one-time state grant awarded specifically for grape diseases and was not viewed as a regular part of the station workers' duties.[4] Plant disease investigation was often mentioned in experiment station charters, such as that of the University of Minnesota in 1885, but mention in a charter was not a guarantee that station workers could devote much time or energy to plant disease research.[5]

In the 1880s, the same forces that urged the USDA personnel toward more scientific research and advocated the inclusion of science as a requisite part of the land-grant college mission demanded that state experiment station workers shift from strictly practical service and into more fundamental science. Land-grant college officials, as well as scientific professionals and educators, particularly those working in the extant state experiment stations, were the most vocal advocates of this new emphasis. "The plea must ever be made for work more scientific in its character," wrote Edward L. Sturtevant, director of the New York

State Experiment Station at Geneva, to Charles Bessey in 1886, "and every Station must abandon the crudities forced upon it by the popular prejudice."[6]

In response to the numerous calls, on March 2, 1887, Congress passed the Hatch Act, which created a federally funded system of state agricultural experiment stations. The Hatch Act was named for William H. Hatch of Missouri, chairman of the House Committee on Agriculture and one of the primary supporters of federal funding for state experiment stations. The Hatch Act was the culmination of a long battle by the proponents of agricultural science to bring the financial power of the federal government to the aid of the nation's farming interests at the state and local level. Under the act, the number of experiment stations across the United States grew quickly. Prior to the passage of the act, there were seventeen experiment stations in fourteen states. By early 1888, there were forty-six American experiment stations, forty-three of which were funded by appropriations from the Hatch Act.[7]

PLANT DISEASE WORK AT THE EXPERIMENT STATIONS

By the last decade of the nineteenth century, workers at many experiment stations around the country were doing plant disease research. This research was often patterned after the applied work begun by Frank Lamson-Scribner and Beverly Galloway at the USDA. Workers at the stations tested fungicidal sprays on locally troublesome diseases, repeating and revising experiments that had been done elsewhere and reconfirming control practices already proven successful. Although the results were not wholly original, this was an effective, sound method for solving many disease problems on a state-by-state basis, and it fit well into the experiment stations' mandate to provide practical assistance for agriculturists. Confirmation of the effectiveness of specific sprays in a specific geographical region reflected an acknowledgment of the effect of differing weather and cultural conditions among states. It also showed the experiment station scientists' desire to maintain their credibility with growers by allowing them to recommend only those practices and remedies that they had tested and found effective. And although fungicidal testing was a mainstay of the work in the experiment stations, scientists at some stations were branching out into new control measures and research on new diseases.

An excellent example of general plant pathology at a station was the work of Frederick D. Chester of the Agricultural Experiment Station at Delaware College. Like most of his fellow station "plant pathologists," Chester was a competent botanist with mycological experience. He investigated diseases caused by fungi, and he even contributed information on previously undescribed diseases such as tomato anthracnose (*Colletotrichum phomoides*).[8] But, like most of his fellow station botanists, Chester believed that fungicidal spraying experiments were the best measure of the value of the station's plant disease research to growers. He tested a variety of fungicides on different diseases that were disruptive to Delaware crops, such as pear scab, leaf blight of pear and quince, potato late blight, and bitter rot of apple.[9] Chester's experiments were well organized and valuable. They expanded knowledge about the application of chemical controls to plant diseases in Delaware and the mid-Atlantic region.

In some cases, the USDA took an active role in sponsoring plant disease studies at experiment stations. Galloway once said that one of the main reasons he had taken over publication of the *Journal of Mycology* was to provide a means of getting pathological information to experiment station workers, thus encouraging them to move more forcefully into plant pathology. In a few instances, Galloway devised research plans, provided funding, and guided station scientists through initial experiments. For example, in 1889 he planned and funded the research by E. S. Goff, station horticulturist at the Wisconsin station, on apple scab caused by *Venturia inaequalis*.[10] Goff had left the New York State Agricultural Experiment Station at Geneva in 1889 and journeyed to Wisconsin, where, under USDA direction, he and A. L. Hatch, a Wisconsin orchardist, applied copper compounds to apple scab with remarkable success.[11]

Dominated by regional concerns, experiment stations struggled to achieve a balance between practical service and fundamental science. Everyone recognized that basic research was necessary, if for no other reason than to bring clarity to a discipline continually charting new waters. The practicality of plant disease research, however, continued to be a source of debate and strain. Some scientists found a comfortable niche, achieving a mix of "pure" and "applied" research, while others struggled to escape the calls for practicality that taxpayer-funded research stations naturally exerted. They longed to concentrate on biology without worrying about how their research might increase crop yields.

Because the balance between the different missions was often dictated by the interests of the individual scientists, some stations were stronger in one area than the other. Most station plant disease researchers did creditable work disseminating basic information on fungal diseases and control measures. But some, particularly in the East, Midwest, and South, pushed to develop the science, and, along the way, they made major contributions to the practical control of plant disease.

Even though California had been one of the first two states to create an agricultural experiment station, the study of plant disease did not play a major role in western experiment stations in the period between the passage of the Hatch Act and the turn of the twentieth century. It was not until the arrival of Ralph E. Smith at the University of California at Berkeley in 1903 that California and the West began to attract scientists with an interest in exploring new areas of plant disease. In other regions of the country, however, agricultural experiment station scientists were bringing new focus and vigor to plant disease research.

THE EAST

Connecticut was one of the first states to use its Hatch Act appropriation specifically for plant disease research. In 1887 the director of the Connecticut Agricultural Experiment Station, Samuel W. Johnson, authorized the use of the station's new federal money to hire a scientist to study plant diseases, specifically those caused by fungi.[12] The scientist chosen to undertake the work was Roland Thaxter, a gifted mycologist who had been working in Farlow's laboratory at nearby Harvard University.

Roland Thaxter had achieved notable academic success at Harvard, earning a B.A. in natural history and English composition in 1882 and then attending medical school for several years before deciding to specialize in cryptogamic botany under William Farlow. In 1888 he graduated from Harvard with a Ph.D. in natural history. Continuing work begun by his mentor, including the study of heteroecious fungi, notably *Gymnosporangium juniperi-virginianae*, the cause of cedar apple rust, Thaxter published his first mycological paper, "On Certain Cultures of *Gymnosporangium* with Notes on their *Roesteliae*," in 1887.[13]

Thaxter's work at the Connecticut Agricultural Experiment Station would result in the identification of a number of diseases caused by fungi and in suggestions for several significant control measures for diseases that were destroying crops in Connecticut and around the country.[14] His first work for the station, and perhaps his most important, was his study of onion smut in 1888. Onion smut was extremely destructive at the time, affecting as much as 50 percent of the Connecticut onion crop.[15] Thaxter's landmark study, dealing with a disease caused by a soil-borne fungus, offered hope of practical control methods based on theories of pathogen dissemination in conjunction with life-cycle studies.

In his onion smut study Thaxter was again following in the footsteps of his mentor, Farlow, who in the late 1870s had been the first to describe the causal fungus of onion smut.[16] But Thaxter expressed his new theories on the dispersal of the pathogen in clear and unambiguous terms. "The popular impression," he wrote, "that smut is disseminated principally in the planted seed is one which is quite erroneous" and "not to be considered for a moment."[17] Thaxter went on to explain that onion smut was different from the more widely understood smut or "bunt" of wheat, which survived and spread through spores in and on the grain, which was then threshed and sewn along with smut-free wheat grain. In contrast, because the spores of the onion smut fungus lived in the soil, they were ready to infect any onion seedlings planted in the infested field year after year.

Thaxter then discussed fungicidal preventives for the onion smut, beginning with the premise that the "usual external applications would be quite useless."[18] He designed experiments in which he applied fungicide to the soil at planting time to act in the soil when the onion was vulnerable to attack. He tested a wide range of chemicals, including Bordeaux mixture, but found the most useful preventive to be sulfur, which increased a harvest by a ratio of about 5 to 1 over untreated fields. Still, after only one season of results, Thaxter hesitated to endorse the widespread use of sulfur in plantings. Due to the lack of proper equipment to deliver the chemical in the field, his tests had been difficult to carry out. Until better tests could be done, he continued to advocate practical measures such as burning off the remnant vegetation in smutted fields in the autumn, pulling weeds, rotating crops within fields, and carefully washing implements to prevent transfer of soil from smut-infested fields to uninfested fields.[19]

A more useful method of preventing onion smut was to transplant seedlings. Onions could be raised from seeds sown in flats with uninfested soil and then transplanted safely into any field with smut-infested soil. Transplantation was successful because transplants were exposed to the pathogen only after the plant had passed through its early seedling stage, which was its most vulnerable period. This cultural practice was already common with onion farmers, although not as a smut-control measure.[20]

Since Thaxter was searching for a practical chemical treatment, he did not advocate transplantation expressly for control. However, William C. Sturgis, Thaxter's successor at the Connecticut Agricultural Experiment Station, recommended it several years later. The delivery of sulfur to the soil continued to be so difficult that transplantation was actually more effective, although laborious. By 1900, however, F. A. Sirrine and F. C. Stewart at the New York State Agricultural Experiment Station at Geneva, and A. D. Selby at the Ohio Agricultural Experiment Station were able to build on Thaxter's ground-breaking work to make effective soil treatment for onion smut with sulfur and lime commercially feasible.[21]

Although his work held major implications for disease control, Thaxter was a true laboratory scientist who actually had little interest in "applied" science. He was never comfortable with his work at the experiment station and remained at the station only three years before returning to an academic life of research and teaching mycology at Harvard. During his years at the Connecticut station, Thaxter revealed in his private correspondence with Farlow his disdain for the style of work at the experiment station and for applied plant pathology in general.[22] He wrote that he wanted to give "the bucolic constituency a sound ducking in Bordeaux mixture and run away where I can be absolutely impractical or as impractical as I choose."[23] He found Bordeaux mixture, that chemical panacea of applied plant pathology, to be "the vilest compound imaginable." It was also Thaxter who coined and popularized the terms "squirt gun botany" and "pocketbook mycology" to describe the emerging science of plant pathology.

Thaxter did, however, make contributions to the practical knowledge about several fungus diseases of plants. He became a charter member of the American Phytopathological Society (APS) in 1909, and during his career, he worked with many students, including George P. Clinton, Fred C. Stewart, and Howard P. Barss, who would all become noted plant pathologists and serve as presidents of the APS.[24]

Not all scientists preferred academic work to work at the experiment stations. Byron David Halsted of New Jersey, for example, reveled in practical, experiment station work. In the first decade after the Hatch Act, Halsted was one of the most important figures in experiment station research on plant disease. Having spent part of his early career as an agricultural journalist, he was a prolific writer and correspondent. The expansive bulletins he wrote for the New Jersey State Agricultural Experiment Station at Rutgers were models for early plant disease publications. He was also a solid scientist who set standards for investigative technique, particularly for diseases of garden, floral, and ornamental plants.

Receiving his D.Sc. from Harvard in 1879, Byron D. Halsted was William Farlow's first doctoral student to specialize in plant disease. His dissertation was entitled "Classification and Description of the American Species of *Characeae*."[25] In a paper delivered before the American Pomological Society in 1883, he coined the phrase *fungicide*, and he spent several years as editor of *American Agriculturist*.[26]

In 1885 Halsted was named professor of botany at the Iowa Agricultural College, replacing Charles Bessey, who had recently departed for Nebraska. He taught botany for four years in Ames and increasingly turned his attention to economic botany and plant disease caused by fungi, particularly downy and powdery mildews and rusts. He wrote two bulletins during his years at Iowa Agricultural College, each one covering more than forty diseases.[27] In 1889, on Farlow's recommendation, Halsted left Iowa for Rutgers College in New Brunswick, New Jersey, where he was named professor of botany and horticulture and staff botanist of the experiment station.[28] Until about 1900 he dedicated most of his efforts at the New Jersey station to the study of plant disease caused by fungi and bacteria. At that time, his eyesight began to fail to such an extent that it was difficult for him to use a microscope. He switched his research emphasis to plant breeding and had a productive second career in that field.[29]

In New Jersey, Halsted devoted much of his time to the study of the many fruits and vegetables that were important to that state's truck-farming industry. One of his focuses was fungal diseases that occurred during the storage and transit of produce. Every year, he wrote about numerous diseases, sometimes merely mentioning their existence and explaining what, if anything, was known about them. Few diseases of garden, orchard, or nursery escaped Halsted's scrutiny on some level. Fortunately, this broad-based program agreed with him. The sheer amount of work that Halsted accomplished was astounding.[30]

Halsted paid special attention to sweet potato diseases, particularly black rot, caused by the fungus *Ceratostomella fimbriata*, and soft rot, caused by the fungus *Rhizopus nigricans*. In any given year, black rot threatened to destroy as much as 30 percent of the sweet potato crop along the mid-Atlantic seaboard. In 1891 Halsted coauthored an article on black rot of sweet potatoes in the *Journal of*

Mycology with David G. Fairchild, his nephew and an assistant in the Division of Vegetable Pathology. They isolated the causal agent in pure culture, reproduced the symptoms by inoculating healthy sweet potato with the isolates obtained, and thus confirmed a fungal cause for the disease. However, there was little they could offer by way of control aside from clean "seed" selection.[31]

Halsted's work on the control of cranberry disease is particularly interesting in highlighting the potential influence of experiment station research. Nearly one-third of New Jersey's cranberry crop was habitually destroyed by two diseases known as gall, caused by the fungus *Synchytrium vaccinii*, and scald, caused by the fungus *Guignardia vaccinii*. In 1889 Halsted outlined the life histories of these two fungi in a special bulletin.[32] Then, in 1890 he led field experiments in cranberry bogs around the state to prove that chemical fungicides were essentially useless.[33] Impressed by the research, the New Jersey legislature passed a law allowing the experiment station to judge whether destroying scalded vines in a particular area could help stop the spread of the disease.[34] Halsted recommended destroying the scalded cranberries, but he also announced that the best method of prevention was better management of water in the bogs. Halsted's investigations proved to be instrumental in reinvigorating the cranberry industry in New Jersey.[35]

Halsted also played a role in the growing cooperation between state experiment stations and the USDA. In 1891 Samuel Mills Tracy, director of the Mississippi Experiment Station, desperately wanted a capable scientist to study a blight of tomato that was causing alarm among growers in the South. Seeking advice from Beverly Galloway in Washington, he eventually settled on Halsted, who had already made a substantial reputation for himself at Iowa and Rutgers, as the best candidate for the special assignment.[36] Traveling to Mississippi on temporary assignment for Tracy, Halsted discovered that "Southern tomato blight" was caused by a bacterium (now known as *Pseudomonas solanacearum*). He correctly associated the disease with a similar condition in potato, sometimes known as brown rot, and he carried out cross-inoculations with the pathogen between tomato and potato. Spurred by complaints of a melon blight in the area, he also inoculated melon with the pathogen, which produced the same disease. With a little more trepidation, he noted similarities between the Southern tomato blight and a "bacterial melon blight," now known as bacterial wilt of cucurbit, caused by *Erwinia tracheiphila*, that occurred in the northeastern United States. Noting that Bordeaux mixture had proven effective as a control for the Southern blight in potato, he recommended it for any of the crops in danger of this bacterial blight.[37]

Another disease that drew Halsted's attention was clubroot of cabbage. First described scientifically in the pioneering 1878 work by Michael Stepanovitch Woronin in Russia, clubroot, caused by the fungus *Plasmodiophora brassicae*, had a long and destructive history among crucifers such as cabbage and turnip.[38] Clubroot could be quite serious, as Halsted noted, "sometimes incurring almost a total loss, and in the aggregate the destruction for the whole country is doubtless represented by millions of dollars." The disease was of concern to Halsted because it "prevailed extensively in the truck regions around the large cities of New York and Philadelphia."[39]

In his 1893 bulletin, Halsted categorized the clubroot organism as a slime mold, a member of a "family of fungi . . . widely distinct from the mildews, rusts and smuts."[40] He described the life cycle of the organism and added several names to the list of susceptible plants. Explaining that the parasite's method of attacking belowground foiled any hope of treating the plant chemically, he laid out the same traditional preventive cultural measures Woronin had in 1878, that is, burning all refuse from diseased plants, choosing disease-free seedlings from hotbeds, and rotating crops. Halsted, however, also recommended liming the soil, a practice that Woronin had erroneously dismissed.[41]

Like most plant pathologists of the period, Halsted drew most of his advice about disease prevention and control from the day-to-day practices of farmers. On the other hand, he provided a massive amount of new scientific information about causation on a wide range of diseases caused by fungi and bacteria. Many, if not most, of the diseases he studied had not received any previous systematic treatment. Halsted created a large corpus of work from which numerous contemporaries and future plant pathologists would profit.

One eastern plant pathologist who was influenced by Halsted in his early years in plant pathology was Lewis Ralph Jones, originally of the Vermont Agricultural Experiment Station. During the 1890s Jones's work in Vermont centered on the nature and control of certain diseases caused by fungi, mainly those affecting potato and orchard crops. Occasionally, he extended his investigations to cereals and vegetables, and he did significant work on bacterial soft rots caused by *Erwinia carotovora*. It was, however, his control experiments for late blight and scab of potato, together with those on apple and pear, that truly defined his early experiment station work. His research placed him squarely in the mainstream of experiment station phytopathology.

In the late 1880s Jones had been influenced to enter the nascent field of plant pathology by the chance to attend the final Ph.D. examination of Erwin Frink Smith at the University of Michigan. Jones had attended Michigan with the intention of earning a medical degree and had studied botany under Volney Spalding, as had Smith. Jones came to know Smith during the years of the latter's peach yellows work as a special agent for the USDA. As Jones later said of Smith's examination, "The glimpses . . . of the significance of and opportunity for research in the field of plant pathology were most inspiring."[42] Jones graduated from Michigan in 1889 with a Ph.B. in botany and was selected by the University of Vermont at Burlington to be an instructor in natural history and the botanist at the Vermont Agricultural Experiment Station.

Some of the earliest and most important work that Jones began at Vermont was his research on potato late blight. Jones credited Halsted with encouraging him to "make a special study of the diseases of the potato."[43] As further impetus, the potato was "the largest single crop raised" in Vermont and extremely subject to disease.[44] Jones's research on potato late blight came to the attention of Beverly Galloway, who wrote to Jones that "if you can only succeed in getting farmers interested in the work the annual saving on this crop alone in Vermont will amount to far more than the total expenditures of the Experiment Station."[45]

Jones not only verified many of the known facts about *Phytophthora infestans*, the causal agent of late blight, but he also uncovered new information about the disease cycle. Additionally, he took the lead in experiments with Bordeaux mixture as a control, announcing in 1891 that the vines at the experiment station that were not sprayed yielded only 86 bushels per acre of marketable tubers versus 155 bushels for vines sprayed once and more than 165 bushels for vines sprayed twice.[46]

But Jones was not completely satisfied with these superb results. Bordeaux mixture was expensive and tended to clog the sprayer nozzles. Jones decided to examine substitutes and seek improvements.[47] The result was perhaps the most extensive tests for quality control that Bordeaux mixture underwent, outside of Galloway's work at the USDA. Jones tested several alternative fungicides on late blight, some of them on the recommendation of Galloway, with whom Jones was in close communication.[48] His tests, however, proved beyond a doubt that Bordeaux mixture was the superior fungicide.[49] Having settled this question to his satisfaction, he moved on to investigations of such practical issues as the most appropriate times and methods of application as well as the best type and strength of solution. Some of his most interesting investigations involved the efficacy of dry Bordeaux mixture, which was being promoted by chemical dealers.[50] He found that these treatments provided "decidedly inferior protection to the foliage as compared with the usual or wet forms of the mixture."[51]

Although Jones's research on late blight and Bordeaux mixture during the 1890s was a hallmark of his scientific and practical ingenuity, it was his studies of the bacterial soft rot of vegetables that showcased his investigative talents and foreshadowed his developing prominence in the science of phytopathology. In 1899 Jones spent six months in E. F. Smith's laboratory in Washington, D.C., studying bacteriology and the physiology of disease. The results of these investigations were most significant for American plant pathology. During his stay in Washington, Jones described *Bacillus carotovorus*, the causal organism of soft rot of vegetables. Perhaps more important, he identified an enzyme associated with the disease. Over the next few years, Jones used heat, filtration, germicides, diffusion through agar, and precipitation by alcohol to isolate the enzyme from the bacterium. He carefully recorded the rotting action of the enzyme pectinase on the middle lamella of the cell wall of the host plant.[52]

Jones's work on soft rot was the first precise articulation of a principle physiological mechanism of infection for disease of plants. This groundbreaking work in disease physiology was accepted for publication in 1901 by the prestigious German science journal *Centralblatt für Bakteriologie*.[53] His final results appeared in a more customary venue for plant disease research, the Bulletin of the Vermont Agricultural Experiment Station, four years later. For this work, in 1904 Jones was awarded the Ph.D. degree from the University of Michigan. Jones's research on bacterial soft rot of vegetables placed him at the forefront of American plant pathology in the early twentieth century. Furthermore, he had demonstrated an appreciation and understanding for both the practical and scientific, or what he termed the "pocket-book and the laboratory," nature of plant pathology.[54]

While at Vermont, Jones would continue to explore new horizons. After 1900 he cultivated an interest in disease resistance, stimulated no doubt by the successful work of his own student, William A. Orton, on cotton wilt, and that of Henry L. Bolley on flax wilt in North Dakota. He also pursued the growing field of the relationship between the environment and disease development. In 1909 he moved from Vermont to continue an illustrious career at the University of Wisconsin, where he was to found a department of plant pathology, based largely on his vision for research and teaching.

Thaxter, Halsted, and Jones devoted their careers to the wide variety of crops and disease problems created by the intensive agriculture of the eastern United States. Following the example they set, the work on plant disease in the experiment stations of the eastern United States branched out to cover a range of subjects. Some agricultural experiment station scientists focused primarily on maladies of truck garden crops including potato, grape, and orchard fruit culture. At some of the stations, including those in Vermont and Geneva, New York, work was done on diseases of cereals. However, the majority of the research on the important diseases of cereals fell to the scientists at stations in grain-growing regions of the American Midwest and Great Plains.

The midwest and great plains

In the late nineteenth century, the states from the Ohio Valley in the East to the high plains under the Rocky Mountains in the West were rapidly becoming the breadbasket of the nation. Wheat, corn, oat, barley, and crops of lesser acreage, such as sorghum and flax, were planted in abundance. New cultivars were tried to find those best adapted to the harsh winters and dry summers of the plains.[55]

Because of the central role of wheat and other grains in people's diets, diseases such as smuts and rusts of grains had been important since the earliest scientific studies of plant disease. Therefore, it is not surprising that research on cereal diseases initially dominated the plant disease interests of the agricultural experiment stations of the midwestern and plains states. Many questions concerning the reproduction, germination, and dissemination of the smuts and rusts remained unanswered. There was already a record of successful measures that prevented some grain smuts, notably seed treatment with copper sulfate. But there was much more work to do on preventive measures as well.

Control strategies for cereal diseases were given a tremendous boost in the late 1880s, when a new type of seed treatment came to the United States from Denmark—hot water. A version of this treatment had been devised originally in the early 1880s to combat late blight in seed potato by Jens Ludwig Jensen, a teacher and agricultural scientist. Outside of Denmark it did not receive a great amount of attention as a control for potato blight, but when Jensen applied the hot-water treatment to seed-borne diseases of grains, the results were noted widely. Jensen's paper "The Propagation and Prevention of Smut in Oats and Barley" was published in the United States in 1889 in the *Journal of Mycology*.

The hot-water process could prevent both covered and loose smuts. Covered smut, also known as bunt or stinking smut when infecting wheat, had been controlled with copper sulfate, which killed the smut spores that overwintered by adhering to the outside of the wheat seeds. Loose smut, on the other hand, had resisted chemical treatment because the pathogen overwintered as mycelium inside the seed and chemicals that were applied to the seed surface could not reach the fungus. Jensen reasoned that the mycelium could be killed with heat, just like the late blight fungus inside seed potato. The only difference was that the potato tuber needed dry heat, but the seed grains could be immersed in hot water.[56]

Jensen's procedure caused an immediate stir in the United States. Experiment station workers in grain-growing states scrambled to apply the hot-water treatment to a variety of diseases. The hot-water treatment was tested by many experiment station scientists, including William A. Kellerman and Walter T. Swingle in Kansas, H. L. Bolley in North Dakota, E. S. Goff in Wisconsin, Louis H. Pammel in Iowa, and even L. R. Jones in Vermont.

At first, the procedure seemed to promise a new era in disease control for previously unstoppable smuts such as that of barley (*Ustilago nuda*), just as Bordeaux mixture had for many other diseases caused by fungi. Additionally, it seemed to promise a simpler treatment for the already controllable smuts such as that of oat, caused by *Ustilago kolleri*, and stinking smut of wheat, caused by *Tilletia caries* and *Tilletia foetida*. But the hot-water treatment, while simple in theory, proved to be cumbersome in practice. Given the available technology, maintaining hot water at an exact temperature was troublesome. In addition, hot water did not help control several other important cereal diseases such as corn smut and a wide variety of rusts. Hot water did prove successful against oat smut and bunt of wheat, but it was no more successful than copper sulfate. Thus, the expected promise of a new era of disease control was not wholly realized.

Nevertheless, scientists at agricultural experiment stations were kept busy uncovering new facts about the life cycles of various grain diseases and carrying out disease-control experiments with a number of new chemical treatments. Among the most important work on cereal diseases was that of J. C. Arthur at the Indiana Agricultural Experiment Station at Purdue. Having earlier confirmed the futility of hot-water treatments for corn smut, in the mid-1890s Arthur experimented with fungicides on the corn disease. In 1895 he announced "the first successful spraying to prevent corn smut yet reported for this country."[57]

Arthur's tests at the Indiana station in 1895 and 1896 involved trials with Bordeaux mixture, ammoniacal copper carbonate, chloride of iron, and sulfide of potassium. Bordeaux mixture clearly outperformed the others. Unfortunately, continued experiments over the next several years indicated that spraying would be of only "moderate value." The efficacy of spraying as a practical control method for corn smut was mitigated by high costs associated with spraying large acreages, the frequent occurrence of weather that was highly conducive to smut, and the impossibility of protecting the ear by spraying. Instead, Arthur recommended destroying the immature smut balls before the spores were disseminated.[58]

Scientists had been trying to determine the etiology of corn smut for years, without success. The pathogen's mode of infection had eluded the earlier attention of William Kellerman.[59] In 1895 Arthur and his assistant, William Stuart, originally from the University of Vermont, began several years of comprehensive studies on the disease. They focused on discovering the mode of infection and pathogen dispersal, and on the possibility of using fungicides as a control. Although repeated attempts to infect the plant by applying smut spores to the seed failed, inoculation of "young plants in second leaf . . . was successfully accomplished" in October 1897.[60] Arthur concluded that smut spores could infect any part of the plant, and that they were airborne. He also realized why the hot-water treatment of the seed had failed as a disease control measure: Smut spores did not enter the plant through the germinating seed.[61]

Meanwhile, in 1890, H. L. Bolley, Arthur's student at Purdue, began important work on the diseases of grains and other crops at the North Dakota Government Agricultural Experiment Station. Bolley had been brought to North Dakota because of his work on wheat rust at Purdue, and his primary duty was studying rusts and smuts of cereals.[62] Bolley did his share of work on seed treatment for smuts, including testing Jensen's hot-water treatment.[63] Also, some of his early work on wheat rust looked at the overwintering of spores and the distance that urediniospores could travel and still maintain their viability and infectivity.[64]

Through his work with potato scab, Bolley also made significant strides in chemical controls. He had established the pathogenicity of the potato scab organism while at the Indiana station, and in 1891 he continued that work at North Dakota. He made a breakthrough in scab prevention by treating seed tubers with mercuric chloride, or, as it was generally known, corrosive sublimate.[65] Kellerman and Swingle had previously tested this substance in 1890 for wheat bunt and oat smut, but they had decided it was too destructive to the grain to be of practical use.[66] Bolley found that potato tubers treated with corrosive sublimate produced a crop that had as little as 5 percent scab, whereas untreated tubers nearby resulted in a crop with 85 percent scab.[67] Although he did some corollary testing on the effects of different soils on the growth of scab, he gave most of his time and attention to the chemical prevention of the disease.

Even Bolley, its greatest advocate, had to admit that mercuric chloride, although effective on potato scab, was quite dangerous to use. He ended his bulletins by warning the reader that "corrosive sublimate is a strong *poison* and too great care cannot be exercised in its use." Wooden vessels were required because the solution would eat through metal.[68] In 1895 J. C. Arthur began to search for an alternative. Through field tests on potato scab in 1895 and 1896, he found that formalin was as effective as mercuric chloride in preventing the disease. In 1897, he recommended formalin as the preferred treatment.[69]

Meanwhile, Bolley determined that formalin was a valuable seed treatment in preventing the smuts of oat and barley and the bunt of wheat. These results came from years of experiments aimed at treating smuts with many different chemicals, including corrosive sublimate. In 1893 Bolley reported that mercuric chloride, in a solution of 1.5 parts to 1,000 of water, "destroyed the smut and seemed to leave the wheat unreduced in yield, if the seed was dried at once after dipping," but it had no discernible effect on the smuts of oat and barley.[70] In 1896 he reported that formalin was also successful for both wheat bunt and the smut of oat and barley. It was "said to be non-poisonous, and . . . promises to excell [*sic*] as a medium of efficient and easy smut preventative."[71] He continued chemical-control work over the next few years and, in 1898, experimented with treating wheat and other cereal grains against smuts by fumigating with formaldehyde gas. The treatment was declared successful, but at the time was not economically feasible.[72]

At the turn of the century, Bolley's most important research still lay ahead. By 1900 he had begun to study the problem of flax wilt and "flax sick" soils in North Dakota. He demonstrated that flax wilt was caused by a pathogen (*Fusarium oxysporum* f.sp. *lini*) that built up in the soil, and he repudiated the common perception that flax naturally diminished the soil nutrients that were required for plant subsistence.[73] Research on the pathogen's role in flax wilt had been published in Japan in the early 1890s, but Bolley was unaware of these papers.[74] Several years later, Bolley did even more valuable work in selective breeding by introducing to North Dakota flax cultivars that successfully resisted fusarium wilt.[75]

Workers at the Iowa Agricultural Experiment Station at Ames were also demonstrating a remarkable ability to investigate many different plant disease problems of midwestern crops. In 1889 Louis H. Pammel succeeded Byron Halsted as a professor of botany at Iowa Agricultural College and also became the botanist at the state agricultural experiment station. Pammel established himself quickly as a worker with broad-based interests, in the same mold as Halsted, and he was graced with good assistants. He examined not only the traditional cereal diseases, as did the other station scientists in the Midwest and Great Plains, but he also did a great deal of fungicidal testing on fruit and garden crops, as was more typical in the experiment stations of the eastern United States.

Pammel first made a name for himself in 1888, when, as an assistant to William Trelease at the Shaw School of Botany in St. Louis, he accepted a summer appointment from the Texas Agricultural Experiment Station to study cotton root rot. A serious disease characterized by a sudden wilting of the plant, cotton root rot was costing farmers in Texas an estimated $3 million to $5 million annually. Pammel had been Trelease's student at the University of Wisconsin, graduating with a B.Ag. degree in 1885, and then, after a brief period at Harvard, followed Trelease to St. Louis. Although benefiting from broad training, he had a specific interest in fungal diseases of plants, and he was exposed to the nature of bacteriological studies when Trelease returned to St. Louis from Berlin after studying with Robert Koch.[76]

Through his initial investigations in 1888 and with assistance from the published work of Robert Hartig, Julius Kühn, Frank Lamson-Scribner, Pierre Viala, and others, Pammel came to believe that the cotton root rot was caused by a soil-borne fungus that he identified as *Ozonium auricomun*. During his first summer, he was unable to do inoculation tests to confirm this hypothesis, but he still advocated empirical control strategies such as the removal and destruction of infected plant stalks and crop rotation.[77]

Pammel's initial results were praised by Frank Lamson-Scribner as "original and I doubt not will lead to valuable scientific results and perhaps practical also."[78] Even after his appointment in Iowa, in 1889, Pammel returned to Texas to continue his studies of cotton root rot. Further research proved conclusively that the disease was caused by *Ozonium auriconum* and hinted at the possibility of control with fungicides.[79]

In Iowa, Pammel spent much of his time testing various chemical fungicides on a variety of diseases. He also tried to improve fungicides by, for instance, studying ways of improving adherence to leaves through additives such as molasses.[80] Unlike most of his colleagues on the Great Plains, Pammel also focused a great deal of attention on the control of several foliar diseases of fruits. Between 1891 and 1896, with the help of assistants P. H. Rolfs, Fred Carleton Stewart, and George Washington Carver, Pammel performed fungicide investigations at the Iowa station on the comparative efficacy of Bordeaux mixture and ammoniacal carbonate of copper on leaf spot of cherry, caused by *Cylindrosporium padi*, and on spot diseases of currant, caused by *Septoria ribis* and *Cercospora angulata*.[81] Carver, the first full-time assistant botanist at the station, also worked with Pammel on important taxonomic and control experiments on powdery mildew of apples, caused by *Sphaerotheca mali*. For the most part, Bordeaux mixture remained the most effective treatment for all these diseases.[82]

Naturally, the scientists at the Iowa station also had a stake in the control of cereal diseases. Damage from rusts in the state amounted "to millions of dollars" each year. In 1891 Pammel and Rolfs tested Bordeaux mixture and ammoniacal carbonate of copper on wheat rusts, caused by *Puccinia graminis* and *Puccinia recondita*, and, like Kellerman and Swingle in Kansas, failed completely to control the rusts.[83] The next year, Pammel tried again on oat rusts, caused by *Puccinia graminis* and *Puccinia coronata*, once again with no success.[84] He also considered the possibilities of disease-resistant cultivars of wheat when fungicides failed to control rusts.[85] And in 1894, he anticipated disease forecasting with an inquiry into the association between weather and disease.[86]

Around the turn of the century, due to the increasing importance of the Midwest and Great Plains to agricultural production in the United States, scientists at the agricultural experiment stations of these regions were extremely active in plant disease research. Although they conducted studies on fruit and garden crops, they devoted most of their attention to diseases of cereals and forages, crops of great economic importance. The search for practical control methods for cereal diseases continued to dominate the experiment stations' agendas into the twentieth century. Eventually, however, the stations of the Midwest and Great Plains would become leaders in the aggressive introduction of new cultivars and the field of breeding for disease resistance.

The south

In the first ten years after the passage of the Hatch Act in 1887, many experiment stations were established across the southern United States. In general, however, these stations devoted less time to studying plant disease than did stations in the East and Midwest, possibly because there was little tradition of plant disease research at southern colleges and universities. For the most part, plant disease research in the South was infrequent and scattered.

There were, however, a few notable exceptions. When he left the USDA in 1888, Frank Lamson-Scribner became the botanist at the University of Tennessee Experiment Station, where he continued his work on diseases of grape and potato. H. Garman, at the Kentucky Agricultural Experiment Station, did excellent descriptive work on brown rot of stone fruits, caused by *Monilinia fructigena*, and was the first to describe black rot of cabbage, caused by *Xanthomonas campestris*.[87]

Another notable exception was the program developed by George Francis Atkinson at the Agricultural and Mechanical College of the Alabama Polytechnic Institute at Auburn. Atkinson had earned a doctorate from Cornell in 1885 before becoming professor of general zoology at the University of North Carolina. In 1888 he moved to the University of South Carolina as professor of botany and zoology. There, he also worked in the state experiment station, producing a bulletin on clubroot of cabbage. In 1889 Atkinson was appointed professor of biology and botany at Alabama Polytechnic Institute at Auburn.[88] In addition to teaching, he was the "biologist" at the Agricultural Experiment Station, with duties that included the investigation of the "diseases of plants caused by parasitic fungi and insects." Plant disease work was normally assigned to a staff botanist or horticulturist, but at the Alabama station, the botanist, P. H. Mell, was too busy juggling responsibilities in his other position as station meteorologist to undertake any work with plant disease.[89]

Atkinson's scientific character was like Roland Thaxter's; he was driven by science, not by a desire for practical disease control. He wanted to devote himself to systematic work and, while at the Alabama station, he found time for excellent mycological studies on the species of *Cercospora* and *Ravenalia* of Alabama and *Erysiphe* of Alabama and the Carolinas.[90] His interest in entomology and his experience with zoology gave Atkinson the background necessary to produce his first plant disease paper, "Nematode Root-Galls," in 1889.[91] This was one of the first major studies in the United States on parasitic nematodes, superseding and correcting similar research done that same year by J. C. Neal of Florida, which had been directed by Charles V. Riley, the USDA entomologist.[92] When Atkinson's study was published, Scribner had just begun studies of a nematode disease of potato, the first plant disease work to be done at the Tennessee station.[93] Atkinson's report was a solid descriptive contribution on the nematodes around Auburn, providing an extensive summary of the existing European publications on nematodes. He concluded by advising farmers to rotate crops and select disease-free seedlings as control strategies.

Atkinson's nematode paper created a noticeable stir among American plant pathologists, many of whom understood the value of the work even if they were not, as Louis H. Pammel of the Iowa Experiment Station wrote, "sufficiently familiar with the zoology."[94] Clearly, Atkinson had shown the way and exposed the need for further research. His paper was reviewed in *Agricultural Science* as "one of the notable bulletins of the year '89, and is the most comprehensive document upon nematodes that has been published in this country."[95] Beverly Galloway wrote Atkinson that "such publications as this raises a station greatly in . . . estimation." Galloway also expressed his hope that Atkinson would "continue the work until it is completed in every detail."[96]

Atkinson did continue his nematode work, but by accident. In 1890 he was sent a sample of rotted cotton roots from a local farm. Expecting to find the root rot fungus that had been described by Louis Pammel the year before in a Texas study, Atkinson began his investigation of the diseased plants and was surprised to find nematodes instead.[97] He could, however, add little to the practical, cultural methods of control that he had outlined in his earlier root gall bulletin.

Atkinson's investigations in 1890 proved to be his last major work with nematodes, but it was the beginning of his research on cotton diseases. Little work had been done on cotton diseases when Atkinson began his studies, and his work in this field earned him a reputation as the premier plant pathologist in the South. There was a great deal of confusion surrounding various cotton diseases. In Alabama, a plethora of common names were used to describe a multitude of symptoms without a scientific knowledge of the pathogens. Beginning in 1891, Atkinson tried to clear up the confusion. While studying a cotton disease locally known as black rust, he observed four species of fungi and one bacterium, many of which were already associated with different cotton diseases.[98] By 1892 he had sorted out the multiple pathogens and the diseases they caused. He discarded all of the confusing and contradictory local disease names and created his own cotton disease nomenclature, based on scientific research. Black rust now became yellow and red leaf blights, which Atkinson recognized as a physiological problem, "being one of imperfect nutrition or assimilation," asserting that "organisms do not initiate the disease, they only aggravate it."[99] Through 1890 and 1891, he did excellent, original, descriptive work on cotton leaf blight, caused by *Sphoerella gossypina*; areolate mildew, caused by *Ramularia areola*; and cotton wilt, caused by *Fusarium oxysporum* f.sp. *vasinfectum*.[100] He was also one of the first to describe a bacterial condition later known as angular leaf spot, caused by *Xanthomonas malvacearum*.[101]

Atkinson's most notable work, however, was on cotton anthracnose, caused by *Colletotrichum gossypii*.[102] He shared credit for the observation and description of this fungus with Effie Southworth of the Division of Vegetable Pathology at the USDA, but Southworth's work was limited to laboratory observation of the pathogen as it existed on the cotton boll. Atkinson studied it in all its forms, prompting Southworth to admit that "your work is much more complete than mine."[103] Once again, Galloway lauded Atkinson's research, informing him that his anthracnose paper "will be valuable for cotton growers throughout the south."[104] Although Atkinson clearly valued the scientific role of experiment stations over the practical, he pointed the way, preparing the ground for further scientific and applied study of cotton diseases.

In 1892 Atkinson left Alabama for Cornell, where he wrote a series of descriptive publications on celery early blight caused by *Cercospora apii*; oedema of apple trees caused by *Fusarium arcuatum*; powdery mildew of crucifers caused by *Erysiphe polygoni*; and a long report on leaf curl and plum pockets.[105] He also reported on the "damping-off" diseases of seedlings in potting beds, drawing special attention to soil moisture, humidity, and light.[106] In 1893 his important bulletin on "Oedema of the Tomato" brought more attention to physiological diseases.[107] At Cornell, Atkinson's freedom from pressure to do applied work was in part due to the fact that the New York agricultural experiment station at Geneva had a tradition of applied plant pathology. In addition, western New York was the area of frequent sojourns for USDA plant pathologists such as Merton B. Waite and David G. Fairchild.

AROUND THE NATION

The state agricultural experiment stations that sponsored plant disease research in the 1880s and 1890s made, by necessity, a strong commitment to the practical aspects of disease control. The Hatch Act, on which the stations had been founded, had been passed on the promise of immediate improvement to agricultural conditions. In exploration of practical solutions to disease problems, station scientists sprayed fungicides, particularly copper sulfate and Bordeaux mixture, on the plant diseases that affected important crops in their states. They continually tried new chemicals and other types of treatments, including Jensen's hot-water procedure. When this small arsenal of "chemical" controls would not serve, they studied and affirmed the value of practices such as crop rotation, destruction of infected plants, liming fields, transplantation of disease-free seedlings, and planting new cultivars. Some of these experiments were actually nothing more than verification of existing, empirical agricultural practices, but the value of understanding the scientific rationale behind the effectiveness of such practices was inestimable. Furthermore, station scientists sometimes dismissed and discouraged common farm practices that proved worthless or even harmful, such as using waste plants from diseased fields for compost.[108] Thus, experiment stations played a significant role in systematizing agricultural activity and occasionally played a part in developing and implementing public policy.

Still, many scientists working in stations and teaching in the associated colleges and universities were determined to advance science in fundamental ways.

They were aware that the developing field of plant pathology could stagnate if the understanding of the biological principles of disease were not advanced. Moreover, plant pathologists were beginning to expand the breadth of their vision beyond the traditional lines of work. A few considered new areas of research, including the physiology of host-pathogen interactions, the effects of environment on disease, and the comparative susceptibility of cultivars to pathogens. These few, who included among their ranks Roland Thaxter, Lewis Ralph Jones, and George Francis Atkinson, were carving a more acceptable place for "pure" research in agricultural experiment stations and elevating the science associated with plant pathology into the realm of the main currents of biological research of the late nineteenth and twentieth centuries.

The decade of the 1890s was an explosive period for plant pathology. Both the science and the institutions developed rapidly. From a handful of teaching and research scientists located at colleges and universities and a small core of plant researchers at the USDA, the science of plant pathology expanded into laboratories for studying plant disease across the entire United States. By 1900 there were fifty-seven state experiment stations, fifty-two funded by the Hatch Act, with a total station employment of nearly 600 people.[109] But there were problems. Each laboratory tended to focus on regional problems, and, with no central authority or administrative body to oversee the work, there was considerable overlap and redundancy. Scientists at different experiment stations often studied the same disease problem without knowledge of each other's work. However, personal and professional connections were bolstering the lines of communication and cooperation. By the turn of the century, the beginnings of a national system could be recognized.

This national system would increasingly be influenced by the centralizing power of the federal government. In 1888 the federal government created the Office of Experiment Stations, a branch of the USDA, to be a coordinating body for experiment station research.[110] The Office of Experiment Stations promoted "unfettered research" over practical service. It also interpreted the Hatch Act as giving the USDA the power to determine proper use of Hatch Act funds by stations. This, in essence, gave the federal government the authority to define research agendas.[111] Under this system, state agricultural experiment stations were increasingly brought into the sphere of influence of the federal government.

CHAPTER 9

Recognition of U.S. Plant Pathology

The USDA's late-1880s discoveries about disease-causing microorganisms in the laboratory and its successes with fungicide treatments in the field had demonstrated the usefulness of plant pathology, laying the groundwork for an extension of federal involvement. As Beverly Galloway later observed, "The fiscal year 1890 was the beginning of a new era for the pathological work."[1]

When Galloway took over as chief of the Section of Vegetable Pathology in 1888, the agricultural budget for that fiscal year included just $5,000 for work on plant diseases in both the field and the laboratory. In 1890 congressional appropriations for these investigations had tripled, to $15,000. Moreover, in the 1888 appropriation, the emphasis centered on "diseases . . . due to parasitic fungi."[2] Perhaps as a result of the initiation of the study of bacterial pathogens, the 1890 appropriations bill broadened the mission to the investigation of "diseases injurious to fruits, fruit-trees, grain, cotton, and other useful plants."[3]

The 1890 appropriations act also instituted an organization change in the USDA: The Section of Vegetable Pathology became a division. The separation of pathological work from the Division of Botany had major implications for federal plant disease research. Expansion had previously been restricted "by the limited appropriations for the Division of Botany. As a separate division, USDA plant pathology was better positioned to undertake additional and expanded investigations on a number of America's most pressing agricultural problems."[4] As it moved into the 1890s, the Division of Vegetable Pathology had a skilled leader in Beverly Galloway, a cadre of competent scientists, a record of success, crucial funding, and a new freedom of direction.

THE RISE OF GOVERNMENT LABORATORIES

Since Frank Lamson-Scribner's first use of "special agents" in his fieldwork with grape diseases in 1886, the USDA had undertaken disease investigations in many important agricultural regions of the United States. Galloway expanded and improved Scribner's program. In 1889 the Section of Vegetable Pathology had agents in many states around the country. Some of these were growers; others were affiliated with agricultural colleges and experiment stations. As Scribner had done with Erwin F. Smith on peach yellows, Galloway also sent assistants from the section into the field to do special reports on diseases.

As long as the Section of Vegetable Pathology remained relatively small, the use of grower-agents, experiment station personnel, and traveling assistants worked reasonably well, particularly in testing fungicides. But the system had weaknesses, most notably the lack of scientific training among grower-agents. With their limited knowledge of an increasingly complicated biological science, growers could not be depended on to make all of the requisite observations or to do rigorous field tests over an extended period of time.

Effective fieldwork increasingly required the careful supervision of a specialist. It called for a commitment to investigations over the course of several growing seasons, especially when the evaluation of control strategies was involved. Although a growing network of college and state experiment station scientists were beginning to deal with state and regional agricultural concerns, the most capable scientists were not always located in regions where acute plant disease problems were occurring. Thus, the Section of Vegetable Pathology had to rely on its own small force of scientists to visit trouble spots. But Galloway's assistants, although successful in gathering firsthand evidence of disease conditions, supervising fungicide tests, and building a good rapport with farmers during their "reconnaissance" missions of the 1880s, were unable to carry out long-term investigations or to provide support for all of the major centers of agricultural production.

In the final decade of the nineteenth century, the USDA faced a barrage of demands from viticulturists and horticulturists in California, citrus growers in Florida, and nurserymen and fruit growers in New York to investigate destructive plant diseases in their regions. The increasing influence of specialized producers in the nation's market economy gave growers a forceful voice in the federal government.[5] Thus, while continuing to persuade experiment station scientists around the country to undertake more plant disease research, Galloway devised a plan in 1891 that would establish research laboratories in Santa Ana, California; Eustis, Florida; and Geneva, New York. This plan allowed federal scientists in the new Division of Vegetable Pathology to respond to the urgent pleas of growers for assistance with extended, on-site research operations.

VINE DISEASE IN CALIFORNIA

On June 15, 1891, Newton B. Pierce submitted his preliminary report on the California, or, Anaheim, vine disease. In the works since November 1890, Pierce's report encompassed the results of his first ten months of fieldwork in California and a five-month trip to Europe. He gave a brief survey of the viticultural industry in southern California, provided an extremely detailed description of the symptoms of the disease, and reported on its spread and impact on Orange County, California. He then related the evidence he had collected on the relation of soil, shade, rainfall, temperature, insects, and fungi to the disease. In a chapter on control methods, he mentioned cultivation, pruning, and cutting back seedlings, and followed this with a discussion on the relative hardiness of cultivars and methods of grafting resistant stock. He also described a number of vine diseases he had observed during a trip to Europe and showed how each differed from the California vine malady.

Pierce suspected that a bacterium was the cause of the California vine disease. "That bacteria are present in the diseased vines is," he wrote, "quite clearly established. They have been isolated and cultivated in various media." Nevertheless, none of the inoculation experiments had resulted in producing the disease: "That the bacteria, isolated from the inner tissues of the vine, bear any causal relation to the disease," he was forced to admit, "is not, however, established and may be very justly questioned."[6] Galloway and Pierce realized that "it [was] too early to state positively the cause of the disease or to make many recommendations as to prevention or cure."[7] Publication of the report was delayed for almost a year—perhaps in hopes that further inoculation experiments and control tests would be successful.

In July 1891, after filing his preliminary report, Pierce returned to California with a new mandate: He was to establish a laboratory and settle in for long-term fieldwork and experimentation. California vine disease was by many accounts a potentially devastating threat to California agriculture. One careful observer claimed in late 1891 that because of the disease, "in the Southern counties the viticultural industry has changed more during the past four years than any other portion of the State. In 1886 the counties of Los Angeles and San Bernardino produced 4,000,000 gallons of wine while during the past season but 500,000 gallons were made. This great reduction was due to the Anaheim disease, which for the past four years has destroyed all the vineyards around Orange, Santa Ana, and Anaheim and a large portion of the vineyards in other parts of Los Angeles county. It is estimated that at least 20,000 acres of vines have been destroyed."[8]

Pierce spent July 1891 visiting different agricultural regions of California, talking to growers and looking for a site for his laboratory. Growers in Orange County attempted to persuade Pierce to locate his lab there, but he chose Santa Ana "as the most favorable point for the continued study of the vine disease." On August 24, 1891, he informed Galloway that "good quarters have been obtained and I am fitting up a laboratory for immediate and future use."[9] Three days later, he was requesting that headquarters send laboratory supplies of agar, peptone, gelatin, test tubes, and glass rods.[10]

Although California vine disease was the primary reason for Pierce's return to California, before long he found himself investigating a number of other regional agricultural problems. His presence was appreciated in a state where horticultural crops were gaining ascendancy. As he told Galloway in November 1891, "I know at this time of several fungus diseases which are gaining a marked foothold here and which will certainly attract the attention of the public before many seasons. . . . It looks as if there would be no rest for the wicked."[11] Over the next few years, Pierce devoted considerable attention to diseases affecting many of the region's major economic crops such as almond, prune, peach, cherry, apricot, apple, olive, lemon, orange, fig, and walnut. Several of his studies were particularly significant, both from the standpoint of advancing fundamental knowledge in plant pathology and of developing disease-control measures. Beginning in late 1891 he took up the investigation of an almond disease caused by the fungus *Cercospora circumscissa*, which attacked the leaves and young twigs early in the season and subsequently harmed the fruit. By 1892 he was able to describe and illustrate the fungus in addition to discovering the overwintering spores and the initial point of infection. This basic knowledge of the life habits of the fungus played an integral role in control experiments conducted late in the year. By the end of 1892 he was reporting encouraging results from the timely application of fungicidal sprays such as an ammoniacal solution of copper carbonate and modified eau celeste.[12]

In 1893 Pierce reported the satisfactory spray treatment of a rust of prune caused by *Puccinia pruni-spinosae* and also the successful prevention of the leaf curl of the peach caused by *Taphrina deformans* in an experimental orchard under his supervision.[13] Showing once again his penchant for quantifying USDA results in economic terms, Galloway delightedly trumpeted, "In the peach orchard where the work was carried on, it was found that the treated trees averaged 160 pounds more fruit per tree than the untreated, although the latter were exactly the same size as the former, and the soil conditions in both cases were practically identical. On this basis the yield of the entire orchard of 20 acres, if treated, would have been increased 345,600 pounds, or more than seventeen car loads."[14]

Another productive area of work looked at the threat of severe postharvest losses due to rapidly expanding opportunities for transporting agricultural commodities from west to east. One of Pierce's first studies in this area involved the attacks of *Penicillium digitatum* on lemons during packing and transfer of the fruit from cold storage to market.[15] As Pierce wrote to E. F. Smith in 1893, "There is hardly a day that I am not called upon to work on some other disease. . . . It is all work that should be done," he insisted, "but it almost prevents me from working on my special subject."[16]

Even with a heavy workload, Pierce continued his research on the vine disease that would later bear his name. His work in the laboratory involved further histological, bacteriological, and physiological studies; his field tests were designed to gather additional evidence on the use of disease-free cuttings.[17] In 1892, still unable to isolate and identify the pathogen, he gave growers their first ray of hope when he recommended the practical control of taking "great pains" to procure "cuttings for planting in a region where the disease does not already exist."[18] Later, his control work centered mostly on the development of disease-resistant cultivars.[19]

Pierce failed to conquer the California vine disease. He did the best science available to ascertain the causal agent, but came up short. He demonstrated the contagious nature of the disease, but was unable to pinpoint the nature of the contagion or the causal agent. Only much later would it be shown that a fastidious prokaryote, *Xylella fastidiosa*, caused the disease that was the subject of Pierce's labor.

During his investigation of California vine disease, however, Pierce had made important strides in establishing the worth of federal plant pathology. He had expanded his laboratory facilities as well as the range of investigations and, perhaps most important, gained acceptance from California growers through his many scientific and applied accomplishments. With pride, he wrote in 1893 that "I am getting a pretty solid hold in southern California and in fact all along the coast, as is evidenced by the letters of inquiry being received, and I cannot help feeling pleased at the situation as it now appears."[20] In 1974, two later students of the California vine disease, M. W. Gardner and W. B. Hewitt, said of Pierce, "It is worthy of note that there was one good result of the Anaheim disease. It brought to California a competent plant pathologist and a laboratory for plant disease research fourteen years before the university and agricultural experiment station were prepared to start work in this branch of science."[21]

CITRUS DISEASE IN FLORIDA

The California vine disease was only one of several geographically important plant disease problems in the 1890s. In his year-end report to Congress for 1891, Secretary of Agriculture Rusk disclosed that for more than three years citrus growers in Florida had issued urgent calls to the USDA for assistance with plant disease problems, but "practically nothing in the way of investigating the many serious maladies of this important group of plants has been undertaken in this country."[22] Earlier that year, however, the Division of Vegetable Pathology had begun to investigate the problem. Galloway had sent special agent L. M. Underwood to Florida to collect preliminary information, and in June had dispatched two additional agents on a fact-finding mission, involving a two-month tour of important citrus-growing regions of the state.[23] These two agents were Walter T. Swingle and Erwin F. Smith.

Walter Swingle had graduated from Kansas State with a B.S. degree in 1890 and had come to the USDA from the Kansas Agricultural Experiment Station, where he worked on plant diseases with Kellerman. He had proved himself an able plant taxonomist, had published in mycology and plant pathology, and had shown an interest in plant breeding.[24] He and Smith examined groves, interviewed growers, and found a public eager for USDA help.[25] This was no surprise, given the growing economic importance of Florida's sizable and thriving citrus industry. In 1870 Florida's orchard products had been valued at $21,259; twenty years later the crop was valued at $7 million.[26] But the investment in citrus groves was severely threatened by a host of major diseases, including orange "blight," "foot rot," and "die back."[27] On the basis of their survey and evaluation, Smith and Swingle reported to Washington that plant diseases threatened economic catastrophe for Florida's citrus growers.[28]

The possibility of a major failure of the citrus crop was exacerbated by the lack of an existing plant pathology program in Florida. During their travels, Smith and Swingle met J. C. Neal, of the Florida Agricultural Experiment Station at Lake City, who had done some pioneering work on nematodes. Because of political difficulties at the Florida experiment station, however, he had been unable to consider adequately many practical matters such as plant disease.[29] At the end of 1891, Galloway announced that "it may be truthfully said that the orange industry . . . is in danger, and a careful and thorough investigation by the Department is most urgently needed."[30] But lacking the funds to commit the division to an extended field project, he chose, instead, to send Swingle back to Florida for another short visit in early 1892.

The limitations of such short visits were obvious. Little could be accomplished in the field during a small segment of one growing season. Inoculation experiments and fungicide tests, for example, required several years of study. Moreover, as Galloway complained, "the long distance from Washington prevented the laboratory studies from being carried on at this place, as it was found impossible, on account of the warm climate and imperfect mail facilities, to obtain good material from Florida."[31]

At a meeting in May 1892, during which citrus growers vented their growing frustration, Swingle promised that "the department is anxious and willing to investigate."[32] Indeed, after further petitions and resolutions from growers, the Orange Growers Convention, and the State Horticultural Society, Galloway decided to take advantage of increased appropriations from Congress and establish a USDA Florida station for research on citrus diseases.[33] In October 1892 Swingle began operations in Florida with his new assistant, Herbert John Webber, hired by the USDA specifically for the Florida project.[34] After completing his B.S. degree under Charles Bessey at the University of Nebraska in 1889, Webber became an instructor of botany at that institution. Later, he became an instructor of botany and did some graduate work with William Trelease at the Shaw School of Botany in St. Louis.[35]

Swingle and Webber chose Eustis, located in the lake region of central Florida, as the site for the station. Their choice was due in part to its proximity to major orange-production areas, but also because the citizens of Eustis agreed to donate a laboratory building and experimental plots for the station.[36] Without the expense of a physical plant, the division was able to furnish Swingle and Webber with the most modern equipment for physiological, anatomical, and pathological research, as well as an impressive library containing some of the world's most authoritative literature on citrus fruit. Galloway declared that the work at the laboratory would encompass "the study of the diseases of all economic plants of the far South, as well as the discovery and trial of remedies and preventives."[37] But this ambitious agenda would have to wait. Certain destructive diseases immediately affecting the orange and lemon industries took precedence.

Over the next few years, the laboratory at Eustis pursued several lines of investigation on orange blight. These various approaches to the research demonstrated once again the increasing sophistication of plant pathology in the United States in the 1890s, particularly in the areas of physiology and bacteriology. Anatomical studies compared healthy and diseased tissue, physiological investigations focused on water loss from wilting tissue as well as carbon assimilation of diseased and healthy leaves, while bacteriological techniques were applied to the search for a pathogen.[38]

But etiological explanations for orange blight remained incomplete. Swingle and Webber found themselves in a predicament similar to Pierce's in California. As Swingle reported sadly to the Florida State Horticultural Society in April 1894, "For three years, [I have been] telling a continued story about blight and I am sorry to say that discussion will have to be continued longer."[39] Unable to cure the crop malady most responsible for bringing them to Florida, Swingle and Webber continued their experiments while turning their attention to other problems.

Another baffling plant disease of citrus was foot rot. Swingle and Webber maintained that foot rot was "probably" contagious and caused by "a parasitic organism," but they lacked evidence. Their knowledge of plant breeding paid dividends, however, when experiments confirmed that foot rot could "be prevented by using immune or resistant stocks." Unfortunately, incompatibility between stocks and soil types hindered breeding experiments and compelled Swingle and Webber to look to other management strategies, particularly cultural schemes such as removing the soil from around the crown roots and cutting away diseased bark and wood.[40]

But Swingle and Webber could claim successes in other areas of their work in Florida, particularly with scab. Aware that a fungus (*Cladosporium* sp.) caused the scab, the USDA plant pathologists began to consider treatment. Extensive field tests were done in 1894 in the Lake Weir region of Florida.[41] Swingle and Webber later reported that both Bordeaux mixture and the ammoniacal solution of copper carbonate, "if properly applied," had performed well.[42]

Another productive area of applied plant pathology at the Eustis Laboratory in the early-to-mid-1890s was research on the cause and control of sooty mold. Caused by a dark-colored, superficial fungus that followed the attacks of insects like the mealy wing or white fly, sooty mold frequently resulted in serious damage to numerous citrus crops. The Florida team made successful experiments with various sprays and fumigants, but perhaps their most interesting experiments were those involving biological control of sooty mold through the use of several entomaphagous fungi.[43] Working with a fungus (*Aschersonia aleyrodis*) parasitic to the insect that was responsible for depositing the "honey dew" on which the sooty mold was nourished, the pathologists searched for ways to artificially infect the insects during the larva and pupa stages. Preliminary, but unsuccessful, experiments involved spraying leaves with spores mixed in water and hanging branches containing the fungus in trees, expecting that the rain would wash the spores down on the larva and pupa. Later, however, the scientists were successful at introducing *A. aleyrodis* into orange groves by transplanting young trees already containing the mealy wing larvae infected with the fungus.[44]

Experiments in biological control for sooty mold demonstrated the scientific talents and innovations of Swingle and Webber and the usefulness of the USDA satellite laboratory in Florida. Unfortunately, however, an event beyond their control abruptly halted these and other investigations at Eustis: In 1894–95 a major freeze wiped out the orange crop. In 1896 both plant pathologists were recalled to Washington. Their untimely departure naturally left incomplete much of their research on major citrus diseases like orange blight, foot rot, and dieback. Nevertheless, Swingle and Webber had carried out good work in Florida in the less than three years since the creation of the laboratory. They had made significant original contributions to the understanding of a number of citrus diseases, had discovered effective control measures for scab and sooty mold, and had begun to diversify their lines of research into other subtropical crops such as pineapple. Furthermore, their investigations revealed an increasing degree of sophistication on the part of a new generation of plant pathologists. For the first time, plant pathologists were turning their attention to physiology in relation to disease and were beginning to recommend breeding for resistance. In future years, Swingle and Webber would apply the insights they had gained in Florida to illustrious careers in plant breeding.

Fruit and orchard interests in New York

Unlike Santa Ana and Eustis, by 1891, Geneva, New York, had a well-established agricultural experiment station. Investigations there in the 1880s by J. C. Arthur and E. S. Goff had launched a tradition of applied plant pathology in central New York, and F. C. Stewart would continue to emphasize practical disease work when he arrived in 1893. In the interim, the Geneva region, which in the 1890s was "one of the largest nursery centers east of the Mississippi," hosted a USDA research laboratory.[45] B. T. Galloway and station director Peter Collier merged interests so that the USDA efforts to test and enhance copper fungicides might work to the advantage of fruit and orchard interests in New York. Galloway's responsiveness to another segment of America's increasingly regionalized agriculture, in this case with particular attention to diseases of nursery stock, further expanded the power and influence of federal plant pathology.

Since David Fairchild's arrival at the USDA in 1888, he and Galloway had collaborated closely on a wide range of chemical tests involving grape rot, pear scab, and leaf blights of numerous fruit and orchard crops. When the "urgent request of a large number of western New York nurserymen and fruit-growers" arrived at the division in early 1891, Galloway immediately chose Fairchild to mount a federal response to a number of severe leaf diseases of nursery stock.[46]

Fairchild's assignment to the Geneva station allowed him to test fungicides under varied conditions of soil and climate. He was pleased with the opportunity to follow up earlier investigations that had looked promising for protecting nursery stock from the attacks of different pathogenic fungi. His only concerns were that he be given sufficient library and laboratory facilities to carry out the work and sufficient time to be effective. Before leaving in March 1891, he remarked that "it does not seem to me that just a month or two will pay. . . . I ought to stay long enough to make my work felt and to carry out *fully* the experiments."[47]

When he arrived in New York, Fairchild's anxiety about facilities was quickly put to rest. Station director Collier and New York nurserymen enthusiastically welcomed him to Geneva on April 17, providing laboratory and library space as well as a special field site. Fairchild was particularly delighted with the location of the test plots, enthusiastically describing them to Galloway as "within stones throw of the station and very conspicuously situated where all can see the results."[48] Nurserymen also furnished stock for the USDA experiments.[49] Soon after his arrival, Fairchild characterized Geneva as "a *most* favorable place for the study of nursery diseases."[50]

Over the next three growing seasons, Fairchild focused his attention on fungicide experiments involving leaf blights of pear, apple, cherry, quince, and plum. He considered three main questions: First, could various leaf blights be prevented by the use of Bordeaux mixture or an ammoniacal solution of copper carbonate? Second, what effect would these fungicides have on the growth of nursery stock? And finally, did the variety of stock have any effect on its resistance to leaf blight? Of these three, Fairchild contributed answers to the first two. The question of disease-resistant stock required more time than three growing seasons.[51]

The work at Geneva between 1891 and 1893 supplied the USDA with reliable data on the use of fungicides and other treatment strategies for many nursery diseases. Fairchild was delighted to report that the leaf blight caused by *Entomosporium maculatum*, "perhaps the greatest obstacle to the profitable production of pear stocks," was highly susceptible to control with copper sprays. His research had also led him to suggest that "diseased seedlings should be headed back to within 1 or 2 inches of the ground and all side shoots likely to harbor the parasite removed." In addition, he suggested employing windbreaks to protect seedlings.[52]

Combined with correlated experiments done at Mullikin, Maryland, the positive results at Geneva gave the USDA greater authority to advocate the use of chemicals.[53] Tabulating the results of the New York investigations, Galloway wrote Secretary Morton that "as a result of this work it was shown that trees in the nursery sprayed three seasons at a total expense of one dollar per thousand were worth, according to the estimates of . . . nurserymen, . . . thirteen dollars more per thousand than those not sprayed."[54] Once again, as in California and Florida, the USDA in New York had demonstrated its ability to respond to a regional agricultural problem and come away with positive results.

With his research at Geneva nearly completed, Fairchild looked beyond the laboratory and fields where he had spent most of his time since joining the USDA in 1888. In 1893 he accepted a commission from USDA Secretary Morton "to proceed to Europe to prosecute studies in connection with plant diseases and other matters pertaining to scientific agriculture."[55] That November Fairchild set out for Europe, initiating what would become a productive and distinguished career of plant exploration and introduction for the USDA, including a search for disease-resistant plants.[56]

A NEW EMPHASIS ON PHYSIOLOGY

The year that Fairchild left for Europe, the international Columbian Exposition opened in Chicago, celebrating the four-hundredth anniversary of the voyages of Columbus to the New World. Opening on May 1, 1893, the exposition lasted for six months. The Division of Vegetable Pathology was there, along with other sections of the USDA, capitalizing on an opportunity to display its applied science to a diverse audience. Galloway was pleased with the public's reception of the plant pathology exhibit: "On the whole," he concluded in his 1893 annual report, "it may be stated that the exhibit was the means of awakening much interest in the work of the division."[57]

A focus of the USDA's exhibit at the Columbian Exposition was its recent achievements in plant disease control. Galloway was quick to point to the increase in understanding and control of grape downy mildew and black rot, potato late blight, peach leaf curl, and diseases of nursery stock, among others. At a time of intense pressure for science to deliver increased agricultural production, as Galloway noted, Bordeaux mixture made his efforts to secure funding much easier.

The diagnosis of fungal diseases and the development of chemical treatments remained the mainstays of federal plant pathology throughout the 1890s. Reports from field laboratories in California, Florida, and New York were encouraging, and investigations in Washington continued to yield positive results. Since 1892 Galloway had focused his own research on a macrosporium leaf spot disease of potato. Reports in 1892 of serious injury to potato crops in Maryland

and surrounding states had stirred the USDA into action. Microscopic investigations linked this most recent outbreak to *Macrosporium solani*. After sending out circulars to gauge the range and the extent of losses, the USDA launched experiments at Garrett Park, Maryland, which showed that the leaf spot disease could be controlled successfully with Bordeaux mixture.[58]

A further demonstration of continuing USDA success in applied plant pathology was the treatment of pear leaf blight in orchards and nurseries. In the early 1890s M. B. Waite, the resident expert on pome fruit diseases, while continuing his studies of fire blight and pollination in pear and apple, directed a series of Bordeaux experiments designed to judge the cost and value of copper treatments on pear leaf blight. In 1894 Galloway reported that the results of using Bordeaux mixture on pear leaf blight proved undeniably the "clear profit" justification for the use of fungicides.[59]

Not all of the applied disease work at the USDA, however, matched the success achieved with the leaf blights of potato and pear. The Division of Vegetable Pathology's first attempts to find practical solutions to cereal rusts proved especially challenging. Galloway set a high priority in the early 1890s on treatments that might check the damage from rusts that "have unquestionably played an important part in keeping the average yield down to a ridiculously low figure." But the results of experiments beginning in 1892 in Maryland and Kansas were not encouraging. Soil and seed treatments showed little promise, and fungicide spraying provided inconclusive results while proving "exceedingly laborious and expensive." Galloway had not expected miracles, but he had hoped that control strategies that were yielding spectacular results against other diseases might prove useful against cereal rusts. "It is not to be supposed that rust could have been entirely prevented," Galloway confessed in 1892, "but even if half the damage had been arrested, our farmers, on their wheat crops alone, would have saved in the neighborhood of $34,000,000."[60] Such savings, however, were not to be realized.

Despite all of the early accomplishments in fungal etiology and control, there were limits to what USDA applied science could achieve in the 1890s. Understanding and conquering cereal rusts, peach yellows, California vine disease, and orange blight proved to be beyond the grasp of USDA plant pathologists. Such limitations posed a severe threat to an agency that depended on public patronage. The Division of Vegetable Pathology was constantly pressured to expand its research to include an ever-growing array of agricultural problems brought to its attention by a demanding constituency. The alternative to expanding its research was extinction. Yet undertaking new projects exposed the new division to the risk of pushing the boundaries of applied science too far.

Thus, the USDA began to consider new approaches to plant disease problems. Although the division continued to focus on disease-causing fungi and copper sulfate, it began to extend its sights beyond the application of chemical fungicides to different plant diseases. A major new component of USDA plant pathology, and one linked directly to recent attention to plant improvement by breeding, was the new scientific field of plant physiology.

Plant physiology was introduced to the United States with the publication of Julius von Sach's *Textbook* in 1865. Interest in the subject at first was lackluster, but it rose sharply in the 1880s, due in large part to advances in the emerging botanical sciences and the publication of important treatises from American scientists such as G. L. Goodale of Harvard. The subject was of particular interest to some USDA plant pathologists who thought that too much attention was being paid to the pathogen and not enough to the host plant. At the very least, they thought, taking account of plant physiology would allow them to broaden their discipline to encompass the advances of the botanical sciences. But plant physiology would prove to be much more than an enhancement of plant pathology; its influence proved to be both fundamental and practical.

Galloway began to explore the idea of incorporating plant physiology into the Division of Vegetable Pathology in 1892.[61] He solicited the views of his assistants and found Swingle and Webber particularly amenable. The two were undoubtedly influenced by their backgrounds in plant breeding, but they also appear to have been motivated by a desire for the division to take action before this potential scientific plum went elsewhere. Swingle assured Galloway that "unless we take it up I believe there will soon be a purely Plant Physiological and Anatomical Division which would be unfortunate for the Division studying physiology should be in position to apply the discoveries made to pathology."[62]

By early 1893 Galloway was working to make plant physiology a new focus of the division. He wrote to Swingle in March of that year announcing that he planned "to make a start of the physiological work pure and simple after the first of July." Galloway had a scientist in mind to lead the work, A. F. Woods of the University of Nebraska. "From what Dr. Bessey says," Galloway continued,

"I believe Woods would be a good man to build up a work such as we have in view."[63] Swingle replied that "this is absolutely necessary to insure true progress in the study of plant pathology."[64]

Filling the position left vacant by Fairchild's departure for Europe, Woods was appointed officially to the USDA at the end of 1893. He brought with him a background in plant physiology and ecology nurtured by Charles Bessey's "new" botany seminar at the University of Nebraska. Woods received both his B.S. (1890) and A.M. (1892) degrees at Nebraska, and served as assistant botanist under Bessey until he joined the USDA.[65]

In 1895 the Division of Vegetable Pathology underwent a name change that reflected its new interests: It was reorganized as the Division of Vegetable Physiology and Pathology. To support its expanding lines of investigation, the new division acquired new quarters, which included laboratory and greenhouse space "better lighted, better heated, and better adapted in every way for the work than those formerly used."[66] Speaking directly to the name change and plant physiology's relation to pathology, Galloway told American agricultural interests that "to make the work of [the division] the greatest practical value it must be based upon a knowledge of the way plants behave in health . . . . Its aims," he explained, "are to broaden our knowledge of how plants grow, how the different conditions to which they are subjected affect their usefulness, and how and to what extent these conditions may be controlled to the advantage of the husbandman."[67]

With the arrival of A. F. Woods, the research divisions of the USDA began to organize themselves along specialty lines. Galloway referred to the new organization as "a proper division of labor" whereby "each member of the force is enabled to concentrate his efforts on a comparatively few subjects, and he is thereby in position to become authority on them."[68] As a foundation for investigating disease resistance and the selective breeding of resistant cultivars, plant physiology became a focus of USDA plant pathology.

The division's response to the wheat rust failure was an example of this focus. Galloway recruited Mark A. Carleton from the Kansas Agricultural Experiment Station in 1894 to head the USDA's new cereal disease investigations. Carleton had trained under Kellerman, receiving his undergraduate and graduate degrees at Kansas State. After a brief teaching assignment at Garfield University in Wichita, he returned to Kansas State in 1892 and participated in experiment station studies of cereal diseases with A. S. Hitchcock. When he joined the USDA in 1894, Carleton's job immediately became finding new control strategies for cereal diseases, particularly rusts.

Carleton's duties at the USDA involved intensive studies on the life-histories of rust fungi, and he would do impressive work in this area as well as in carefully controlled inoculation studies. His applied research, however, focused on the possibility of controlling rusts by breeding host cultivars for resistance. Galloway knew that the best hope of managing rusts lay with the selection of resistant cultivars.[69] Because "no experiments to determine the rust resistance of different varieties" had "been made in this country," Galloway decided to develop a large-scale program.[70] By 1895 Carleton was considering the distribution and severity of cereal rusts throughout the principal wheat-growing regions of the country in order to determine the relative susceptibility of different cultivars to the rust pathogen. On the fertile promises of disease resistance, Carleton had begun a line of research in the mid-1890s that would soon take him not only across the United States but also around the world in search of disease-resistant cultivars adaptable to American agriculture. Although it would be several years before the results of this work became evident, the idea of breeding for resistance as a means of plant disease control had found a home in USDA plant pathology.

Though linked most closely with breeding for resistance and improvement, plant physiology also filtered into other areas of federal plant disease research. There were experiments to test the effect of Bordeaux mixture on plant growth, and irrigation studies to examine the connection between water supply, plant growth, and root disease. Moreover, USDA plant pathologists began to consider "disorders" of plants that were suspected to be of nonparasitic origin. Shortly after the reorganization of the division in 1895, several new lines of inquiry were initiated, including a study of the "effects of gases and fumes from smelting and other similar works" and a series of investigations on the diseases of shade and ornamental trees. The latter study included diseases caused by insects and fungi, but focused on factors such as nutrition and climate. An ongoing area of USDA research was diseases of plants grown in greenhouses, particularly the Easter lily, violet, rose, and carnation. Such studies explored nutrient supply, light, heat, water, and soil in relation to plant health.[71]

But plant physiology was not the only attempt to augment and strengthen the discipline of plant pathology. One of the most innovative and forward-thinking scientists at the USDA in the 1890s was Erwin F. Smith. Smith's investigations of plant disease–causing fungi and bacteria, combined with an impulse to create a world-class scientific research institution at the USDA, propelled federal plant pathology in new directions. For Smith, international recognition of the merits of scientific plant-disease research at the USDA was essential to his field's rightful place among the major biological sciences.

PLANT BACTERIOLOGY

Another feature of the expansion and specialization of work in the Division of Vegetable Pathology in the 1890s was a new focus on bacteriology. Actually, the study of bacteriology had begun in 1888, when M. B. Waite borrowed equipment from the Bureau of Animal Industry to set up the first laboratory in the department for the study of bacterial diseases of plants.[72] In this laboratory, Waite made studies of fire blight of pear and apple that eventually revealed the role of insects in the transmission of the disease. Waite's bacteriology laboratory proved valuable in 1891, when N. B. Pierce sent samples from California to test his preliminary identification of a bacterium as the causal agent of the vine disease.

As Waite was setting up the new bacteriology laboratory in the late 1880s, Erwin F. Smith was still engrossed in the seemingly unsolvable problem of peach yellows. Smith took great interest in Waite's new laboratory, expressing a desire to use bacteriological techniques to continue the search for the cause of peach yellows. He worked up culture media, including some neutral peach-twig infusions, but obtained negative results. He had, however, taken a first exploratory step in the methodology of plant bacteriology, a field in which he would soon excel.[73]

Smith's association with bacteriology had actually begun in the mid-1880s in Michigan, when he was working in the field of human hygiene and sanitation. Assigned to produce a review of the literature on the subject, Smith waded through a great deal of material, including the most recent work by Europeans such as Louis Pasteur, Robert Koch, and Joseph Lister. This work brought him into contact with men such as Frederick V. Novy, who was destined to become one of the United States' leading medical bacteriologists.[74]

Coincidentally, in 1889, just as Smith was beginning to take an interest in

Waite's new lab, Theobald Smith and Veranus A. Moore, of the Bureau of Animal Industry, published a landmark study on hog cholera. This study heightened Smith's excitement about the scope of the work of human and animal bacteriologists and his curiosity about their techniques.[75] Smith acquainted himself with the hog-cholera study, as well as with the studies of others including William H. Welch and Harry Luman Russell at Johns Hopkins. Russell, formerly a student of Koch's in Germany, was at the time a student of Welch's at Johns Hopkins University, where he wrote his thesis in 1892, "Bacteria in Their Relation to Vegetable Tissue."[76]

As he had before with his work on human hygiene and peach yellows, Smith took pains to review the literature on plant bacteriology to understand thoroughly the background of the science. He had enough familiarity with the subject to have some firm opinions. In 1889, Smith had reviewed the second edition of Robert Hartig's *Lehrbuch der Baumkrankheiten* for the *Journal of Mycology* and criticized the noted German forest pathologist for his lack of attention to the groundbreaking phytobacteriological research done by Burrill and Arthur in the United States.[77] Smith did not share the opinion of Hartig and many other botanists that bacteria caused few, if any, diseases in plants. This school of thought held that bacteria were likely to be opportunistic infectious agents taking advantage of wounds or existing damage done by fungus diseases. Even Anton De Bary had stated in 1885 that the case for bacterial plant diseases simply had not been proved through proper experimentation.[78] Although Smith believed that bacterial diseases were more common than generally thought, he did share De Bary's opinion that the experimental proof was less than convincing, with the exception of the work by Thomas J. Burrill, Joseph C. Arthur, Edouard Prillieux, and Jon Hendrik Wakker.

To bring his science onto an even footing with human and animal pathology, Smith spent considerable time studying and disseminating information on the methodology of plant bacteriology. He stressed the point that "pathogenesis must be worked out conclusively." This, Smith claimed, could be accomplished only if the scientist followed four steps, the first three derived from Koch's postulates and the fourth added by Smith. First, the scientist must establish a "constant association of the germ with the disease." Second, the scientist must isolate that germ from the diseased plant and culture it. Third, the scientist must reintroduce the cultured germ, causing characteristic symptoms of the disease under investigation. And finally, the scientist must reisolate the bacteria from the inoculated diseased tissue and establish that it is identical to the original germ.[79]

This process depended on flawless, pure-culture techniques, which Smith also investigated. He developed practical ideas on bacterial tolerance to heat and light in the laboratory as well as their reactions to alkaline or acid environments. He wrote about various poured-plate media and the problems of culture contamination. He also sought to adapt the microscopic staining techniques of plant pathologists who were more accustomed to working on fungi.[80]

In 1893 Smith applied his increasing expertise in bacteriological technique to the first of a series of specific bacterial plant diseases, cucumber wilt.[81] In 1892, a bacterial wilt of cucurbits had struck cucumbers, muskmelons, and squash around Washington, D.C. Smith may have been drawn to investigate the wilt because he had already seen a disease of muskmelons in Michigan caused by an *Alternaria* sp. In 1893 he studied the wilt, observing the advance of infection through the hosts. In 1894 he demonstrated that the disease was caused by a bacterium, which he described and named *Bacillus tracheiphilus.* Smith devoted no time to prophylactic experiments, instead assuring himself that cucumber beetles were the culprits in spreading the disease. At this stage, other researchers may have concluded their work in victory, but Smith was just beginning. He launched several years of detailed experiments to observe the bacillus's growth and development in different culture media and to chart its effect on the host plant's physiology.[82]

Smith's work with bacterial wilt of cucurbit encouraged him to study similar wilt diseases of watermelon, cotton, and cowpea that were found across the United States, but particularly in the South. He thought these wilts might be other bacterial diseases, but through field studies in South Carolina with watermelon in 1894, he discovered the presence of a *Fusarium* sp. He studied the mechanisms of the wilt, that is, the clogging of vessels, as he had with *B. tracheiphilus.*[83] This work too would stretch over many years, and Smith did not publish a definitive bulletin on the subject until 1899.[84] He made important breakthroughs on the soil-borne nature of the pathogen and contributed significant information on fusarium diseases of many staple crops such as potato and tomato, including preliminary studies on breeding for resistance.

In 1895 Smith began a study of southern bacterial wilt, which was similar to the fusarium wilt he had been examining, both in the choice of hosts and the damage to them. Wilt was a serious impediment to the cultivation of tomato, potato, and eggplant in the southern United States, and as Smith later found, it also affected tobacco. In 1890 Burrill had published some notes on the disease as it applied to potatoes.[85] In 1891–92, Byron D. Halsted had done important experiments on "southern tomato blight" and made associations between it and "brown rot of potatoes" and also more tenuous ties to a "melon blight," which was later shown to be the bacterial wilt of cucurbits.[86] Halsted's work was a good beginning and Smith gave him credit for exposing the southern wilt and perceiving the similarities between these vascular-damaging diseases.

In 1895 S. M. Tracy, the director of the Mississippi Agricultural Experiment Station, asked Smith to study the disease, following up on Halsted's work.[87] As the first step in what would become twenty-five years of investigation on the subject, Smith described and named the pathogen *Bacillus solanacearum*. L. R. Jones of the University of Vermont wrote in 1896 that Smith's examination of southern bacterial wilt "stands unrivalled as a thorough piece of work on bacterial diseases of plants."[88]

In 1897 Smith continued to expand the knowledge base on bacterial diseases with his work on black rot of crucifers. Black rot had been first noted in cabbage in 1891 by H. Garman in Kentucky. In 1895 L. H. Pammel of Iowa State described and named the pathogen as it appeared in rutabaga.[89] Smith read Pammel's work, and when he first saw black rot in cabbage fields, he wrote to A. F. Woods that "the disease is likely to attract wide attention in this country before many years and the Division ought to lead the way."[90] H. L. Russell, a bacteriologist at the University of Wisconsin, had already spent several years studying the disease, but Smith published first, much to Russell's dismay, securing the credit for himself and for the Division of Vegetable Physiology and Pathology.[91]

Smith's study of black rot of crucifers revealed that the initial point of infection was water pores on the leaves. This was the first time such a point of infection had been demonstrated for a bacterial plant disease. Previously, the path of infection for bacterial plant disease had been assumed to be wounds and insect transmission. Smith described the black rot bacterium more fully and renamed it *Pseudomonas campestris*. He also extended the study of its pathogenicity to cauliflower, kale, mustard, and other cruciferous hosts.[92]

Although Smith had successfully identified a number of bacterial pathogens by the late 1890s, he found that plant bacteriology had not gained full acceptance in the scientific community, particularly in Europe, which was still considered the world leader in plant pathology. In fact, in 1897, the same year that Smith was studying black rot of crucifers, Alfred Fischer, a distinguished professor of botany at the University of Leipzig, published *Vorlesungen über Bakterien*, a well-respected collection of lectures in which Fischer stated that "no single example is yet known of bacteria which can insinuate themselves into the closed living cells of a plant."[93]

Fischer did not intend his statement as a bombshell; he was simply repeating the opinion held by many, if not most, botanists, particularly in Europe. American scientists had already begun to recognize that in the field of bacteriology, the United States, not Europe, was leading the way. In 1891, for example, H. L. Russell decided to abandon the University of Berlin for Johns Hopkins University to follow his interests in phytobacteriology.[94] However, to Smith, Fischer's text was both an egregious error of fact that demanded clarification and a cavalier slight to the scientific work done by himself and others, revealing either an ignorance of the current literature or an unwillingness to accept its implications. Either possibility was intolerable to Smith, who felt that the noted German botanist was perpetuating a scientific misperception that had already been refuted.

In 1898 Smith prepared a reply to Fischer's *Vorlesungen über Bakterien*. Aimed at a German academic audience and titled "Are There Bacterial Diseases of Plants?" the article was published in April 1899 in the prestigious journal *Centralblatt für Bakteriologie*.[95] Smith's tone, and the tenor of the debate to follow, was determined in the article's introduction: "It is seldom in a genuinely scientific book that one finds so many unwarranted assumptions and serious misstatements in the space of a single page." He went on to name eight plant diseases that he thought had been shown definitively to be of bacterial origin: cucurbit wilt, brown rot of potato, black rot of cabbage, bacteriosis of bean, yellows of hyacinth, pear blight, soft rot of hyacinth, and olive tuberculosis. Smith had personally worked on the first six of these and was instrumental in determining the causal agent of the first four. In closing, he chastised Fischer for speaking on a subject without being "familiar even with the readily accessible literature in his own language."[96]

In the same number of the *Centralblatt für Bakteriologie*, Alfred Fischer published "The Bacterial Diseases of Plants: Answer to Dr. Erwin F. Smith." Clearly angered by Smith's attack, Fischer argued that his original presentation on bacterial diseases of plants "was completely adequate for a book of the moderate range of my lectures." He had not cited, in his lectures, the literature mentioned by Smith because he did not think it worthy. The experiments were not rigorous enough to prove indisputably that the bacteria that had been found in the diseased plants were, in fact, the cause of the symptoms rather than a researcher's misinterpretation or contamination due to poor laboratory technique. Fischer felt particularly "obliged to show why I hold Smith's own work to be unsatisfactory. As long as Smith does not submit to carry out and prepare his experiments satisfactorily with all the necessities of experimental pathology," Fischer argued, "he has no subsequent claim that his work will be evaluated as well guaranteed and certain achievements of science." Responding specifically to Smith's nationalistic tone, Fischer commented that after studying elements of J. C. Arthur's famous fire blight studies "one loses the desire to read American works."[97]

Smith's counterattack was not long in coming. Before the end of the year, he had dispatched another article to the *Centralblatt* titled, suggestively, "Dr. Alfred Fischer in the Rôle of Pathologist." Smith's article was an emotional statement intended primarily to rebut the criticism of American work on fire blight and to question Fischer's credibility to pass judgment on the work of plant pathologists. "The positive statements of a pathologist," Smith wrote, "who has spent five years industriously studying these special diseases ought to be worth considerably more than the negations of a man who is not a pathologist, and who has not seen a single one of these diseases, much less critically studied one of them."[98]

In early 1901, Smith published the final salvo against Fischer and, more important, against the general hesitance by many scientists to accept bacterial diseases of plants. He prepared a detailed discussion, published in three parts, of his own bacteriological work, refuting the criticisms leveled by Fischer point by point. A focus of this discussion was a defense of his laboratory techniques. He began by outlining the criteria for proving the pathogenicity of any causal organism and then proceeded to demonstrate how his experiments met those conditions. He used numerous photographs to illustrate his points, mindful of the fact that the lack of such evidence had been one of Fischer's leading criticisms.

Smith's three-part defense was persuasive and decisive. It was also less abrasive than earlier pieces in the exchange. "My final answer is ready," Smith wrote. "It is dignified in tone. It is entirely free from personalities, and best of all it is *unanswerable!* I wish to use it both for the sake of my own reputation in Europe and for the sake of the dignity and honor of the U.S. Department of Agriculture, and of American workers generally."[99]

Smith's final defense in *Centralblatt* in 1901 closed the dialogue and settled the matter. The debate had overstepped academic niceties and had taken on personal and national overtones. Fischer never forgave Smith for the whole affair. Smith later regretted the rancor of the exchange, but he took comfort in the fact that other scientists, both in the United States and Europe, supported him.

Many bacteriologists and plant pathologists had been dismayed by the opinions expressed in Fischer's *Vorlesungen über Bakterien* and welcomed Smith's authoritative voice on the issue. In 1897 W. H. Welch of Johns Hopkins University wrote to Smith that he did not "know what to make of A. Fischer's position. . . . He goes so far as to deny the possibility even of such diseases . . . being indeed quite sarcastic on the subject."[100] In 1899 Professor Emerich Ráthay, director of the Enological Pomological School of Klosterneuberg in Austria, proclaimed to Smith that "I am quite convinced, that there are Phytobacterioses and that the objections of Prof. Dr. Alfred Fischer concerning the existence of them are not at all founded."[101] L. R. Jones, who had contributed significant experimental evidence for a physiological mechanism of disease through the actions of bacterial enzymes, congratulated Smith on the debate in 1901: "I doubt if [Fischer] ventures to say anything more and so thoroughly convincing a presentation of the matter must close the mouths of other critics as well."[102]

In terms of science, however, the exchange brought out nothing new. The heart of the matter was not the existence of bacterial diseases of plants, but rather the scientific quality of experimental proof offered by the field of phytobacteriology. Fischer's comments in *Vorlesungen über Bakterien* placed him squarely in the middle of an argument for which he was neither prepared nor particularly qualified. He was not thoroughly acquainted with the current literature on phytobacteriology and had simply followed the assumptions about bacterial diseases of plants that prevailed in Europe.[103]

The ramifications of the Fischer-Smith controversy went far beyond the actual subject. Although Smith did not present new information on phytobacteriology in his articles in the *Centralblatt,* he summarized the state of knowledge brilliantly and displayed it in the glaring light of a scholarly exchange that attracted great attention. The debate has become a central part of the story of the foundations of plant pathology. Moreover, the exchange gave American research a new place in the forefront of plant pathology. The glow of victory shone particularly on the scientists of the USDA and most notably on Erwin Frink Smith.

By the end of the century, the Division of Vegetable Physiology and Pathology had become a very different organization than the Section of Mycology had been in 1886. USDA plant pathology was no longer confined to research, largely applied, on fungi and fungicides. The drive for practical disease-control measures was now balanced with the pursuit of fundamental science. Although the scientist's view through the microscope was still connected ultimately to crop yield, it was now acknowledged that the interconnections between the laboratory and the field were complex and likely to become more so as plant pathologists confronted new issues in the twentieth century.

Part Four

THE MATURATION AND PROFESSIONALIZATION OF PLANT PATHOLOGY

Agriculturists and horticulturists all over the world have of late years directed a great deal of attention toward the study of plant diseases and the methods of combating them. The Department of Agriculture of the United States has done vastly more than any other institution in the world along this line and the results have been well worth the efforts, for many efficient methods of fighting these pests have been devised.
—Edward M. Freeman

Pathology is not only one division of botany, it is by far the largest division, it is growing very rapidly and must continue to grow rapidly in the future. As a result most pathologists are young, with the zeal and enthusiasm of youth and of expanding opportunity.
—George R. Lyman

The turn of the twentieth century was an extraordinary time in the United States. Having filled its geographical limits on the North American continent, the nation turned from expansion to consolidation. It was commonly assumed that the key to the nation's continued development was not the further acquisition of resources, which the country seemed to possess in limitless abundance, but rather the rational organization and exploitation of those resources—natural, human, and intellectual.

This period, known as the Progressive Era, was characterized by a fierce debate over how to marshal most efficiently the massive resources at the nation's disposal in order to create the most improvements in the lives of the most citizens—in other words, how best to organize progress. The debate over the proper courses of action, and the questing spirit that spawned it, crossed boundaries from politics and the role of government to industry and agriculture to education and science. Because the most efficient means of delivering the best results with the least waste was assumed to be division of labor, the Progressive Era became the era of the specialist.

During the Progressive Era, the greatest source of expertise and intervention in the name of progress came from government, particularly the federal government. It was accepted and even expected that the government had a significant role to play in the lives of the average citizens to support those activities that were generally agreed to be necessary to keep the nation flourishing. One of the most important and groundbreaking roles that government workers acquired was overseeing the activities of business and industry, whether as regulators or as central agents for guiding the economy. Progressivism advocated the control of economic cycles, like those that had caused the Gilded Age of the late 1800s to struggle through periods of boom and bust. The theory ran that, although corrective policies of government might dampen explosive growth, they would also mitigate the disastrous effects of depressions.[1]

The first quarter of the twentieth century was a golden age for American agriculture. The growing role of the United States in international commerce led to more agricultural exports. Farm prices increased, and so did yields, bolstered both by scientific farming methods and by the increasing mechanization of farm labor.[2]

The federal government was eager to assist in bringing progressivism to agriculture. Congress increased spending to help farmers in many ways. It created regulatory legislation to be administered by the United States Department of Agriculture, supported various forms of research to increase production, and founded government-supported agencies charged with delivering the improvements in agricultural science and practice to even the most isolated farmer in the country.[3] To meet its growing mission, the USDA was enlarged and reorganized several times between 1900 and 1920. The expansion was ambitious, but judging from the resulting increase in agricultural production, it was also successful.

The Progressive Era's brand of rational and efficient organization facilitated by the federal government became a central theme in the development of the field of plant pathology. Under the direction of Beverly Galloway, USDA research programs dedicated to plants, including plant diseases, were organized into a single, massive bureau. In theory, the bureau would oversee the progress of the nation's plant scientists, directing resources where needed and keeping redundancy and waste to a minimum. Galloway encouraged cooperation between his staff of federal scientists and researchers at state agricultural experiment stations and professors at colleges and universities across the nation. Since the federal government was providing much of the funding for plant disease studies, Galloway and his successors had great power to enforce a system of nationwide research.[4]

But plant pathologists wanted to organize themselves beyond the network created by federal funding. The maturing science was finding that it could now stand alone, free from disciplines such as botany and mycology that had nurtured it over the previous quarter-century. The general trend in the sciences was an emphasis on specialization. Plant pathologists, following this trend, created a specialists' association to further the goals of communication and cooperation.

They also founded a specialized journal dedicated to publishing the best research in the many areas of inquiry that made up plant pathology.

Plant pathologists also sought to organize in the arena of education. Inspired by the German model, the university system was finally gaining widespread acceptance. A university education, which among the general population had long been considered as a trophy for dilettantes, was becoming a necessary ingredient for social and career advancement. The tenets of progressivism demanded proof of competence and a strict definition of qualification that a university degree was designed to confer. Whereas the "self-made man" remained an American icon, the college-educated professional of the growing middle class quickly filled the ranks of "experts" in both government and private industry.

The growth of the university system was critically important to the developing field of plant pathology. The creation of specialized departments of plant pathology freed the discipline from the possession of dedicated, but scattered, professors of botany or mycology who had established and dominated the teaching of the science at the college level over the previous quarter-century. By establishing discrete university departments, plant pathologists defined the criteria by which a professional in the discipline could be recognized and created the means to generate needed numbers of additional plant pathologists to fill the ranks of the growing science.

Plant pathology was developing into a complex and multifaceted science. With every year of study, the intricacies of pathogens and their interactions with their host plants were being revealed more clearly. In many cases, the new discoveries raised more questions than they answered. For example, the discovery of races in some fungi was a startling result of one line of research, but it served mainly to create more questions and open new avenues of research for the future. Meanwhile, several devastating plant epidemics created additional complications for the science, emphasizing the need to study new disease-causing organisms.

In the late nineteenth century, plant pathologists had achieved some highly visible practical successes against some diseases by spraying fungicides and using simple cultural control practices. Unfortunately, such practices proved insufficient to handle the staggering disease problems that were beginning to confront American agriculture at the beginning of the twentieth century. Although practical research on spraying and basic control continued, the leaders of plant pathology agreed that their efforts must turn to more fundamental science to pave the way for improved control measures in the future. The emphasis on innovative areas of research resulted in the establishment of several new branches of plant pathology, such as breeding for disease resistance, forest pathology, and nematology.

On another front, several devastating epidemics that were caused by exotic pathogens caused plant pathologists to become more active in the political realm. Many plant pathologists joined the movement for laws controlling the importation of plants from overseas and regulating the movement of plants within the United States. Such activities placed plant pathologists squarely in the Progressive Era, expert advocates of an activist role in controlling plant disease.

Fig. 4.1.   Laboratory of Erwin F. Smith, January 1905, (L–R: Adeline Ames, Alice Haskins, J. F. Brewer, all assistants to Smith), courtesy of the National Archives.

Fig. 4.2.   William A. Orton, courtesy of
the National Archives.

Fig. 4.3.   Sea Island cotton plants resistant to wilt grown from seed
selected from resistant plants by William A. Orton, 1900, from USDA
Division of Vegetable Physiology and Pathology Bulletin No. 27.

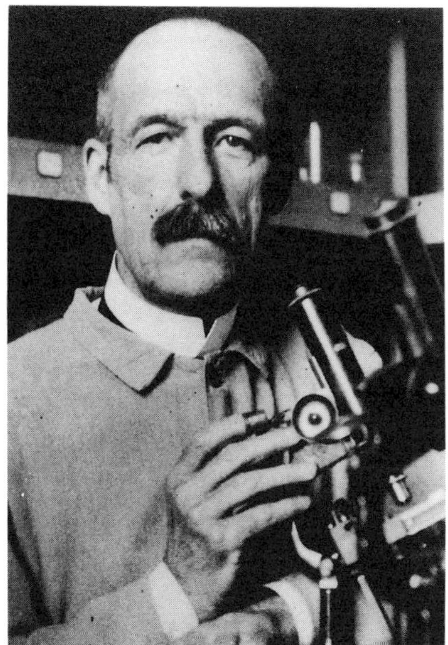

Fig. 4.4. Nathan A. Cobb, courtesy of the National Archives.

Fig. 4.5. Mark A. Carleton crossing wheat at Garrett Park, Maryland, 1894, courtesy of Special Collections, National Agricultural Library, Beltsville, Maryland.

Fig. 4.6. Haven Metcalf, courtesy of the
National Archives.

Fig. 4.7. Chestnut tree suffering from chestnut blight, from the
Pennsylvania Chestnut Blight Conference, Harrisburg, February
20 – 21, 1912.

Fig. 4.8.   Herbert H. Whetzel, courtesy of the Department of Plant Pathology, Cornell University.

Fig. 4.9.   Cornell Plant Pathology staff, 1909, standing (L–R): Gertrude Whetzel, Herbert H. Whetzel, Jensen, J. J. Taubenhaus, Mortier F. Barrus, Donald Reddick; seated (L–R): Mrs. Whetzel, Agnes McAllister Vandermark, Erret Wallace), courtesy of the Department of Plant Pathology, Cornell University.

Fig. 4.10.   Plant Pathology laboratory at Cornell, 1907, courtesy of the Department of Plant Pathology, Cornell University.

Fig. 4.11.   Bailey Hall, courtesy of the Department of Plant Pathology, Cornell University.

Fig. 4.12.   Edward M. Freeman, courtesy of the
University Archives, University of Minnesota.

Fig. 4.13.   Elvin C. Stakman examining wheat plants in the greenhouse, 1915, courtesy of the University Archives, University of Minnesota.

Fig. 4.14.   Plant Pathology building, known formerly as the Old Drill Hall and known affectionately as the "Tottering Tower," courtesy of the University Archives, University of Minnesota.

Fig. 4.15. Lewis R. Jones, courtesy of the
Department of Plant Pathology, University
of Wisconsin.

Fig. 4.16.   Agricultural (Science) Hall,
site of Lewis R. Jones's office in 1910,
courtesy of the Department of Plant
Pathology, University of Wisconsin.

Fig. 4.17.   John C. Walker, courtesy of the Department of
Plant Pathology, University of Wisconsin.

Fig. 4.18.   Cabbage plants resistant to yellows disease, 1913, courtesy of the Department of Plant Pathology, University of Wisconsin.

Fig. 4.19.   Cornelius Lott Shear, author's personal collection.

PROGRAM

OF

THE FIRST ANNUAL MEETING

OF

THE AMERICAN PHYTOPATHOLOGICAL SOCIETY

———

BOSTON, MASS.

December 30th and 31st, 1909

———

L. R. JONES, *President.*

A. D. SELBY, *Vice-President.*

C. L. SHEAR, *Secretary-Treasurer.*

B. M. DUGGAR and J. B. S. NORTON, *Members of Council.*

———

The society will meet in connection with the American Association for the Advancement of Science. The sessions will be held at the Harvard Medical School, Building A, Room 201.

———

Thursday, December 30th—2 P. M.

Joint Program with Section G, A. A. A. S.

The Morphology and Life-History of *Puccinia Malvacearum* Mont. Illustrated. (10 min.)	J. Tabenhaus.
A Spinach Disease caused by *Heterosporium variabile*. Illustrated. (10 min.)	H. S. Reed.
Common Names for Plant Diseases. (8 min.)	F. L. Stevens.
A New Species of *Endomyces*. (10 min.)	C. E. Lewis.
Malnutrition Diseases of Cabbage, Spinach and other Vegetables. (10 min.)	L. L. Harter.
Three Species of the Type of *Aecidium cornutum* Pers. Illustrated. (8 min.)	F. D. Kern.
Contributions to the Life-History and Structure of Certain Smuts. (10 min.)	B. F. Lutman.
Present Status of the Cotton Anthracnose Investigation at the S. C. Experiment Station. (10 min.)	H. W. Barre.
Life-History of *Melanops Quercuum* (Schw.) Rehm from the Grape. (10 min.)	C. L. Shear.

Fig. 4.20. Program of the First Annual Meeting of the American Phytopathological Society, 1909.

PHYTOPATHOLOGY

OFFICIAL ORGAN OF THE

AMERICAN PHYTOPATHOLOGICAL SOCIETY

EDITORS

L. R. JONES

C. L. SHEAR   H. H. WHETZEL

ASSOCIATE EDITORS

G. P. CLINTON	HAVEN METCALF	ERWIN F. SMITH
E. M. FREEMAN	W. A. ORTON	RALPH E. SMITH
H. T. GÜSSOW	W. M. SCOTT	F. L. STEVENS
F. D. HEALD	A. D. SELBY	ROLAND THAXTER

BUSINESS MANAGER

DONALD REDDICK

VOLUME I

JANUARY-DECEMBER, 1911

WITH 30 PLATES, AND 14 TEXT FIGURES

PUBLISHED FOR THE SOCIETY
AT ITHACA, NEW YORK
PRESS OF ANDRUS & CHURCH

Fig. 4.21.   Title page of the first issue of *Phytopathology*.

CHAPTER 10

Growth and Development at the Federal Level

The field of plant pathology developed even more rapidly in the early years of the twentieth century than it had during the 1890s. There was continued growth in the areas of research that had sustained the science for two decades, most notably the study of fungal and bacterial pathogens and the search for chemical and cultural methods of disease control. In addition, the science broadened in ways that had barely been imagined a decade earlier. New phytopathological subdisciplines included breeding and selecting plants for disease resistance, studying nematodes as plant pathogens, and applying existing knowledge to forest pathology. As the science of plant pathology expanded, so did its influence on public policy. The early twentieth century saw several devastating plant epidemics that would alter the natural landscape of America, and plant pathologists came to wield increasing political influence. Meanwhile, the institutions where plant pathology was done, notably the USDA and the state agricultural experiment stations, experienced continuing centralization and growth.

A NEW BUREAU AND CONTINUING EXCELLENCE

In the early 1890s, Beverly Galloway had begun to lobby for a centralized and coordinated system of USDA plant research similar and equal in status to the Bureau of Animal Industry.[1] In the late 1890s, several factors contributed to a new climate that was receptive to such a change in the USDA. First, the science of plant pathology was able to point to a growing record of successes. The USDA Division of Vegetable Physiology and Pathology, though barely fifteen years old, had made notable strides in fundamental research. This research had led to the control of a number of America's most destructive crop diseases through the use of chemical sprays and improved cultural practices. Second, James "Tama Jim" Wilson was installed in 1897 as secretary of agriculture. Unlike his predecessor, J. Sterling Morton, Wilson favored the expansion of scientific activity in the department.[2]

An opportunity to expand plant pathology's influence in the USDA came on September 11, 1900, with the death of William Saunders, chief of the USDA Division of Gardens and Grounds. For nearly forty years, Saunders had jealously guarded his division's facilities, only reluctantly granting needed greenhouse and ground space to USDA plant pathologists. With his death, Beverly Galloway saw a way not only to gain experimental facilities for plant pathology and physiology but also to expand his own sphere of influence in the USDA. He proposed to Secretary Wilson that he take control of the Division of Gardens and Grounds and that his deputy, A. F. Woods, become head of the Division of Vegetable Physiology and Pathology. Thus, two men with strong interests in plant disease came to control the operations of two major divisions of the USDA.[3] Galloway recognized that the existing, decentralized structure among the divisions concerned with the plant sciences resulted in harsh and, at times, acrimonious competition for funds, duplication of efforts, and discord among researchers. He believed that a unified Bureau of Plant Industry, which would centralize administration and coordinate research, could solve many of these problems.

Galloway's concept of a unified bureau was not without opponents. Frederick V. Coville, USDA botanist, "objected violently to such a plan being forced on him by 'squirt-gun pathologists' and others whom he did not consider real botanists, and who clearly had no intention of asking him to head the proposed new organization."[4] Coville's animosity probably had less to do with his personal temperament than it had to do with his uneasiness about the movement toward increasing specialization in botany in the late nineteenth century. His opposition, however, could not stand up to Galloway, who successfully garnered the support of the chiefs of the other divisions that would make up the Bureau of Plant Industry.

Galloway and Erwin F. Smith helped draft the legislation that created the Bureau of Plant Industry (BPI), which commenced operations on July 1, 1901, under the direction of a new chief, Beverly Galloway. Five divisions involved with plant research were consolidated into the new bureau: Gardens and Grounds, Vegetable Physiology and Pathology, Pomology, Agrostology, and Botany. During the BPI's first year of operation its scope was enlarged to add seed and plant introduction, congressional seed distribution, experimental work with tea, and the administration of the Arlington Farm.

Galloway, as the bureau's chief, had political acumen and an ability to lead others to share his enthusiasm for excellence in federal plant science research. These qualities allowed him to succeed in securing increased salaries for his scientists and greater overall appropriations for research. He also strengthened scientific federalism by encouraging even closer cooperation between USDA plant scientists and their counterparts at the state experiment stations and agricultural colleges. In 1913 he would be named assistant secretary of agriculture.

Plant pathology was "one of the fundamental lines of work" of the new BPI.[5] The BPI's Division of Vegetable Physiology and Pathology, under the direction of A. F. Woods, vigorously maintained and expanded the research and practice in plant pathology that had earned it an exemplary record of successes. It aimed primarily at understanding and ameliorating diseases caused either by fungi or by bacteria. In general, the division kept to Frank Lamson-Scribner's model: Study the pathogen's life cycle and then create or fine-tune control measures based on that information.

Plant pathology work within the division fell to several laboratories headed by senior scientists: E. F. Smith was in charge of the primary pathology laboratory in Washington; M. B. Waite headed up work on orchard fruit diseases; N. B. Pierce continued at the Pacific Coast Laboratory in California; and Peter H. Rolfs reestablished the Florida-based subtropical research begun by H. J. Webber and W. T. Swingle in the 1890s. In addition, over the next ten to fifteen years several new laboratories and offices were established in specific regions or to handle specific groups of commodities or pathogens. Research on plant disease control, highlighted by pathogen life-history studies and chemical testing, remained a priority. Examples of this work included breakthrough life-history studies of the fungus that caused apple bitter rot and W. M. Scott's rediscovery of self-boiled lime sulfur for use on some apple and peach diseases. Scott's work helped create a resurgence in the use of sulfur fungicides.[6]

But activities like discovering new fungicides were only part of federal plant pathology. As a service-based institution, the BPI also had an obligation to disseminate information by publication and, increasingly, through demonstration. Because some farmers resisted the work of mixing and applying compounds, scientific advances in controlling plant diseases with chemical treatments did not always help agriculture as much as it might have. Holdouts remained even when the chemical industry made progress in providing chemicals that were increasingly ready to use. As Galloway lamented, "Although the treatment for apple scab was worked out by this Department some fifteen years ago, in many sections of the country orchardists are still permitting this serious fungous disease to destroy their crops of apples."[7]

Demonstration work very quickly became a routine aspect of the bureau's duties, centered particularly on the control of fruit disease, for which fungicidal spraying had proven particularly useful.[8] To overcome the growers' hesitation to try new procedures, in 1906 W. M. Scott was dispatched to Nebraska, Missouri, and Arkansas to demonstrate the value of fungicides for apple scab and other diseases. Galloway noted that "further work along these lines is looked upon as an important part of our services."[9]

Demonstration and extension work became a necessary part of plant pathology in part because the complexity and scope of the science continued to increase as the twentieth century proceeded, becoming even more of a challenge for nonspecialists. Those who were shaping plant pathology were increasingly more likely to spend a greater portion of their time with a microscope and laboratory apparatus than with a sprayer. As E. F. Smith wrote in 1902, "The so-called 'practical man' has gone about as far as he can go and must have help from the technical and laboratory man."[10] Although this was a natural progression for a developing science, it was a difficult path for a profession dedicated to its service-based tradition. In 1913, L. R. Jones, of the University of Wisconsin, said, "I recognize clearly that the highest duty in plant pathology is service, and that the chief aim in that service is to lessen the loss from plant diseases." But to accomplish this, he thought that "as fast as conditions permit, we must be moving on to the attack on the more fundamental problems, the performance of the more enduring service."[11] By 1912, a bureau report confidently asserted that "our leading pathologists have developed lines of work which have been epoch making in their nature."[12]

Federal plant pathology's finest example of the pure scientist was probably E. F. Smith. Best known for his continuing research in plant bacteriology, Smith broke additional new ground with his studies of crown gall in 1904. When he had first looked at crown gall in 1892, his investigation was incidental to his other research.[13] During the 1890s James W. Toumey of the University of Arizona Agricultural Experiment Station had studied crown galls of almond, peach, and cherry trees, among other plants, indicating in 1900 that he suspected the causal agent to be a myxomycete.[14] In 1902 the USDA reported that crown gall was "one of the most serious enemies to fruit growers . . . from California to the Alleghenies."[15] Investigations by George Hedgcock at the USDA's Mississippi Valley Laboratory resulted in publications for growers that described the symptomology of crown gall and suggested wrapping grafts for protection against the disease.[16] The cause of the disease, however, remained a mystery.

It was Smith and his assistants, C. O. Townsend and Nellie A. Brown, who finally discovered the cause of crown gall. They began their work in 1904 while studying gall on Paris daisy. As usual, Smith first familiarized himself thoroughly with the literature available on the subject, which in this case meant the work in Italy and France on various gall diseases, including the symptomatically similar olive knot caused by *Bacillus oleae-tuberculosis*. Smith credited Fridiano Cavara of Italy with the first firm proof of the bacterial origin of crown gall based on pure culture and reinoculation studies in 1897. Smith's work, however, was the first study to be complete in its methodology and description of the organism. In 1907 he named and described the causal agent, *Bacterium tumefaciens*.[17]

From this start, Smith developed an intense interest in plant galls and the histology of tumors. He began making comparisons between the growth and structure of plant tumors and human cancer tumors, and he advanced the concept that human cancer might be parasitic in nature. At first Smith's ideas were greeted coolly by cancer researchers. But as he developed his work on crown gall, he discovered "a tumor strand and . . . a stem structure in secondary tumors." Meanwhile, some research in animal tumors seemed to demonstrate the inoculability of certain sarcomas. Smith concluded that everything "tended strongly to unsettle the crystallizing belief in the nonparasitic origin of cancer."[18]

Plant bacteriologists praised Smith's work, and some members of the medical profession began to take an interest. Smith was invited to attend conferences and lecture on the relationship between crown gall and human cancer, and he received several awards from medical societies in recognition of his research.[19] In the end, however, Smith's cancer research was inconclusive. He eventually found that he could induce tumors in plants in the absence of the bacterium through purely chemical means or even by oxygen deprivation. He continued his intense study of crown gall and *Bacterium tumefaciens*, but tended to deemphasize the connection with animal cancers, a theory that found fewer supporters over the next decade and lost many of those it once had.[20]

Plant breeding for resistance to disease

By the beginning years of the twentieth century, the breeding of plants for disease resistance was beginning to take shape as a major new branch of plant pathology—a tool with exciting potential in the arena of disease control. The USDA possessed several visionary scientists who perceived the importance of this area of research and quickly became its leaders. In 1902 Erwin F. Smith wrote that "in my judgment, the treatment of diseases by spraying with copper fungicides has reached its climax and is now on the wane. We shall have to devise other methods for dealing with many plant diseases. Plant breeding is one of the most hopeful."[21]

Agriculturists and horticulturists had been selecting and breeding better plants long before the USDA began research in this area. As the nineteenth century advanced, the trend toward plant breeding in the United States was bolstered by a new, commercial movement to introduce foreign plants. This movement stressed the expansion of agriculture by the introduction of new commodities to the American market, concentrating on new crops that would thrive in unique and often hostile environments. New introductions were often crossed with native cultivars. By the turn of the century, American plant breeders had introduced many important crops, improving them in categories such as productivity, succulence, and environmental adaptation through selection and hybridization.[22]

As early as the mid-nineteenth century scientists and farmers began to observe systematically that some plants were susceptible to diseases and others were not. During the potato late blight epidemics of the 1840s, for example, discussions filled the agricultural press about which cultivars of potato, if any, remained relatively free of the rot. But this interest in the relative disease resistance of crop cultivars did not lead immediately to activity by agriculturists to select or breed new cultivars with the specific goal of disease control. Rather it signaled an increased awareness of plant diseases, resulting in a conscious selection of the best and healthiest plants or seeds from such plants to provide the next generation of crops.

Starting in the 1880s, however, plant disease workers and some agriculturists began to pursue the acquisition of disease resistance in crops as a goal. Charles Bessey, for example, was already considering the concept of growing rust-free wheat. At the USDA, efforts to use resistant stock to combat diseases began in the 1890s with H. J. Webber and W. T. Swingle's advocacy of the use of "immune or resistant stocks" to prevent citrus foot rot and their work to breed a cold-resistant orange.[23] In California, N. B. Pierce investigated the possibility of controlling the grape disease that would later bear his name by importing vine cuttings from other areas in the hope that they would exhibit resistance to the mysterious disease.[24]

At the turn of the century, the potential for actively breeding plants for disease resistance received a tremendous boost from a reexamination of the work of Gregor Mendel. In the 1860s Mendel had done experiments with the garden pea, demonstrating that inheritable traits could be charted through mathematical probabilities. Plant breeders found Mendel's methodology to be a key to efficient and predictable work.[25]

Among the earliest examples of this new direction in plant breeding came from the work of William Allen Orton, a pathologist with the BPI. Orton was born in North Fairfax, Vermont, in 1877. At age sixteen he entered the agricultural school of the University of Vermont, and, at age twenty, he received his bachelor's degree with Phi Beta Kappa honors. Orton was a man of extraordinary will and accomplishment, particularly given the fact that he was partially deaf from birth and was troubled throughout his life by ill health due to diabetes. After graduation, he stayed at the university for two more years, pursuing a master's degree and working with L. R. Jones on pathogenic fungi. When Jones visited E. F. Smith's Washington laboratory in early 1899, he recommended Orton as the most appropriate person to take up work that Smith was contemplating on cotton wilt. Orton was immediately called to Washington and then hurried off to South Carolina.[26]

Because cotton was the South's most valuable commodity, a threat to the cotton crop was a threat to the economic stability of the entire region. Cotton growers in the South and their political representatives were bemoaning the destructiveness of diseases and pests even prior to the devastation caused by the boll weevil.[27] Cotton wilt, which E. F. Smith had determined only recently to be caused by a *Fusarium* species, was one of the South's most troubling diseases.

This was because the *Fusarium* species infested the soil and persisted, thus rendering the land incapable of sustaining the production of cotton. By the waning years of the nineteenth century, the pathogen had already spread across most of the Carolinas and was making inroads into the Deep South.[28]

In 1899, when Orton arrived in eastern South Carolina, he had never seen a cotton plant. He claimed that the wilt was "not now the cause of great financial loss," but he admitted that the "possibilities of future injury are very great . . . on account of the peculiar and almost unique conditions attending the culture of the Sea Island cotton."[29] Imported originally from British plantations in the Bahamas, so-called Sea Island cotton grew only in the coastal plains. One of the most prized cottons in the United States, it was proving to be very susceptible to fusarium wilt, and its spread meant that the coastal land might be rendered useless for cotton production.[30]

Orton began work in the usual way, investigating "hygienic or preventative measures" including "preliminary experiments with chemicals."[31] However, because he suspected that these chemical experiments might be futile, he simultaneously did more intense studies of the pathogen, succeeding for the first time in inoculating cotton with the pathogen from a pure culture and reproducing the disease, a feat that had eluded both G. F. Atkinson and E. F. Smith.[32] He also preserved seeds from plants that thrived in the fields even when they were surrounded by badly wilted plants.

Planters and scientists had been working on cotton breeding for years, with some success in improving factors such as productivity, size of boll, size and strength of fiber, and even earliness of ripening. This success came through intercrossing numerous American-grown cultivars and hybridizing American cultivars with introduced Egyptian cottons. However, this work had not considered disease resistance as a desirable factor for incorporation into the new cultivars. In fact, it had not yet been demonstrated that disease resistance could be inherited.

During the winter of 1899–1900 Orton spent time in Washington, surrounded by men such as Webber, Swingle, and Carleton, who were fascinated by the use of breeding for plant improvement.[33] Even Orton's direct supervisor, E. F. Smith, immersed as he was in bacterial studies, found plant breeding to be the most interesting work in the department, "next to my own."[34] By the growing season of 1900, Orton was sure that breeding wilt-resistant cotton was the only solution worth seeking.

Returning to South Carolina, Orton observed and experimented with many cultivars of cotton. He planted the seeds from plants that had survived the previous year in fields with soil infested by the wilt pathogen, hoping that they would give him a wilt-resistant cultivar. The plants from the selected seed thrived while the other cotton plants around them wilted and died. This demonstration was truly significant: It showed that selection for wilt-resistant cotton was a practical possibility and that resistance was a heritable trait.[35] Although the quality of the cotton was not quite as good as that of the parent plant, Orton noted that "it is believed, however, that by continued cross breeding and selection in succeeding years the quality of the cotton may be improved without loss of resistance to the wilt disease. Work along this line has already been started in a small way by the Department, which it is hoped may be enlarged."[36]

After barely two years of work, Orton's results were phenomenal—if not in actual returns, certainly in their potential for the control of an economically important disease. Orton's first departmental publication, Bulletin No. 27, "The Wilt Disease of Cotton and Its Control," was referred to by his mentor, L. R. Jones, as a "historic classic."[37]

Orton continued for the next several years to work with growers, developing and stabilizing the resistant cultivars and increasing the stock of seeds for continued research by himself and others.[38] The new Sea Island cotton even returned to its tropical roots, when seed supplied from Orton's South Carolina experimental grounds was sent to Barbados for experimentation. The commissioner of the Imperial Department of Agriculture at first viewed this seed with suspicion, but he was soon won over by Orton's expertise.[39]

Orton next extended the scientific and economic value of his cotton-breeding work with an investigation into the fusarium wilt of cowpea. He spent two years testing different cultivars of cowpea and other plants such as soybean as possible options for rotation crops. By 1902 he found a cowpea cultivar called Iron that "proved to be the solution to the wilt problem wherever it has been cultivated on 'pea sick' lands in South Carolina, though more extended trials, now in progress by the Department, will be necessary to determine its value for other localities."[40]

H. J. Webber, now the USDA's chief plant breeder, was quick to exploit these breeding successes for their potential political gain for the department. He traveled to the South Carolina home of the powerful Senator Tillman, who "seemed to warm up and really became rather enthusiastic over some of the new cottons and this resistant cowpea." Webber then requested an increased budget for breeding work, writing to A. F. Woods that "the results obtained this year already justify the request" and suggesting that Woods should turn to Senator Tillman to support the "many important breeding problems that must be pushed."[41]

Much as the use of copper fungicides had in the previous century, the practice of breeding for disease resistance settled into a period of hard work and debate. Orton's next project, fusarium wilt of watermelon, was not an immediate success. His problems with breeding a successful, wilt-resistant watermelon demonstrated one of the great difficulties that scientists would encounter in breeding for disease resistance—time contraints. Plant breeding in general took time, and breeding for resistance to a specific pathogen could take even more time, particularly when the scientists were also trying to include other essential traits such as taste or productivity. E. F. Smith warned that breeding "is a slow process, and the man in the field will sometimes become impatient unless he is a philosopher as well as a farmer."[42]

Meanwhile, it became clear that disseminating and maintaining resistant seed would be another challenge to the success of breeding programs. Without a clear answer to this challenge, the benefits derived from the breakthroughs in breeding disease-resistant cultivars could not be translated into long-lasting, crop-saving techniques. Much of the new disease-resistant seed was disseminated through the congressional seed-distribution system, through which seed was sent to growers who were willing to cooperate with the USDA in experimentation and demonstration.[43] In the first decade of its existence, the congressional seed-distribution system had disseminated more than 7,000 tons of seed. In 1912 the federal government, with regional cooperation from colleges and state agricultural experiment stations, as well as some private firms, began to increase its capacity for creating and certifying improved seed.[44] This system, however, was not sufficient to overcome widespread, traditional adherence to old cultivars or ensure a commercially viable distribution of resistant seed. As W. A. Orton noted, "The Government cannot put into the hands of every farmer a supply of resistant seed, and, even if it could, experience has shown that this would not permanently relieve the situation, for special care is needed to prevent the resistant cultivars from deteriorating through crossing with nonresistant cotton in neighboring fields, through lack of careful selection, and through the mixing of seed at the gin."[45]

Turmoil in the science of breeding for disease resistance extended beyond practical logistics; the theoretical foundations of the science was undergoing great flux. New concepts were announced frequently, some of which threatened to topple the discipline's fragile theoretical foundations. One of the most influential, but erroneous, was the so-called bridging-host theory offered in 1902 by Harry Marshall Ward of Cambridge University. In a study of rust on bromegrass, Ward theorized that a type of rust that was pathogenic to one cultivar of grass, but not a second, could adapt itself to the second by moving first onto a third, intermediate cultivar.[46] Other significant work that raised questions about breeding for disease resistance was that of Ernest Salmon of the Southeastern Agricultural College at Wye, Kent. Salmon announced that although barley mildew could not usually parasitize wheat, it could do so if the wheat was injured. This led some noted plant pathologists, including the prominent J. C. Arthur, to proclaim that the imperfect knowledge base about the "whole meaning of parasitism" called into question the value of breeding for disease resistance. As Arthur asked, "What assurance have we that after years of work in establishing an immune variety of grain or vegetable some mishap . . . will give just the right opportunity for the fungus, which we have taken so much pains to circumvent, to gain a foothold in our supposed immune crop, and . . . become after a short period of adaptation a greater pest than it was to old varieties?"[47] The concept, if not the specifics, of Arthur's caveat was well founded. The difficulties inherent in developing disease resistance were rapidly becoming evident.

Despite concerns about both the economic value and the scientific bases of breeding for resistance, the field continued to grow. The BPI took the lead with continued breeding work on cotton, tobacco, and many other crops. J. B. Norton's research on breeding asparagus resistant to rust succeeded not only in conquering asparagus rust, but in totally revising the commercial cultivars of asparagus grown in the United States. At the Tennessee Agricultural Experiment Station successful breeding was also done by S. M. Bain and S. H. Essary to provide anthracnose-resistant clover and by Essary independently on tomato resistant to fusarium wilt. At the Colorado Agricultural Experiment Station, P. K. Blinn worked on a leaf blight–resistant cantaloupe, and in North Dakota H. L.

Bolley bred wilt-resistant flax. In addition, L. R. Jones and William Stuart of Vermont demonstrated a level of resistance to late blight in certain cultivars of potato.[48]

At the USDA, the work on breeding for disease resistance associated with most important garden and truck crops fell under the Office of Cotton and Truck Diseases and Sugar Plant Investigations. Orton went on to direct and staff this office with a remarkable selection of scientists. He himself moved on to focus on the diseases of tropical plants and on the organization of plant pathology on a national and international level.

THE CREATION OF NEMATOLOGY

Another new subdiscipline in plant pathology in the early twentieth century was the systematic study of nematodes. As with so many other areas of plant disease research, the study of nematodes in the United States was advocated most passionately by scientists associated with the USDA. Thus, although USDA scientists did not create the science of nematology or the need for it, they did cultivate and encourage this nascent science.

Nematodes certainly were not unknown organisms at the turn of the twentieth century. As early as the seventeenth century, European microscopists had pioneered the study of free-living nematodes, particularly those living in vinegar, earning them the common name "vinegar eels." Then in the next century, the role of plant-parasitic nematodes as the causal agent of some diseases of cereals including "cockles" of wheat was noted by John Needham (1743), Carolus Linnaeus (1767), and Giovanni Scopoli (1777), among others.[49] In the nineteenth century significant work on plant-parasitic nematodes occurred in Germany, where a nematode-caused disease threatened the important sugar beet industry. H. Schacht first wrote about the disease in 1859, and in 1871 A. Schmidt named the causal agent *Heterodera schachtii*. Also in the 1870s, the famous German agriculturist Julius Kühn had some success controlling the sugar beet nematode through crop rotation. Although he found that soil fumigation using carbon disulfide and other chemicals was also effective, he determined that these chemicals were too difficult and expensive to apply to be useful as a control method.[50]

Another control strategy proposed by Kühn and others was the use of *Fangpflanzen*, or "trap crops." The idea was to plant a susceptible, rapidly growing crop such as rape (canola) in a field infested with nematodes. The theory was that the nematodes would be attracted to the trap crop instead of the base crop. At the right time, this trap crop would be removed carefully and destroyed with the nematodes clinging to the roots. If this was repeated several times over a growing season, theoretically, most of the nematodes could be eliminated from the field. Although the trap-crop control method was still being discussed in the early twentieth century, it never proved to be effective on a practical basis.[51]

The study of nematodes inhabited a gray area between the disciplines of plant pathology and entomology. Scientists were not sure which discipline had "responsibility" for them. Plant disease studies were dominated by mycologists, who tended to ignore nematodes. Entomologists, on the other hand, did not count nematodes as insects. This confusion as to who should study nematodes did not, however, lessen the damage done to crops by these organisms. In the United States, diseases caused by nematodes affected a wide variety of crops and were disruptive particularly in the Southern states and in greenhouse agriculture.[52] Attempts to explore the nature of nematodes as plant pathogens were initiated by entomologist J. C. Neal of Florida, in 1888, and by George Atkinson of Alabama, a botanist with zoological training, in 1889.

Neal first confronted the problem of nematodes in the course of a survey of root-knot diseases of various Florida crops under the auspices of the USDA Division of Entomology. His unfamiliarity with the subject of nematodes was obvious. As USDA entomologist C. V. Riley later noted, Neal had not had access to the current literature on the subject, and although his "work is not without scientific interest . . . the bulletin makes no pretense to be a scientific treatise on the life history of these worms." Riley suggested that "the technical and classificatory treatment of the subject should be undertaken by some competent helminthologist."[53]

Meanwhile, at the Alabama Agricultural Experiment Station, George Atkinson was undertaking a study of "nematode root-galls."[54] Atkinson was an excellent descriptive physiologist and morphologist, and perhaps because of his earlier training in zoology, he was also more disposed to study nematodes than many of his plant pathology colleagues were. His work revealed a familiarity with the European literature.

The question of who should study nematodes was still unresolved a decade later. In an 1898 bulletin on nematodes, George E. Stone and Ralph E. Smith, botanists for the Hatch Experiment Station of the Massachusetts Agricultural

College, felt compelled to "explain why we have undertaken a work which is zoölogical rather than botanical in nature." The Massachusetts station and other stations around the country had been receiving complaints about crop damage due to nematodes for years, but the work continued to fall between the cracks between plant pathology and entomology. "As a consequence," wrote Stone and Smith, "very little has been done in investigating the pest in this country and nothing at all in this section, though the necessity for such investigation has been continually increasing. . . . Realizing the impossibility of making definite recommendations to those seeking advice in the matter and feeling that the subject was one of great importance to the gardeners of Massachusetts, we finally undertook investigations."

Because their concern was primarily economic, however, Stone and Smith ignored much of the life-history work, leaving it for others who were "practiced along the lines of modern zoölogical technique."[55] They preferred to discuss the symptomology of root-knot, and they concentrated on controlling root-knot on plants grown in greenhouses rather than in the orchard or field. This concentration meant that treating the soil with chemicals or sterilizing it with steam might be a practical control strategy. Such strategies were largely irrelevant to Atkinson's work in fields, where such measures were prohibitively expensive.

The breakthrough in the control of nematodes came when scientists began selecting and breeding cultivars for disease resistance. A very early suggestion of possible resistance to root-knot came in Neal's 1889 entomology bulletin, when he recommended the use of certain cultivars of orange trees to combat nematode infestation. In 1897 a clearer observation on nematode resistance came from outside the United States, when Albrecht Zimmerman determined that *Coffea arabica* was more susceptible to root-knot than *C. liberica* and he recommended grafting the more valuable *C. arabica* onto resistant root stock of *C. liberica*.[56]

But the earliest practical work in the United States occurred in 1901, when USDA plant pathologists W. A. Orton and H. J. Webber, in the course of their experiments with wilt-resistant cowpea, discovered that Iron cowpea resisted not only wilt, but also root-knot nematodes.[57] Webber called this "a genuine out and out discovery . . . which . . . opens up important new fields" since "so far as my knowledge of this class of diseases goes no resistant strain of any plants effected [*sic*] have ever been known."[58]

Several years later, A. D. Shamel and W. W. Cobey investigated strains of to-bacco that seemed to show resistance to root-knot nematodes.[59] Ernst Bessey, son of the famous botanist Charles Bessey and a pathologist with the BPI, also observed that certain grape cultivars were resistant to nematodes, particularly those American cultivars that were also resistant to phylloxera, and that Orton's wilt-resistant watermelon cultivars seemed remarkably free from root-knot. He recommended that "simple selection can be and ought to be practiced by every-one who grows his own seed; more complicated breeding work, unless per-formed by men who can devote considerable time to it, hardly pays for the time and expense required."[60]

Despite these moderate advances in nematode control, fundamental research on nematode biology continued to stall. The USDA made a halting attempt to address plant pathogenic nematodes in a systematic way when Ernst Bessey was assigned to the work in 1900. But this work was "interrupted for a period of years," resumed in 1905, and "pursued with various interruptions until its com-pletion," marked by the publication of Bessey's bulletin on root-knot in 1911.

Bessey was not a nematode expert; he divided his time between nematodes and many other projects. His bulletin was largely an excellent summary of the work from around the world, supplemented with some original research. He at-tempted to simplify the morass of root-knot nematode taxonomy, claiming that the variety of names for the organism was due to lack of coordination within the field. He lumped all of the previously named species together into one species—*Heterodera radicicola*. His taxonomic simplification would eventually be proved incorrect, but was justifiable at the time. Due to the continuing lack of detailed physiological and morphological studies, the vast diversity of nema-todes simply was not well understood. Furthermore, there is no denying the truth of Bessey's underlying point—the need for more coordination among the scientists who were studying nematodes.[61]

The character of the study of nematodes began to change in a fundamental way with the studies of Nathan Augustus Cobb. Cobb was one of the first and most vocal advocates for a specialized science called nematology. His contribu-tions to the study of nematodes, as well as his efforts to popularize such studies and to train the first generation of true nematologists, established his reputation as one of the most important figures in plant pathology of the early twentieth century.[62]

Nathan Cobb was born on June 30, 1859, in Spencer, Massachusetts. He had a broad, practical knowledge of handicraft and manufacturing plus a fascination with tinkering that would serve him well throughout his career. His acquaintance with the noted landscape painter Joseph Greenwood helped develop his excellent artistic skills.[63] He attended college at Worcester Polytechnique Institute and taught for six years at the Williston Seminary in Easthampton, Massachusetts. In 1887, after losing a scholarship to Johns Hopkins University, he uprooted his wife and three children and went to the University of Jena in Germany, where it took him only ten months to earn a doctorate. At the University of Jena he was initiated into the world of nematodes, specifically marine nematodes, and he wrote a thesis titled "Beiträge zur Anatomie und Ontogenie der Nematoden."[64] Nematodes clearly fascinated Cobb, and their study eventually came to dominate his career.

After a brief stint at the prestigious Zoological Research Station at Naples, Italy, Cobb took his family to Australia, where in 1890 the Department of Agriculture of New South Wales made him Australia's first full-time government-funded "vegetable pathologist." At the time of his appointment, however, Cobb was not a plant pathologist, even by the standards of the day, and he scrambled to retrain himself. He spent the first several years in New South Wales acquainting himself with basic agriculture and economic botany. Soon he was writing informed bulletins on fungal diseases of plants, with a focus on wheat rust and fruit diseases. He also did important early studies on a bacterial gumming disease of sugarcane, which subsequently was named Cobb's disease.

Cobb's work in Australia attracted the attention of E. F. Smith, who recommended Cobb for a USDA position in 1905. Cobb spent two years in Hawaii as the first chief of the Hawaiian Sugar Planters' Association Experiment Station, where he studied diseases of sugarcane. He finally arrived in Washington as an agricultural technologist in 1907. By 1915, his work was exclusively dedicated to research on nematodes.[65]

Since leaving the University of Jena, Cobb had maintained his interest in nematodes, particularly the search for new species. He named his first genus, the marine nematode *Tricoma*, in Naples.[66] His desire to add to the list of known species placed him firmly in the middle of the confusing world of nematode taxonomy. For instance, in Cobb's 1890 paper on root-knot nematodes for the *Agricultural Gazette of New South Wales*, his first on a plant-parasitic species, he placed the species he observed under *Tylenchus*, although by that time the names *Heterodera* or *Meloidogyne* had already been proposed and were generally accepted.[67] Taxonomy continued to play a prominent role in the new discipline. Over the course of his career, Cobb named and described well over 1,000 species of nematodes.[68]

Cobb was not, however, just concerned with describing and classifying nematodes; he also did research on nematode morphology and matters of economic importance. In Australia, he wrote numerous economic papers on nematodes as well as other parasites in farm animals.[69] He even delved briefly into human-parasitic nematodes, developing a special capsule containing eggs of *Oxyuris* spp. and going so far as to swallow it. The capsule was permeable to liquids, exposing the eggs to stomach acids as it passed through his digestive tract, but preventing the release of the eggs into the stomach or intestine. When examined, the capsule held thriving larvae, demonstrating that these nematodes could live in humans.[70]

Another of Cobb's early breakthroughs occurred in Hawaii in 1906, when he observed *Dorylaimus bulbiferus* and made one of the earliest, if not the first, mentions of parasites on nematodes.[71] He also made numerous suggestions for new morphological terms such as *amphid*, *phasmid*, and *deirid*, which found widespread acceptance. Other terms that he tried to inject into the science, for instance, shortening *nematode* to *nema*, were not as popular.[72]

Cobb was aided throughout his research career by his knack for tinkering. "He loved equipment; loved to design and make it, to remake and improve it . . . and to have it everywhere in abundance. He was an ingenious inventor, and held a number of patents for photographic devices and microscope accessories."[73] After he joined the USDA, he made or enhanced several devices, including a soil-sampling tube and microscope slide, often referred to as a Cobb slide, with an aluminum frame and dual cover slips that allowed the subject to be viewed from both sides. For counting nematodes in soil samples he recommended using a modified Syracuse, or "Beltsville," dish, and he also created special microscope eyepieces complete with crosshairs made from spider webs.[74] Many of Cobb's creations became standards in nematode research.

Although the research, the procedures, and the instruments that originated with Cobb were great contributions, it was his insistence on the necessity of establishing a separate branch of study for nematodes that marked his true value to the developing science of phytopathology. In 1914, he started *Contributions to a Science of Nematology*, a periodical that he published at his own expense. In the first issue, Cobb called for the removal of nematode research from helminthology and for the creation of a new subdiscipline of science called nematology.[75] The next year Cobb published "Nematodes and Their Relationships," a feature article in the *Yearbook of the Department of Agriculture*, the seminal piece in his campaign for a science of nematology. In it, he discussed the understanding of nematodes as it then stood and as it should be, "the last great organic group worthy of a separate branch of biological science comparable with entomology."[76]

Cobb's goal in his yearbook piece was nothing less than unveiling a new field of science and convincing the reader of its value. He began by admitting that there was little obvious reason to study nematodes since they "do not furnish hides, horns, tallow, or wool. They are not fit for food, nor do they produce anything fit to eat; neither do they sing or amuse us in any way; nor are they ornamental—in fact, when they are displayed in museums the public votes them hideous . . . they fail even in the furnishing of any moral or praiseworthy example; they are not known to be industrious like the ant, or provident like the bee . . . or frugal, or honest, or anything else that is admirable." However, Cobb informed his readers, because nematodes infest humans, animals, and plants, "beyond doubt a complete knowledge of nematodes, if properly applied, would enable us to save a vast amount of life and treasure and prevent a vast amount of suffering."[77]

This appeal to the practical worth of scientific research on nematodes was a potent weapon, but Cobb went beyond it to argue that nematodes needed to be studied for their own sake. He portrayed these organisms as omnipresent in nature, yet mysterious and unknown. He pointed to the value of the nematode as a vehicle for scientific discovery because of the "unusually clear view they give us of the various processes relating to life."[78] He also associated them with the most progressive biological science of his day, genetics and the study of heredity, by noting that "it was in a nematode egg that it was first observed that the animal male and female nuclei actually coalesce to produce the new compound nucleus, or pronucleus," which divides and "sends into each of the two new cells a definite portion of the matter it derived from each parent."[79]

Cobb's article was full of figures illustrating the diversity of nematodes, most of them drawn by Cobb's longtime partner, W. E. Chambers, who was probably the finest illustrator of nematodes in the history of the science. Many of these drawings showed basic examples; others rendered the nematodes dynamically, complementing some of Cobb's more lurid descriptions, including "savage," "rapacious," and a "monster which feeds upon others of its kind."[80] Cobb's article, on the whole, was masterful, and it was the first comprehensive statement in the United States of the value and need for studies in both fundamental and applied nematology.

In 1918, when the USDA created the Division of Nematology in the BPI, it made Cobb its head. During his tenure as head of the division, Cobb continued to push the young science from its new institutional base. He built a staff of eager and talented scientists. The best of the next generation of nematologists in the United States came largely from Cobb's laboratory and were trained and inspired under his tutelage. Cobb's imprint was evident on all aspects of the science of nematology in the United States well into the twentieth century.

CHESTNUT BLIGHT AND FOREST PATHOLOGY

In 1904 forester Herman W. Merkel noticed that chestnut trees in the Bronx Zoological Park in New York City were dying. The foliage of the affected trees yellowed and wilted, and eventually became brown and shriveled. On the bark, rusty brown, sunken cankers circled the swollen stem. Merkel sent specimens to mycologist W. A. Murrill of the New York Botanical Garden, who studied the problem, publishing the results of his studies in 1906. Murrill identified the cause of the disease as the fungus *Diaporthe parasitica*.[81]

In the meantime, other specimens from the vicinity were sent to Flora W. Patterson, mycologist of the USDA Bureau of Plant Industry. These samples indicated that by 1907 the disease had reached as far north as Poughkeepsie, New York, and as far south as Trenton, New Jersey.[82] No one realized at the time that one of the most devastating of all plant disease epidemics, one that would virtually eliminate the American chestnut as a forest tree throughout its natural range in the eastern United States, had just begun.

The chestnut held important economic and aesthetic value for Americans. In the central Appalachian forests, one out of four trees was of this species. In terms of forest products, chestnut lumber was second only to oak in usefulness and durability. Its wood was beautifully grained and excellent for house trim, furniture, and flooring. It was extremely resistant to weathering and dry rot and was used for telephone poles, railroad ties, and fence posts and rails. In the Appalachian Mountains, the economic well-being of whole communities depended on the chestnut as a source of tannin for making leather. Americans had become enamored of the chestnut's plentiful nuts, which wildlife also depended on for food. The chestnut had also become a popular shade tree in parks and yards.[83]

In 1907 investigation of chestnut blight headed the list of tree-disease research at the USDA Laboratory of Forest Pathology. The chief investigator was Haven Metcalf, who brought in forestry experts J. Franklin Collins, professor of botany at Brown University, and W. Howard Rankin of Cornell University.[84] Metcalf was among the few who knew that a potential disaster was lurking. "It is no exaggeration to say," he wrote, "that it is at present the most threatening forest-tree disease in America. Unless something now unforeseen occurs to check its spread," Metcalf warned, "the complete destruction of the chestnut orchards and forests of the country, or at least of the Atlantic States, is only a question of a few years' time."[85]

Haven Metcalf was born at Winthrop, Maine, on August 6, 1875. He attended Colby College and Brown University, where he received the B.S. in 1896 and A.M. in 1897, respectively. After further graduate study in botany, bacteriology, and mycology at Brown in 1897–98, Harvard in 1899, and Indiana in 1900, he earned his Ph.D degree in 1903 with Charles Bessey at the University of Nebraska. Metcalf held teaching positions at Brown, Nebraska, Tabor College, and Clemson College, before becoming a special agent with the BPI in 1904.[86]

Prior to joining the BPI, Metcalf had made important contributions to the understanding of plant diseases of economic significance. He had worked with G. G. Hedgcock in Nebraska on soft rot of the sugar beet caused by *Bacterium teutlium*.[87] In a cooperative venture between the South Carolina Experiment Station and the BPI in 1904 and 1905, Metcalf did important studies on rice blast. Concluding that rice blast was caused by a fungus, he suggested an association between the disease and the use of nitrogenous fertilizers.[88] Later, as a USDA pathologist, he was involved with introducing resistant rice cultivars to control the disease.

Although forest pathology was not a new undertaking for the USDA, Metcalf, as pathologist-in-charge, faced some real challenges. Forest pathology suffered from an identity problem. Although contributions had been made over the last decade, forest pathology had yet to prove itself a viable, practical subdiscipline. Most often it was viewed as an interesting branch of mycology.

Forest pathology had begun in the USDA with tentative first steps in the 1890s, primarily on diseases of shade and ornamental trees.[89] With the dual, and sometimes conflicting, influences of the booming timber industry and the growing conservation movement that valued woodlands for their own sake, work in forest pathology expanded late in the 1890s. In 1899 Hermann Von Schrenk, a botanist at the Shaw School of Botany in St. Louis, Missouri, was appointed as a special agent to investigate diseases of forest and shade trees. A few years later, the new Mississippi Valley Laboratory was formed around him. With a staff of two pathologists, he began to widen his investigations on the diseases of trees. Given the economic nature of plant pathology, it was not unusual that Von Schrenk's research concentrated on the diseases that affected timber stands and also on postharvest problems such as methods of preserving lumber. It was Von Schrenk who introduced the use of creosote for the preservation of railroad ties.

But this was a time when plant pathology was moving in the direction of fundamental research, and Beverly Galloway was dissatisfied with Von Schrenk's focus on engineering and business and with his close relationship with the timber industry. Galloway also wanted to keep forest pathology out of the hands of the Division of Forestry, which administered the Mississippi Valley Laboratory jointly with the BPI. To accomplish several of his goals, in 1907 Galloway dissolved the Mississippi Valley Laboratory and set up a new forest pathology laboratory in Washington, easing out Von Schrenk at the same time.

Under the leadership of Haven Metcalf, and with the skills of the pathologists who staffed it, the new laboratory and, later, the Office of Forest Pathology (1909), gained considerable exposure and influence. Service to state and federal nurseries as well as studies in the national forests were high priorities. The staff of the fledgling Office of Forest Pathology had prestigious credentials. Perley Spaulding, who earned an undergraduate degree under L. R. Jones at Vermont before earning his Ph.D. degree at Washington University in St. Louis while working for Von Schrenk, was responsible for the eastern United States, as well as nursery and plantation diseases. George Hedgcock, a student of Bessey's at Nebraska and another Washington University Ph.D., focused on the West and was detailed to survey the diseases of the national forests. Before long, however, inadequacies of manpower became evident, and a number of new scientists were called to fill new positions. Most notably, in 1909, C. J. Humphrey came to investigate the microbiological deterioration of wood. A year later, E. P. Meinecke, with experience in Robert Hartig's laboratory in Germany, was appointed to investigate forest-management practices with the aim of decreasing the prevalence of heart rot and other diseases in the national forests. And in 1913, James R. Weir, another European-trained scientist, arrived after studying under J. C. Arthur at Purdue University and working on forest pathology at the University of Munich with Karl von Tubeuf.[90]

With greater resources, federal forest pathology began to score the practical successes that still were the essential measure of value in pathological work. During the chestnut blight epidemic, the American plant disease epidemic that attracted the most public attention in the early twentieth century, the Office of Forest Pathology played a central role in plant pathology's clearest and most lasting effect on public policy in the United States to date.

In 1909 Metcalf and J. F. Collins wrote the first general USDA bulletin on the chestnut blight, describing its history, etiology, and symptoms. They urged states to adopt control measures, including the inspection of nursery stock to detect disease, the prompt eradication of diseased trees, and the treatment of diseased trees in ornamental plantings.[91] Two years later, in the USDA's Farmers' Bulletin No. 467 Metcalf and Collins recommended more sweeping attempts at controlling the disease.[92] In addition to the strategies mentioned above, they recommended quarantines and the establishment of a blight-free zone along the border of major infected areas.

Four years passed between the publication of Metcalf's first bulletin on chestnut blight and his control recommendations in the 1911 Farmers' Bulletin. Since its discovery in 1904, the disease had spread rapidly across New Jersey and parts of New York and Connecticut before being reported in Massachusetts, Rhode Island, Delaware, Pennsylvania, Virginia, and West Virginia.[93] One possible explanation for Metcalf's seemingly belated response was a continuing debate over the origin of the disease and the practicality of quarantine and eradication-control procedures.

There were two schools of thought on the origin of the causal fungus, *Diaporthe parasitica.* One side argued that the chestnut blight fungus was native to the United States. This group maintained that because relatives of the fungus grew as saprophytes on dead and dying oak and beech roots in the south Atlantic region, it was logical to assume that one of these saprophytes had become pathogenic on chestnut. In terms of control, if the fungus was indeed a native species, quarantine efforts and attempts to eradicate it would be in vain. The other side argued that the chestnut blight fungus was an introduced pathogen. Metcalf had hinted at this possibility as early as 1908, when he reported his observations of immunity to the blight in Japanese chestnut cultivars.[94] To this second group, quarantine and eradication were logical and essential maneuvers to check the spread of the pathogen.

Although Metcalf had carried out eradication experiments in a thirty-five-mile section near Washington, D.C., in 1908, little else had been done to study the disease in the increasing number of states affected by the blight. Metcalf was well aware that quarantine and eradication could be truly effective only if a concerted effort was made among the different states. Without a federal law "to cope with the emergency . . . each State must act on its own initiative and control the disease or let it go as its officers and legislative bodies see fit."[95] Realizing the difficulties involved with state cooperation in the absence of a national policy, Metcalf continued to advocate quarantine and eradication. Meanwhile he looked for conclusive evidence that the blight organism was not a native species, but instead had been introduced into the United States. Once he accomplished this, he believed, he would be able to muster more aggressive cooperative action on the state and federal levels.

But the disease was progressing too rapidly to wait for scientific confirmation of its origin. In 1910, the citizens of Philadelphia, alarmed by the damage wrought in their city by chestnut blight, decided to take action. Seven members of the Main Line Citizen's Association helped secure passage of a 1911 Pennsylvania law that authorized a special commission to do whatever was practical to thwart the spread of the blight. The Commission for the Investigation and Control of the Chestnut Tree Blight Disease in Pennsylvania had the power to invoke strong legal measures to enforce inspection, quarantine, and eradication, if they determined it was necessary. The commission was granted an appropriation of $275,000 to control the blight, a huge amount by the standards of the day. By comparison, the federal appropriation for fighting the chestnut blight in 1911 was only $5,000 (although it did grow to $80,000 the next year).

In 1912 the Pennsylvania Chestnut Blight Commission held an unusual conference. Attended by state officials and businessmen together with scientists from many states, the federal government, and Canada, the main purpose of the meeting was to determine the most appropriate strategy for the elimination of the blight. The issue of the effectiveness of quarantine, inspection, and eradication measures quickly took center stage. A small group led by F. C. Stewart and W. A. Murrill of New York, Mel Cook of New Jersey, Westley Webb of Delaware, and G. P. Clinton of Connecticut argued that any attempt to arrest the disease would be futile. They asserted that eradication efforts, particularly against a pathogen with such a capacity to spread, were a waste of time and money. Stewart declared, "It is better to attempt nothing than to waste a large amount of public money on a method of control which there is every reason to believe cannot succeed altogether. . . .What will be the future course of the disease can only be conjectured, but it can be safely predicted that nothing that man can now do will materially affect its course."[96]

The majority of the meeting's attendees, however, solidly favored taking strong action against the blight. Some of the proponents of quarantine and eradication based their arguments on scientific evidence, suggesting that the pathogen had been introduced to the United States and was not native. Eradication, they maintained, combined with inspection and quarantine, would prevent the disease's spread into regions where the chestnut still thrived.

It was sentimentality, with little consideration for science, that got the better of other supporters. R. A. Pearson, a former commissioner of agriculture for New York and chairman of the conference, tugged at emotions and patriotism when he told the audience:

It has been suggested that we should do nothing to counteract the ravages of the chestnut tree disease, because we are not fully informed as to how to proceed. That is un-American. It is not the spirit of the Keystone State, nor the Empire State, nor the New England States, nor the many other great states that are represented here, to sit down and do nothing, when catastrophes are upon us. It has been suggested that we should wait patiently until the scientists have succeeded in working out these questions in all their minutae: that thus we may be able to accomplish our results more quickly. But that is not the way that great questions are solved. If we had waited until the application of steam should be thoroughly understood, we would be still waiting for our great trains and steamboats, which are the marvel of the Age [applause].[97]

Pennsylvania opted against waiting for scientists to provide more details on the disease. Emotions were running high, and enough information had been presented by state and federal experts, Pennsylvania officials insisted, to justify moving forward. Some federal scientists had pushed for an aggressive eradication program in the state. Thus, over the next two years, the state of Pennsylvania mounted a massive control campaign. This campaign proceeded with the cooperation of the federal government and under the direction of Mark A. Carleton, chief of the successful wheat importation work and the USDA expert on foreign-plant introduction. Carleton was optimistic about the chances of success against the spread of the disease in Pennsylvania. If the state legislature provided enough funding, he asserted, "at the end of two more years we shall have the chestnut blight disease practically under control."[98]

Over the next two years, the Pennsylvania legislature delivered on Carleton's call for financial resources, providing more than $500,000 to fight the blight. Meanwhile, workers began scouting for infections, cutting and burning trees, treating individual trees, and promoting the use of dead or diseased trees for timber. But Carleton's forecast of success in stopping the spread of the disease proved to be ill-fated. At the end of two years, the chestnut blight was still sweeping through the state rapidly, largely unimpeded by the control campaign. Not surprisingly, the optimism of earlier years was difficult to find. In his final report to the governor in 1914, the chairman of the Pennsylvania commission wrote, "In conclusion, it seems necessary to call sharp attention to the real lesson to be learned from the chestnut blight epidemic—*viz*: the necessity of more scientific research upon problems of this character; to be undertaken early enough to be of some value in comprehending, if not controlling, the situation."[99]

The Pennsylvania effort failed mostly because of inadequate knowledge about the distribution of the disease in the United States. Given the widespread range of the chestnut blight, by 1911 eradication efforts, particularly on a state level, were simply too little, too late. Scattered attempts at eradication could never hope to catch up with new infections.

The Pennsylvania campaign was not, however, a total failure. The program was somewhat successful in delaying the spread of blight and in providing growers with valuable time to harvest timber.[100] Perhaps even more beneficial was the research that came out of the cooperative venture between the Pennsylvania commission and the USDA. Paul J. Anderson, for example, did important studies of the morphology and life-history of the blight fungus that led to its placement in the genus *Endothia*.[101]

By 1911 it was known that the causal agent was disseminated by spores formed in cavities of diseased bark. Both summer and winter spores had been discovered. Nonetheless, important questions remained. In 1914, Frederick D. Heald and R. A. Studhalter showed that spores were easily disseminated by birds, insects, other animals, and rainwater.[102] A year later, Heald, M. M. Gardner, and Studhalter found that spores were propelled into the air and disseminated by wind from pustules in the cankers.[103] In other cooperative studies, N. Rex Hunt and J. Russell Smith demonstrated that the chestnut blight could not be controlled by spraying, while Roy Pierce demonstrated that tree surgery was ineffective. Later, Caroline Rumbold showed that chemical injections were worthless.[104]

PUBLIC POLICY AND QUARANTINE

The chestnut blight demonstrated to plant pathologists the alarming consequences of their lack of influence in the political sphere. In general, the early twentieth century was a period of increased application of science to public policy, evidenced particularly by the rise of federal regulatory power. Many plant pathologists were eager to use their expertise to influence laws regarding the importation and traffic in plants and plant material, which, they argued, was an important tool in their fight to control plant disease.

Chestnut blight may have been the most dramatic disease to be introduced in the early twentieth century, but it certainly was not the only one. American plant pathologists were aware of the dangers of other foreign pathogens. They realized the destructive potential of other exotic diseases knocking at the door, particularly potato wart and white pine blister rust. Both of these plant diseases were of serious economic importance in Europe and were discussed extensively in the phytopathological literature. By 1912, USDA scientists were warning American growers of the potential for major losses due to these diseases. "Within the past few years," they told farmers, "very serious European plant diseases have been brought to North America upon imported plant material. These diseases are of more than ordinary economic importance to us," they cautioned, "as one is apparently the worst enemy of the white pine in Europe, while the other is practically certain to become one of the worst enemies of the potato wherever it occurs."[105]

Although they perceived the dangers inherent with introduced diseases, plant pathologists felt they lacked a useful means to prevent epidemics. American plant pathologists lacked the legal mechanisms to react to strong biological evidence of potential disaster. Unlike many European countries, America had no quarantine system, relying instead on a loosely organized and haphazardly maintained practice of inspection after importation. Often nurserymen, in an obvious conflict of interest, held the final say over the movement of plant material.

Compounding the problem were the difficulties inherent in the actual detection of diseases such as white pine blister rust and potato wart by inspection. USDA scientists lamented that these two diseases were "being brought into America regardless of the danger of their becoming permanently established in this country." Further, they warned that their control could "not be efficiently handled by inspection." "The only practical method of dealing with these two diseases," scientists maintained, "is that of total prohibition of the importation of white pines and potatoes from certain localities." A national quarantine law, they insisted, was the only way to ensure the complete exclusion that was so vital to preventing or at least minimizing the plant disease caused by exotic pathogens.[106]

The legal exclusion of plants suspected of harboring dangerous pests was a fairly recent concept. The European countries had led the way in the second half of the nineteenth century with restrictions and regulations against plant pests, stimulated in part by the importation from America of both the Colorado potato beetle and the grape insect *Phylloxera*. In the United States, no attempt at legal exclusion occurred at the national level in the late 1800s, although a number of state laws were enacted that reflected an "international character."[107] In several of these early state laws, legal restrictions were placed on the transportation of nursery stock suspected of carrying diseases such as peach yellows, fire blight of pear and apple, and black knot of plum.[108] It rapidly became evident, however, that state laws would not be adequate to thwart the introduction and spread of exotic pests. While a rigorous law in one state might prove effective in impeding the spread of diseased plant material, another state without legal restrictions or with a weak law in place could very easily provide a home for the imported pests. Crop losses told the real story. By the turn of the century, imported insect pests and plant diseases were costing Americans nearly $1 billion annually.[109]

In 1905 two events occurred that bode well for attempts to regulate plant disease: the passage of the Insect Pests Act and quarantine legislation on contagious livestock diseases.[110] Both pieces of legislation reflected the activist nature of the USDA under Secretary Wilson, who believed in the application of science to the regulation of agriculture. In the absence of federal quarantine legislation regarding plant pathogens, the BPI, probably the largest importer of new plants in the United States, was maintaining its own strict inspection of all of its imports and either fumigated or destroyed suspect plant material.[111]

But USDA scientists promoted stronger legal restrictions, arguing that inspections alone were insufficient to prevent the importation of plant pests. BPI plant pathologists played a significant role in encouraging lawmakers to consider quarantine legislation. Testifying before Congress in 1910 on the threat of white pine blister rust, Haven Metcalf insisted, "The only way in which this can be kept out of the country further is by definitely prohibiting the entry of white-pine nursery stock, not related to nursery stock, but simply to white pine from those particular regions where we know the disease occurs and has occurred."[112] William Orton testified that "in the case of diseases, as opposed to insects, it is much more difficult to enforce an inspection for the reason that the presence of the disease is not always apparent on the imported stock. Some of these fungi may gain entrance to the plant before its maturity, and not develop to the visible point until after the stock is widely distributed in this country."[113]

Opposition to legal exclusion developed in the Senate Agricultural Committee. Bolstered by the Legislative Committee of the American Association of Nurserymen, the opponents sought a bill that was weaker with respect to restrictions on imports and interstate commerce. As one critic later described the issue, "In general, the attitude of the nurserymen was to have the bill so changed as to place practically no restrictions on imports and particularly to provide for the acceptance of foreign certification in lieu of inspection on this side, a system which was recognized to be practically useless as far as excluding pests was concerned."[114]

Gradually, however, the expert testimony of the scientists began to win converts. On May 3, 1912, Rep. James S. Simmons of New York introduced to Congress a bill intended "to regulate the importation of nursery stock and other plants and plant products; to enable the Secretary of Agriculture to establish and maintain quarantine districts for plant diseases and insect pests; to permit and regulate the movement of fruits, plants, and vegetables."[115] The Plant Quarantine Act was finally signed into law on August 20, 1912. The act placed restrictions on incoming nursery stock and allowed for the use of similar regulations on imported fruits, vegetables, seeds, and certain plant products. It created the authority for the appointment of a federal Horticultural Board, composed of five members from the bureaus of Plant Industry, Entomology, and Forestry, and gave wide discretionary powers to the secretary of agriculture. Congress's decision was obviously influenced by the threat of imported plant disease. The act contained three comprehensive quarantines, two of which pertained directly to plant diseases, white pine blister rust and potato wart.

A NEW ERA

The Plant Quarantine Act of 1912 signaled that plant pathology was entering an era of new and unique goals and responsibilities. One source of the science's significant growth in the United States was the Bureau of Plant Industry, the new institutional platform in the USDA created primarily through the political genius of Beverly Galloway. During his tenure as chief of the BPI, Galloway, a strong advocate of scientific federalism, guided the rapid expansion of the BPI and its associated laboratories and crafted an unbreakable bond of funding and cooperative research between the bureau and the state agricultural experiment stations across the country.

The activities of the BPI coincided with a period of general growth in funding for agricultural science and a continuation of the work begun in the last decades of the previous century, particularly in the area of the chemical control of fungal diseases and the understanding of diseases caused by bacteria. In time, the bureau added skilled scientists and created numerous new administrative divisions devoted to specific commodities or types of research. The growth of the bureau reflected the expansion of plant pathology as a field.

Meanwhile, plant pathologists were also building and developing the conceptual structure of their scientific discipline. The first decade of the twentieth century saw the reorganization of the field into new subdisciplines that had not even existed as recently as the 1890s. Such subdisciplines as breeding for disease resistance, nematology, and forest pathology represented the incorporation of new scientific concepts, such as the Mendelian theories of heredity, to plant disease research and the application of both old and new concepts to new areas of study, such as trees and associated diseases in their natural habitats. These new areas of research reflected the rapid growth of the biological sciences in the late nineteenth and early twentieth centuries and signaled the eagerness of plant pathologists to adopt new theories and methods.[116] The corpus of knowledge in plant pathology was expanding at a phenomenal rate.

But despite these tools for understanding and control, diseases still occurred that seemed to defy even the best of the professionals in the vigorous, young discipline of plant pathology. This brought about a call for yet another weapon in the fight against plant disease, quarantine. Quarantine and plant-import restrictions were certainly not new concepts. They had been practiced on state and local levels for decades. However, it took a devastating epidemic of chestnut blight to provide the basis for a very compelling argument for national quarantine laws to protect forest trees and crop plants from imported pathogens. The role that plant pathologists played in affecting this public-policy argument was just another sign that the science of plant pathology was starting to mature.

❧ CHAPTER 11 ❧

The Creation of University Departments

During the first decade of the twentieth century, plant pathology began to be recognized and accepted as a unique scientific discipline at many universities across the United States. Universities in California, Delaware, Michigan, Illinois, Iowa, and North Carolina not only had research in progress on plant disease at experiment stations, but also offered one or more courses on plant diseases and their control. A logical next step was to collect the instruction and research on plant disease from the various academic departments such as botany, horticulture, or even general agriculture, where plant pathology had existed through the last quarter of the nineteenth century, into a single identifiable collegiate unit, a separate university department. The institutional leaders in experimenting with this kind of administrative reorganization were Cornell University, the University of Minnesota, and the University of Wisconsin.

Acceptance of plant pathology as a separate academic discipline was pivotal in the continued development of the science. Separate university plant pathology departments played a significant role in the professionalization of the science. Along with assuming responsibility for training a new generation of plant pathologists, they established and maintained intellectual standards for scientists and practitioners entering the discipline. By providing a central location for related interests, they helped the science to grow into new and increasingly more sophisticated research programs. They also provided the resources for extension activities that applied the practical knowledge of the increasingly complex science to the benefit of agriculturists across the nation.

THE CHANGING EDUCATION SCENE

The late nineteenth and early twentieth centuries was a time of transition in American higher education. The growing American middle class was beginning to see the advantages of a college education for social and professional advancement, and, as a result, college enrollment was spiraling upward. Administratively, the old model of the American college was being put aside in favor of a new German model of a "university" system, one basic element of which was graduate schools for advanced specialized studies, particularly in the sciences. The increasing number of graduate schools in American universities meant that American students could now acquire an education on American soil that they formerly could have obtained only in Europe.[1]

American colleges and universities were a natural and congenial fit for scientific studies. This was particularly true of the land-grant colleges after the 1880s, when applied and fundamental research triumphed over purely vocational instruction.[2] Although government-funded research was still driven by practicality, the university was now seen as an institution of research where academic freedom reigned.

Research funding in the United States continued to grow, particularly for agricultural sciences. Although some of this money came from private groups and families such as the Carnegies, most of it came from the federal government. In an era of rapid agricultural growth, enlightened educators and politicians had unprecedented access to the public purse.[3] For example, the Adams Act of 1906, the brainchild of educators and politicians sympathetic to the promotion of agricultural science, increased funding to state agricultural experiment stations and provided funds for both fundamental and applied research. Enhanced federal funding offered visionary educators and scientists the opportunity to augment experiment station work and to create new university departments for both teaching and research.

The reorganization of American universities sparked vigorous intellectual and political debate, much of which reflected the rapid changes that were occurring in science and in the institutions of science in the early twentieth century. For plant pathology, a central controversy in the debate was the role of science versus service. If plant pathology departments were to be primarily science oriented, some argued, they belonged in the "arts and sciences," areas of the university where chemistry and physics and even some of plant pathology's close relatives such as botany resided. But plant pathology also had strong connections to the tradition of practical science in the agricultural colleges. Colleges of agriculture were working hard at this time to change their utilitarian reputations. A way of doing this was to break down their often yield-oriented disciplines into constituent elements supported by sound, basic scientific research. To the extent that they were successful in this, they could provide a natural home for departments of plant pathology. The outcome of the controversy was the placement of science-based plant pathology departments in agricultural colleges.

Among the earliest efforts to create a university-based program dedicated solely to plant pathology was at the University of California in 1903. More programs would soon follow across the nation. The early departments at Cornell University, the University of Minnesota, and the University of Wisconsin were models that would be emulated at other universities throughout the twentieth century. They would play important roles in developing the science of plant pathology, in professionalizing the science, and in educating a new generation of plant pathologists who would take the science into the complex and challenging years of the twentieth century.

CORNELL UNIVERSITY

Since its establishment in 1867, through the largess of Ezra Cornell and the Morrill Act, Cornell University, in Ithaca, New York, had been a center for agricultural education and science. By the 1880s research and teaching about plant disease was an established presence at Cornell. One significant event was the granting of a doctorate of science to J. C. Arthur in 1886, the first advanced degree in mycology conferred by Cornell. Arthur was recognized for his experimental work on the bacterial origins of fire blight at the New York State Experiment Station at Geneva, a rival research institution, but one that could not confer academic degrees. Granting the degree had the effect of attaching Arthur's notable scientific achievement to Cornell University.

The next instance of plant pathology at Cornell came, not unexpectedly, from the University Agricultural Experiment Station. Bulletins in 1889 and 1890 were researched and, in one case, written by women who were university fellows in botany. Botany was a fashionable and acceptable avocation for women of this era. Miss J. W. Snow was responsible for the research presented in an 1889 bulletin on strawberry leaf blight, and Miss J. K. Howell, another botany fellow, wrote a 1890 bulletin on clover rust.[4] That these bulletins were examples of systematic rather than applied plant pathology was not surprising, given that the two researchers were educated in Cornell's Botany Department, a traditional, systematic program managed through the College of Arts and Sciences rather than the College of Agriculture.

While attention to plant disease at the Cornell experiment station was to be expected, given the funding pressures of the Hatch Act and the practically minded USDA, this attention did not necessarily translate into instruction on pathological subjects in the classroom. For the first two decades of Cornell's existence, instruction in plant science was dominated by A. N. Prentiss, head of the Department of Botany. Prentiss was a mycologist of the old school. Although he had some interest in pathogenic fungi, he was not dedicated to the study of plant disease. He did offer a course at Cornell called "Parasitic fungi," but the study of plant disease was not an important part of botany instruction during his tenure.[5]

Prentiss's student, W. R. Dudley, brought a growing interest in plant disease to Cornell, particularly during his tenure as cryptogamic botanist of the experiment station. Dudley served as Prentiss's assistant before graduating in 1876 to become assistant professor of botany (1876–83) and then assistant professor of cryptogamic botany (1883–92), until he was enticed to join the faculty at Stanford University. While at Cornell he wrote several early bulletins for the Agricultural Experiment Station on anthracnose of currants, strawberry leaf spot, and hollyhock rust. His primary contribution to plant pathology, however, was the group of students he instructed at Cornell. These included W. A. Kellerman, W. A. Trelease, and G. F. Atkinson. Returning to his alma mater from Alabama to assume Prentiss's place on the faculty when the elder professor retired in 1896, Atkinson did research on plant disease and taught elements of parasitic fungi in his mycology course. As with Dudley, however, even more significant to the development of Cornell plant pathology than Atkinson's personal research was the guidance he gave to students such as Benjamin M. Duggar.[6]

Duggar had been a student of Atkinson's in Alabama. He followed him to Cornell in 1896, receiving his Ph.D. there in 1898. He was the first scientist at Cornell specifically trained to work on plant disease. Duggar did research on a number of fungal and physiological diseases of important garden and orchard crops such as celery, sugar beet, pear, and peach. In 1901 he left Cornell to work for Beverly Galloway at the USDA, but he returned to Ithaca in 1907 to become the head of Cornell's new Department of Plant Physiology. In 1909 he wrote the milestone book *Fungous Diseases of Plants*, the first general text on plant pathology published in the United States. Duggar was destined to be a major figure in plant pathology, leaving Cornell again in 1912 for a career that would carry him to the Missouri Botanical Garden and the University of Wisconsin. He is credited for the discovery of aureomycin.[7]

In addition to the work in plant pathology in the Department of Botany under Prentiss, Dudley, and Atkinson through the 1890s, some attention was given to plant disease in Cornell's Department of Horticulture. Originally a subdepartment of Botany, Horticulture became independent in 1888 with the arrival of the flamboyant Liberty Hyde Bailey from Michigan Agricultural College. A student of William J. Beal and a friend of Erwin Frink Smith, Bailey had spent two years after college as Asa Gray's assistant at the Harvard herbarium before coming to Ithaca. Bailey was a man uniquely suited for his position as horticulturist at Cornell. Not only was he a brilliant scientist, but he was motivated to spread the news about the value of horticulture and Cornell's College of Agriculture to a variety of audiences.[8] Bailey served as head of the Department of Horticulture until 1903, when he succeeded the venerable Isaac P. Roberts as dean of the College of Agriculture and director of the experiment station.

Bailey's primary horticultural interest was not pathology, but he wrote several bulletins for the experiment station on plant disease, often in cooperation with botanists such as Duggar. The majority of the disease work in the Department of Horticulture, however, was done by Bailey's student and assistant, E. G. Lodeman, who wrote a number of plant pathology bulletins for the experiment station involving spraying fungicides, particularly in order to control orchard tree diseases. Originally submitted for a M.S. degree, Lodeman's work was eventually expanded into the influential book *The Spraying of Plants*.[9]

Another professor who played a significant, though indirect, role in the development of plant pathology at Cornell was Mason B. Thomas, another student of W. R. Dudley. After graduating from Cornell in 1891, Thomas joined the faculty of Wabash College in Crawfordsville, Indiana, as a replacement for the noted botanist John M. Coulter. At Wabash, Thomas served as instructor and mentor for a long line of superior botanists and mycologists, many of whom turned to plant pathology. Nine of his botanical students traveled to Cornell for graduate study. Seven of the nine would serve for several decades as the nucleus of Cornell plant pathology. And one of them, H. H. Whetzel, would actually create the Plant Pathology Department.[10]

Herbert Hice Whetzel was born on September 5, 1877, near Avilla in northeastern Indiana. He developed an early love of nature that expanded significantly during his high school and college years. At Wabash, Whetzel was drawn to mycology, and during his senior year he presented papers to the Indiana Academy of Science on *Gymnosporangium juniperi-virginianae*, the causal agent for cedar apple rust, and on a local species of *Stemonitis*. After he graduated with an A.B. degree in 1902, Thomas encouraged him to go to Cornell for graduate study, heartily recommending his protégé to Atkinson.[11]

At Cornell, Whetzel worked closely with Atkinson, accompanying him on many fungus-collecting expeditions around Ithaca. In 1903 he undertook the bulk of the plant disease research at the experiment station. Atkinson considered Whetzel "one of the very best research students we have ever had and . . . more precise, rapid and successful I think than any."[12]

Whetzel's first experiment station bulletin was "Onion Blight," published in April 1904. That same year he was appointed to an instructorship in the Botany Department. At the request of Dean Liberty Hyde Bailey, Whetzel taught "farm botany" for two winter sessions. The course was dominated by information on plant disease. In 1906 he was made an assistant professor of botany and was also appointed head of the new Department of Agricultural Botany, which was separate from Atkinson's Botany Department.[13] Whetzel's department was part of the newly created New York State College of Agriculture, which had been established under the leadership of Liberty Hyde Bailey in 1904.

Although Whetzel completed all the requirements for a Ph.D. at Cornell, he was not allowed to receive the degree because of his faculty status. Atkinson fought for his student, but he was informed that "in order to receive his degree Mr. Wetzel [*sic*] must now be examined by a body of which he is a member and in which he has a vote. This is absolutely forbidden. . . . The fact that the work was done before he became a member of the Faculty is of no consequence."[14] Whetzel never did receive his doctorate from Cornell. He was, however, granted an honorary M.A. by Wabash College in 1906, and an honorary D.Sc. by the University of Puerto Rico in 1926.[15]

In the summer of 1907, the ambitious and aggressive Whetzel asked Liberty Hyde Bailey if he could change the Department of Agricultural Botany into a department specializing in plant pathology. According to Whetzel, Bailey was "astonished at my temerity in suggesting an entirely new kind of chair in university faculties."[16] Dean Bailey agreed, however, and went before the board of trustees with Whetzel's request. In October 1907 the Department of Plant Pathology was created, with Whetzel as head. His faculty consisted of Donald Reddick, a Cornell Ph.D. by way of Wabash College. There were only two courses offered in the first full year of operation—Plant Pathology and Methods in Plant Pathology, with thirty-five students in "the beginning course" and six in "the advanced course."

A year later, Whetzel and Reddick were joined by another Wabash/Cornell alumnus, Mortier F. Barrus. Several graduate students assisted with disease research. By 1909 the department was offering seven courses with an enrollment of sixty students, and two classes for the short winter course, also with sixty students.[17]

Throughout his career, Whetzel was first and foremost an educator. He was devoted to ensuring that his students received the best, most rigorous instruction available, even at the temporary expense of his own department. For example, for the first six years of the the Department of Plant Pathology's existence he sent his students to learn mycology from Atkinson in the Department of Botany in College of Arts and Sciences. Only after 1912, when Whetzel hired Harry M. Fitzpatrick, another Wabash student who had earned his Ph.D. under Atkinson, did he keep his students at home for basic mycology. When Atkinson died from influenza while on a trip to California in 1918, his position was never restaffed. Instruction in mycology assumed a permanent home in the College of Agriculture's Department of Plant Pathology.[18]

Whetzel taught on both the undergraduate and graduate levels. He thought that introductory, undergraduate classes were the best places to spot the most promising students and guide them into the science. He also guided many graduate students, serving as a constant source of inspiration. Even when an original idea may have been his own, he was sure to give students credit for their work. Above and beyond his normal duties, he created a special course in German, teaching his advanced students to read scientific German in a year. He also taught a course on the history of plant pathology, using a textbook that he wrote specifically for the course.[19]

Whetzel demanded precision from his students and from the science in general, and he advocated rethinking scientific terminology and its application. Like N. A. Cobb in nematology, Whetzel introduced new terminology or, more often, sharpened the proper applications of old terms. Many of the terms he introduced, such as *necrotic*, *hyperplastic*, and *hypoplastic* became standard. His substitution, however, of *epiphytotic* for *epidemic* and *suscept* for *host* were not as generally accepted. He organized control measures into conceptual categories, including *exclusion*, *eradication*, *protection*, and *immunization*, and classified the action of fungicides as either *protectants*, *disinfectants*, or *disinfestants* under specific conditions.[20]

Whetzel knew that proper instruction could not occur without a proper institutional base. His first battle, and one that would continue throughout his career, was to secure sufficient classroom, laboratory, and office space. Because of "a rapidly increasing demand in colleges of agriculture, experiment stations and elsewhere for thoroughly trained men to take up the work of teaching and investigation in the field of Plant Pathology," enrollment was burgeoning, and could have increased even more with adequate facilities.[21] Whetzel astutely argued that he needed "a substantial increase in the general maintenance fund . . . if we are to meet the legitimate demands made upon us by the growers of the State. The opportunity for increased service to the State is before us; we are ready and anxious to do the work. Only the lack of necessary means with which to do the work hinders."[22] His arguments were successful. In 1913 the department moved into new, expanded quarters in the basement of Bailey Hall.[23]

While fighting the floor-space battle, Whetzel also worked to develop other areas of infrastructure necessary for a mature teaching and research program. He continued the expansion of the university's botanical library that had begun under Prentiss, Dudley, and Atkinson. He stocked his laboratories with the best equipment available. As a naturalist and collector, he knew that mycological specimens were of foremost importance, so he created a herbarium with his own personal collection of about 5,000 entries. Finally, to further streamline operations, he hired a department photographer.[24]

Whetzel also knew that excellent instruction would require an adequate financial base. In the face of insufficient commitment from the University, he began to search for unique sources of funding. In early 1907 Whetzel happened upon an article in the *North American Review* by Robert K. Duncan titled "Temporary Industrial Fellowships," discussing how the University of Kansas supported their chemistry research with funding from industrial sources. Whetzel seized on this idea and approached various businesses in New York whose interests coincided with those of the Department of Plant Pathology. He decided to focus first on a single problem, one with a local angle and a very practical, definite goal—the replacement of Bordeaux mixture with lime-sulfur solution for spraying orchard crops.

When the officers of the Niagara Sprayer Company of Middleport, New York, a "small concern pushing commercial lime-sulfur as a summer spray" expressed an interest in the idea of funding research, Whetzel brought his plan to Dean Bailey, who supported it.[25] In 1909 the Niagara Sprayer Company provided a two-year research grant of $1,500 per annum, and Erret Wallace, Whetzel's first undergraduate specializing in plant pathology, undertook the research. Although it is questionable how much Wallace's research actually influenced the switch in fungicides, a mere three years later, "apple growers of the State had largely shifted to lime-sulfur . . . in the control of apple scab."[26]

The use of industrial fellowships flourished in Cornell's Department of Plant Pathology. "We are now making it a uniform policy," Whetzel wrote in 1910, "to undertake only those investigations which are of sufficient importance to the grower to warrant his financial co-operation with the department in undertaking the work."[27] In 1913, when Liberty Hyde Bailey retired as dean, there were sixteen active industrial grants. When Whetzel resigned as department head in 1922, there were thirty-three. During this period, fellowships provided more than $59,000, which was over and above the normal operating budget.

In addition to providing the department with direct fellowships, interested growers interceded with the New York legislature to provoke special appropriations for investigations of specific disease problems. Growers also provided "land, labor, seed, and machinery necessary for conducting the experiments and demonstrations." But Whetzel believed that industrial fellowships had value beyond mere dollars and cents: They spurred dissemination of research. "Growers who have invested substantially half of the funds out of their own pockets," Whetzel wrote, "are quick to interest themselves in the findings and to put into practice the improved methods that are thus discovered and demonstrated under their own eyes." He contrasted this eager acceptance with "the critical attitude and frequent indifference of growers to the results obtained from research financed entirely by the State."[28]

The list of students produced at Cornell during Whetzel's tenure from 1907 until 1922 included many who would play prominent roles in twentieth-century plant pathology. The first generation were all from Wabash. Donald Reddick was the first, although not a student of Whetzel's. He was followed by M. F. Barrus, the department's first extension faculty member, Walter H. Burkholder, L. R. Hesler, W. H. Rankin, Charles Chupp, and, finally, Louis M. Massey, who succeeded Whetzel as department head in 1922. Most of these scientists spent their long and fruitful careers at Cornell.[29]

In the first decade of its existence, the Cornell Department of Plant Pathology undertook investigations on a wide range of plant diseases. The work on bean diseases begun in 1906 by M. F. Barrus was of special importance because of its contributions to the field of mycology. In 1908 Barrus noted the existence of two strains of the anthracnose fungus (*Colletotrichum lindemuthianum*). This was the first documentation of physiological races in the Fungi Imperfecti or Deuteromycetes. Barrus's discovery of pathogen races paved the way for the development of an anthracnose-resistant strain of red kidney bean. Barrus's colleague, W. H. Burkholder, worked on resistant white marrow bean and, significantly, later discovered a third race of the pathogen.[30]

Cornell also was home to studies with a wide variety of fungicides, including F. M. Blodgett's work on sulfur dusting for apple scab and powdery mildew of hops, as well as the somewhat less successful experiments on the fungicidal properties of sulfate of iron sponsored by a fellowship from the American Steel and Wire Company of Chicago. Blodgett's research helped spark a renewed interest in dusting as a method of fungicide application.[31]

During this period, the Cornell Department of Plant Pathology also established the basis for an extensive program on potato diseases. The program began in 1910 with tests of sulfur for the control of potato scab and investigations on the breeding of an improved potato made with the cooperation of Cornell's Department of Plant Breeding under Herbert J. Webber, a former senior scientist in the USDA's Bureau of Plant Industry. Potato disease work of one form or another would eventually involve nearly all of the first generation of plant pathologists at Cornell and their students. Throughout the 1920s, research on potato diseases would gather momentum and lead to a much improved understanding of late blight caused by *Phytophthora infestans* and virus diseases such as leaf roll, mosaic, and yellow dwarf. Cornell scientists pioneered the creation of seed inspection and certification systems for potato as well as developing widespread spraying circles throughout New York.[32]

In 1922 Whetzel resigned as head of the Cornell Department of Plant Pathology. It was not an amicable separation; over the years Whetzel's innovative and blustery ways had created as much enmity as devotion.[33] As he was stepping down, Whetzel remarked that "I am convinced that with rare exceptions no man should be allowed to head a department for more than fifteen years. . . . I love [the department] as a father loves his child. . . . I could not bear to see it stagnate or die, as not infrequently happens when one retains administrative responsibility too long." He wrote to the dean at the time of his resignation that "I know that I can teach, I wish to do it. I think that I can do research work, and I wish to try it."[34]

Although he owed much to those who came before him, H. H. Whetzel essentially created the Department of Plant Pathology at Cornell. He served as its guiding spirit, both in teaching and research, during his years as department head, and continued as the touchstone long after he had become just another faculty member and, in fact, after his death.

THE UNIVERSITY OF MINNESOTA

Although the University of Minnesota received aid from the Hatch Act of 1887, it took a number of years before federal funding could be translated into productive agricultural education and research. Meanwhile, like most American colleges and universities attempting to make the transition to agricultural education and research, the University of Minnesota, in St. Paul, Minnesota, grappled with small enrollments and the difficulties of altering traditional academics. Out of this struggle arose a unique arrangement of instruction and research.

Within the University of Minnesota's Department of Agriculture, or, as it was commonly known, University Farm, three separate but coordinated units functioned: the School of Agriculture, which offered a two-year practical course for young people to improve farming skills; the College of Agriculture, which provided more advanced training to high-school-level graduates; and the experiment station, which focused on applied research and public service.[35] In time, teaching and research in plant pathology came to represent an integral element of all the parts of this system.

At the University of Minnesota, research on plant disease was first undertaken at the Minnesota Agricultural Experiment Station. For the first two decades of its existence, however, the station operated without a plant pathologist, so plant disease problems fell on the shoulders of station botanists, horticulturists, and agronomists. The resulting investigations were fragmented and pedestrian, generally reflecting the lines of work going on elsewhere around the nation—the testing of fungicides and other promising controls.

The one early case at the Minnesota Experiment Station where original work seemed to hold promise was the investigation of flax wilt by Otto Lugger, the station entomologist, who was in charge of both entomology and botany. In the spring of 1890 Governor W. R. Merriam asked Lugger to look into the cause of an unexplained disease that was threatening the health of the state's entire flax crop. During the 1880s, flax had become significant to the state's agriculture, bolstered by efforts to diversify from wheat and by high prices for linseed oil. Soon after flax culture became established, however, a mysterious ailment, characterized by the extensive wilting and rapidly dying plants, created consternation among growers. Lugger had visited with a number of these growers in 1889 and had a good idea of what the condition looked like in the field.

Lugger's initial experiments centered on the effect of fertilizers and fungicides on the disease. His results led him to believe that he was dealing with an infectious agent that was transmitted by water and by flax straw. To prove his theory, he inoculated flax with a crude extract prepared from diseased plants, but he was unable to isolate a pathogen, forcing him to conclude, somewhat surprisingly, that "we have not to deal with a disease, but that the straw flax itself is the cause of this trouble."[36] In the final analysis, Lugger's final report had little to offer the farmers of his state.

In all fairness to Lugger, one season of experimentation was not enough to answer questions that had dogged flax farmers in Europe as well as America. Lugger was aware that further research was required: "The following report is therefore only a preliminary one," he explained. "It will be necessary to continue these experiments . . . where all the surrounding conditions can be observed or kept under control." Unfortunately, it does not appear that he ever got the opportunity to continue his research.[37]

Station workers, who had myriad constituent services to perform, were largely unable to devote much time or energy to lengthy investigations like those required by the flax wilt problem. Studies on plant disease at the Minnesota station on this and other diseases were intermittent and resulted in few groundbreaking pathological details. They did, however, add to the existing data on the effectiveness of some control methods. Perhaps more important, the station bulletins also provided Minnesota farmers with current information on the causes and controls for some of their worst enemies—diseases including potato late blight, potato scab, and wheat smut.

The need for closer attention to plant disease became painfully obvious to many Minnesotans when in 1904 the state's farmers were jolted by a disastrous epidemic of stem rust of cereals. Mark Carleton of the USDA referred to the epidemic as "the most severe attack of rust ever known in that region, certainly the most severe in the last twenty or twenty-five years."[38] The seriousness of the epidemic was not lost on Edward Freeman, a young assistant professor of botany at the University of Minnesota, who had a developing expertise in plant pathology. Realizing that plant diseases such as stem rust had received inadequate scientific scrutiny in Minnesota, Freeman soon devoted his career to rectifying these shortcomings and improving the agriculture of his native state.

Born in St. Paul on February 12, 1875, Edward Monroe Freeman entered the University of Minnesota in 1894, where he came under the influence of Conway MacMillan, professor of botany, and immediately chose to pursue a career in the plant sciences. At that time, botany was not taught to any appreciable degree at University Farm, so to further his studies, he had to study in the College of Science, Literature and the Arts rather than the agricultural wing at University Farm.

Freeman graduated from the University of Minnesota in 1898 with a B.S. and a year later, with his master's degree. While involved with his studies he taught botany in the College of Pharmacy. During the summer of 1900 he studied at Woods Hole Biological Station in Massachusetts, where he became interested in plant pathology. He spent 1901 at Cambridge University in England, studying plant parasite-host relationships under the guidance of the renowned English plant pathologist Harry Marshall Ward. Freeman was influenced by the experience in Ward's laboratory, particularly with respect to the latter's research on rusts. He returned to the University of Minnesota as an instructor of botany in 1902 and assistant professor in botany between 1903 and 1905, teaching courses in plant pathology and industrial botany. He earned his Ph.D. in botany from the College of Science, Literature, and the Arts in 1905.[39]

The stem rust epidemic of 1904 offered Freeman the opportunity to focus on plant pathology. Sensing that the climate was right for educating his state's farmers as well as politicians and university administrators, Freeman embarked on "a systematic survey of the plant diseases of the state, sufficiently thorough to determine the full extent of the damage due to these diseases."[40] Published in 1905, *Minnesota Plant Diseases* developed out of Freeman's studies on the fungi of Minnesota, done under the auspices of the State Geological and Natural History Survey.

In embarking on his text, Freeman was encouraged by the recent success of Ward's *Diseases in Plants*. After the disastrous 1904 epidemic of stem rust, he felt that there would be not only a requirement for a survey of information but also a need for a plan of action. In the preface to his book, he explained, "The possession of an accurate knowledge of plant diseases and their causes is not only of commercial use to the farmer, . . . but also, by making him an intelligent observer, adds hosts of assistants to the small corps of men who are devoting their time to this study of botanical science . . . . In fact it is only with the intelligent and hearty co-operation of farmers that such work can successfully go forward. . . . Upon such knowledge, widely disseminated, can be built a substantial system of disease prevention. In short, the aim of this work is rather *educational* than *immediately practical*, for in the former feature the author hopes that it will be ultimately most useful."[41] After publishing *Minnesota Plant Diseases*, Freeman served in the USDA Office of Grain Investigations for two years. While on an inspection trip in the West for the USDA in 1907, he received a telegram from the governing committee of the University of Minnesota's College of Agriculture requesting that he return to the university in order to establish a new department of "vegetable pathology."

The decision to create this new department was influenced by the university's having received funds from the Nelson Act. Enacted largely through the efforts of Minnesota's Senator Knute Nelson, the Nelson Act was a 1907 amendment to the Morrill Act. One of a number of similar acts during this period, the Nelson Act provided additional funds for the land-grant colleges, stipulating that the support would be used for "special preparation of instruction" in agricultural education. This additional funding formed the basis for the future of plant pathology at the University of Minnesota.[42]

On August 1, 1907, the Board of Regents of the University of Minnesota established the Division of Vegetable Pathology and Botany in the Department of Agriculture, with Freeman as its head (the title was changed later to Plant Pathology and Botany). When Freeman was given a leave of absence until January 1908 to complete his work at the USDA, the university saved five months of salary funds. Already thinking about preparing his new department, Freeman spent the money on a new high-powered Zeiss microscope.[43]

When Freeman returned to the St. Paul campus in January 1908, he immediately faced a showdown over the location of the new division within the university structure. As at Cornell and Wisconsin, the choices were between a college of arts and sciences and a college of agriculture.[44] At Minnesota, the issue was whether the Division of Vegetable Pathology and Botany should be a part of the Botanical Department of the College of Sciences, Literature, and the Arts or a part of the Department of Agriculture of the College of Agriculture.

Freeman proved quite successful at sorting out this organizational conundrum. Soon after his arrival in 1908, the Board of Regents held a hearing to find a solution to the problem. Freeman argued that plant pathology should be a separate unit in the College of Agriculture, and the board decided in his favor. The Division of Vegetable Pathology and Botany was to be an autonomous division within the Department of Agriculture.[45]

Freeman overcame his next administrative hurdle even more easily. When he took charge of the Division of Vegetable Pathology, Agricultural Botany was a separate department. An opportune circumstance, however, allowed Freeman to bring these two units together. When the existing instructor of agricultural botany, W. L. Oswald, became ill early in 1908, Freeman was asked to teach Oswald's classes. Thus, the merger occurred naturally.[46] Upon Oswald's return, agricultural botany classes were taught as part of the new curriculum of the Division of Vegetable Pathology and Botany.

After sorting out the location of the division in the university structure, Freeman turned his attention to the organization of teaching and research in plant pathology and agricultural botany.[47] For the first two years, Freeman and Oswald, as professor and instructor, respectively, managed the classes alone, with Freeman handling the bulk of the teaching load. The curriculum was "designed to assist the student in correlating the fundamental facts and theories of botany and the practical problems of agricultural and horticultural work." Undergraduate classes were offered in general Plant Pathology, Advanced Pathology, Agricultural Botany, and Wood Technology, with a special Advanced Pathology course offered as graduate work.[48]

The division gained the assistance of a bright and energetic young teacher in 1909, when Elvin Stakman agreed to join the staff. Stakman had accepted an assistantship in the new division with the opportunity to pursue an advanced degree. He thus became the first graduate student in plant pathology at Minnesota. This turned out to be a recruitment coup. Over the next seventy years, Stakman, as a teacher, researcher, and administrator, would become a vital part of the growth of plant pathology on the St. Paul campus and a world leader in plant disease research.

Born on May 17, 1885, on a farm in Wisconsin overlooking the western shore of Lake Michigan, Elvin Charles Stakman moved with his family to Brownton, Minnesota, seventy-five miles west of Minneapolis. He entered the University of Minnesota in 1902, and as an undergraduate, focused on botany, German, and political science. He earned a B.A. in 1906 with high marks. Upon graduation, he was elected to Phi Beta Kappa and offered assistantships to continue his studies in botany and German. Choosing instead to accept a teaching position in a high school in Red Wing, Minnesota, he taught for two years, after which he became the superintendent of the high school at Argyle, Minnesota, where, in addition to administrative duties, he taught courses and coached athletics.

Freeman's offer in 1909 of an instructorship with the opportunity for graduate training steered Stakman back to the university and toward plant pathology. As an undergraduate, Stakman had come into contact with Freeman during his laboratory work in botany. Stakman was impressed with his professor and interested in the practical application of plant pathology to agriculture. He later recalled that he accepted Freeman's offer for two main reasons: "I wanted to study with Dr. Freeman if I could," and "I wanted to be as scholarly as I could but I wanted my scholarship to be productive. In other words, I wanted to combine knowledge and interest with utility as much as possible."

Having grown up on a farm and in a state where farming prevailed, Stakman came to the University of Minnesota with much to learn about plant pathology but with "a lot of experience" in "agriculture from the practical standpoint."[49] Through intensive training he quickly acquired the plant pathology knowledge that he needed. Along with his responsibilities of teaching, research, and extension, he earned two graduate degrees over the next four years. His master's thesis topic was the germination patterns of cereal smut spores. This study was published in 1913 as Minnesota Agricultural Experiment Station Technical Bulletin No. 133.

After completing his master's degree, Stakman followed up on the topic of variation among fungi with research on the physiological specialization in the stem rust of wheat pathogen for his Ph.D. In 1894 the Swedish scientist Jacob Eriksson had shown that the rust *Puccinia graminis* consisted of special forms that were host-specific. Soon afterward, H. M. Ward in England elaborated on this idea of variation with his "bridging-host" hypothesis. Ward hypothesized that the stem rust fungus could alter its pathogenic capabilities on different hosts. If this hypothesis were true, and the fungus could readily adapt to a new cereal host, breeding for resistance would turn out to be an ineffective control strategy.

Stakman's research confirmed Eriksson's work and went on to demonstrate, with the use of differential cultivars, that each special form or variety of *Puccinia graminis* comprised a number of distinctive biologic forms, or races. Thus, the validity of Ward's bridging-host theory looked doubtful, as Stakman would later conclusively demonstrate. For this work, Stakman received the Ph.D. degree in 1913, the first for the Division of Vegetable Pathology and Botany.[50] His dissertation was published as Minnesota Agricultural Experiment Station Bulletin No. 138.

The concept of physiologic specialization was a significant breakthrough in understanding the genetics of fungi. It had major implications for breeding disease-resistant cultivars and for plant pathology at Minnesota. As a Minnesota scientist later recalled, "The discovery and identification of pathogenic races . . . enabled a systematic evaluation of sources of resistance and a rational basis for selecting resistant progeny in a breeding program."[51]

But cereal rusts and breeding for resistance had been a part of plant pathology research at the University of Minnesota even before the creation of the Division of Vegetable Pathology and Botany. Cooperation between the USDA and the university's experiment station in the area of breeding for rust-resistant cereal cultivars had begun when Freeman worked in the Office of Grain Investigations from 1905 to 1907. The earliest published results of this research, co-authored by Freeman, were on cereal smuts and rusts. Little progress was made in the early years, however, because of the need for more information on the pathogens as well as the difficulty posed by attempting to locate effective resistance in cereal crops.

Freeman understood that these problems required more research. While organizing the teaching in his new division, he also undertook coordinating the plant disease work of the experiment station. Plant pathology, he insisted, "should function in improving and insuring agricultural production." Thus, he supported early research at the station on cereal rusts and smuts as well as diseases of fruits and potato. As Stakman later said, Freeman "charted the course of basic investigations for decades to come."[52]

In all likelihood, Freeman encouraged Stakman's interest in cereal diseases and genetic variation. After all, in England, Freeman had studied with Ward, an expert on rusts, and he came home a believer in the bridging-host theory. When Stakman's research destroyed the idea of bridging hosts, Freeman displayed his true merit as a scientist and scholar: He changed his way of thinking about fungi. In part because of Freeman's early leadership, physiologic specialization of plant pathogens and breeding disease-resistant crop cultivars became a principal component of Minnesota's plant pathology research program.

About the time that Stakman was completing his Ph.D. in 1913, plant pathology within the University of Minnesota was once again undergoing a major organizational restructuring. The change was prompted by events that had occurred since A. F. Woods arrived as dean of the Department of Agriculture in 1910. According to one historian, "Dean Albert F. Woods and his right-hand man, Professor Freeman, found themselves in sympathy so complete that the latter was able to testify, 'He and I have worked together as parts of one man.'"[53]

Woods and Freeman's close collaboration became apparent in the reorganization of the Department of Agriculture, which occurred soon after Woods's arrival. Plant pathology had been split between the school, college, experiment station, and extension units. Woods and Freeman joined forces to centralize operations. As a result of the restructuring, the Division of Vegetable Pathology and Botany acquired more autonomy to function across the different administrative units and gained an expanded curriculum and staff as well as an enlarged and improved physical plant.[54]

Under the 1913 reorganization, Freeman remained administrative head of the Division of Vegetable Pathology and Botany, but was also appointed assistant dean of the Department of Agriculture. Within the division, sections were created with Stakman as head of Plant Pathology and Oswald as head of Agricultural Botany. Although Freeman continued to teach plant pathology courses, as the curriculum expanded, others filled in the gaps. Meanwhile, Stakman took on more responsibilities, and the Section of Plant Pathology acquired the assistance of another graduate student, Arne G. Tolaas, who began teaching the general plant pathology course with Freeman. By this time, courses also were offered in industrial mycology (Stakman), plant pest control (Stakman and Tolaas), dendropathology, and seed testing.[55]

Space for teaching and research had long been a problem. In 1908, its first year, the division made due with a small office in the administration building, the agronomy lecture room, and a few laboratories. In 1909, half of the third floor and a lecture room were made available in the horticulture building. In 1913, however, "an unusually large increase in . . . appropriations" provided for significant physical improvements. With an allocation of $10,000 for remodeling, the Old Drill Hall became the Plant Pathology Building, or the Tottering Tower, as it was commonly known because of vibrations from a nearby power plant. The plant pathology staff moved in the following year. They also acquired an additional four acres of land on University Farm, providing them with nearly seven acres for research.[56]

Outside events in 1913 also affected applied research in the Division of Vegetable Pathology and Botany. Freeman and Oswald had formulated a pure-seed law in Minnesota, requiring the labeling of seed as to purity and germination, and in 1913 this law was passed by the state's legislature. In response, the division established a State Seed Laboratory. Laboratory staff provided free seed analyses throughout the state until 1919, when the work was transferred to the Minnesota Department of Agriculture.[57]

A major component of this regulatory effort was the seed potato service. This service was an early example of where demonstration and extension work in plant pathology had a noticeable impact on the state's agriculture. In cooperation with county extension agents in 1914 and 1915, Stakman and Tolaas used demonstration plots to show growers the value of clean seed. Tolaas was responsible for much of the early success of this program, which led directly to the creation by growers of a State Seed Potato Association. In 1916, he became the first extension plant pathologist at Minnesota. Three years later, with Stakman's assistance, Tolaas helped draft a seed certification law, which was enacted by the state's legislature in 1919. Afterward, he resigned from his extension position in the division to head the potato certification program in the Minnesota State Department of Agriculture.[58]

By the mid-1920s, teaching, research, and extension in plant pathology were well established at the University of Minnesota. Perhaps the best evidence for the maturation of the division followed the stem rust epidemic of 1916, during which Minnesota plant pathologists, particularly Stakman, demonstrated leadership in the creation of the cooperative barberry eradication program in 1918. Not only was Stakman a major figure in the establishment of the enormous federal-state program, but the division came to play a leading role in stem rust epidemiology and breeding for rust resistance over the next fifty years.

But even with Stakman's rapidly evolving presence in the division, plant pathology at the University of Minnesota continued to reflect the vision and practice of its founder, Edward Monroe Freeman. As his associates later said, it was Freeman who "laid the foundation for a sound program of instruction, research, experimentation, and public service in his own department and helped significantly in raising standards in the entire institution."[59] Freeman became dean of the College of Agriculture, Forestry, and Home Economics in 1917, a position he held until his retirement in 1943. Throughout his career, he remained an ardent champion of the teaching, research, and extension of plant pathology and agricultural science at the University of Minnesota.

THE UNIVERSITY OF WISCONSIN

When L. R. Jones left the University of Vermont and the Vermont Agricultural Experiment Station in 1910 to answer the call to establish a department of plant pathology at the University of Wisconsin, he probably had an idea of the potential impact he could have as a plant pathologist in the vegetable-growing state of Wisconsin. He may also have envisioned the possibilities for a strong program of graduate education to train a new generation of plant pathologists. What he undoubtedly did not realize, however, was the extent to which he and his department would change the face of vegetable growing in Wisconsin, add to the dimensions of research in the growing discipline of plant pathology, and be responsible for helping to populate the growing ranks of plant pathologists with outstanding scientists.

Although Jones was the first to come to the University of Wisconsin with the formal title of plant pathologist, others, including Dean of Agriculture Harry Lumen Russell, who hired Jones, had done work on plant disease at the university. William Trelease, an instructor in botany from 1883 to 1885, was influenced greatly in his teaching by the work of T. J. Burrill at Illinois, and he included lectures on plant bacteria in his botany courses.[60] Trelease wrote what F. L. Scribner called a "very full account" on the white rot of strawberry in the second annual report of the Wisconsin Agricultural Experiment Station.[61] When he left Wisconsin to fill the professorship of the new Department of Botany at Washington University and to direct the Shaw School of Botany in St. Louis, he recommended Arthur Bliss Seymour as his successor.[62]

Before accepting the position as botanist at Wisconsin in the fall of 1885, Seymour had been a student and assistant to Burrill at Illinois, had spent time as a botanist at the Illinois State Laboratory of Natural History studying smuts and downy mildews, and had served as assistant to W. G. Farlow at Harvard.[63] He turned down an offer at Illinois for a permanent position, deciding instead to take the job in Madison because "the opportunity to be chief botanist of a great university allured me."[64] But Seymour, who wanted to be a pure researcher, soon found that the "routine of teaching large classes was too burdensome." With a desire "to learn and to help the studies of specialists," in the fall of 1886 Seymour left Wisconsin to accept Farlow's offer of an appointment to the staff of the cryptogamic herbarium at Harvard.[65] Early research on plant disease in Wisconsin was also done by Emmett Stull Goff, who moved to Madison from Geneva, New York, in 1889 to accept a professorship in horticulture and to become horticulturist of the Wisconsin Agricultural Experiment Station.[66] Goff's first important research at Wisconsin was an 1890 study in which he and B. T. Galloway carried out a successful series of experiments with fungicides for the control of apple scab.[67]

In the 1890s, H. L. Russell pursued a research program in plant bacteriology that presaged the successes that would occur later in the Department of Plant Pathology. A native of Wisconsin, Russell took a class on bacteriology from zoologist Edward A. Birge, using the equipment that Trelease had ordered prior to his departure. Recognizing the talents of the young Russell, Birge urged him to pursue a master's degree in bacteriology, then a relatively new field. Following the completion of this degree in 1890, Russell traveled to Europe to learn for a year from the masters. He studied in the laboratory of Robert Koch in Berlin, Germany, and then at the Zoological Institute in Naples, Italy, concluding his European trek with a visit to the Pasteur Institute in Paris, after which he attended the Congress of Hygiene and Demography in London as a translator for the Russian scientist, Elie Metchnikoff.[68]

Filled with excitement about the new field of bacteriology, Russell returned to the United States and started doctoral studies at Johns Hopkins University under the direction of medical pathologist William H. Welch. Although not a plant scientist at all, Welch supported Russell's dissertation topic: an examination of the immunity of plants to bacteria.[69] But the facilities in Welch's laboratory were not conducive to growing plants, so Russell appealed to Secretary of Agriculture Jeremiah Rusk, a fellow Wisconsinite, for permission to use greenhouse and laboratory facilities of the Department of Agriculture in Washington. Thus, with his research in Washington, D.C., and course work and an advisor in Baltimore, Russell continued work on his thesis, which would be published in 1892 as *Bacteria in Their Relation to Vegetable Tissue.*

Declining an offer to investigate the bacteriology of citrus fruits at the USDA's Eustis (Florida) laboratory, in 1893, after a year's fellowship at the new University of Chicago, Russell was named an assistant professor and experiment station bacteriologist in the College of Agriculture at the University of Wisconsin. Russell's teaching duties during the winter term were divided between the College of Agriculture and the College of Letters and Science. His new research concentrated on the bacteriology of dairying, and it was in this field that Russell established his professional reputation.[70] He did not, however, abandon research on bacteria in relation to plants.

During the summer of 1895 Dean William A. Henry of the College of Agriculture was notified that a severe rot had been found in the cabbage-growing region of southeastern Wisconsin. He sent Russell to investigate.[71] Preliminary isolation and inoculation experiments convinced Russell that the disease was of bacterial origin. In 1896 the cabbage rot was again severe. Encouraged by the results of his experiments during 1896, Russell was looking forward to continuing his work in 1897. He learned in June 1897, however, that E. F. Smith had described a bacterial disease of cabbage at a May meeting of the Washington Biological Society.[72] In late July, as Russell was completing his experiments, he learned that Smith had already published his studies on the cabbage black rot and had discussed the disease thoroughly in the *Centralblatt für Bacteriologie*.[73] Although Russell subsequently published his results as an experiment station bulletin, he was understandably irritated that, even though his work had actually been completed before Smith's, he would not receive the credit as the pioneering researcher on the black rot disease of cabbage.[74]

When Russell became dean of the College of Agriculture in 1907, he set about building on the firm foundation of science in service to the agricultural interests of Wisconsin that had been established by his predecessor, Dean William A. Henry. Although Russell recognized the need for applied research, he also believed that the answers to many practical problems lay in basic investigations.[75] He emphasized later in life that "unless research is constantly opening up new facts, new explanations, and new principles," agricultural teaching "will soon sink to a sterile repetition of old material and die of dry rot."[76] On this basis, in 1907 Russell moved forward with plans to expand the College of Agriculture by establishing new departments of economic entomology, experimental breeding (later renamed genetics), plant pathology, poultry husbandry, and veterinary science.[77]

By 1909 Russell had received permission to create a Department of Plant Pathology. He set about finding the right scientist to build this new department, writing in July that he had "a strong line out for a first-class man in Plant Pathology—Professor L. R. Jones of Burlington, Vermont. . . . He is the best expert on potato diseases in the United States, and as you know, this is a line which we are very desirous of pushing as hard as we can."[78]

When Jones settled into his new, but small, office in Room 39 Agriculture Hall in early 1910, one of his first tasks was to prepare a revised course syllabus for Plant Pathology 1, the general plant pathology lecture and laboratory course, and Plant Pathology 2, colloquium.[79] Plant Pathology 1 actually continued the course initiated as Botany 25 in 1901 by R. A. Harper in the Botany Department of the College of Letters and Sciences.[80] Another course, Diseases of Trees, had been offered in Botany by C. E. Allen. This course included lectures and laboratory work on "the diseases of economically important trees, and the causes of decay in timber used for building and other purposes."[81]

At Wisconsin, as at Minnesota, a debate was raging over whether research and teaching in plant pathology should reside in the College of Agriculture or the College of Letters and Sciences. A well-known dairyman, William D. Hoard, chairman of the agricultural committee of the Board of Regents, wrote to Russell in late 1907 that he "wanted to see [the department] placed under the direction of the College of Agriculture . . . where it would be made of some practical value to the men who are growing plants."[82] Nevertheless, in 1909, when Russell revealed his plans to expand the College of Agriculture, he faced the opposition of his former mentor and now dean of the College of Letters and Science, Edward A. Birge. Claiming that Russell's plans violated the proper place of pure research in the university, Birge maintained that basic science belonged in his college, not in a vocational institution like the College of Agriculture.

In early 1910 the controversy over the placement of the new Department of Plant Pathology became serious enough that Birge arranged a meeting with Russell and university president Charles R. Van Hise. Intent on securing the new Department of Plant Pathology for the College of Letters and Science, and specifically, on ensuring that L. R. Jones would become a member of his faculty and not of the College of Agriculture, Birge maintained that fundamental work should be undertaken only in pure science laboratories. Van Hise initially agreed with Birge, but Russell stood firm: "Well," he said, "the only thing for you to do then, is to get another dean."[83] Russell's firmness carried the day. Plant Pathology and L. R. Jones remained in the College of Agriculture.

Jones developed a strong teaching program. At the end of his first semester of teaching in Madison, he proposed and won approval to divide the basic plant pathology course into two courses. In 1910–11 the department offered Plant Pathology 1: Diseases of Plants in the fall semester and Plant Pathology 2: Methods in Plant Pathology during the spring semester. Jones was the instructor of both courses, with Irving E. Melhus assisting. Additionally, students could participate in a thesis course, a seminar in plant pathology, and research throughout the year.[84]

Reflecting the speed of development in plant pathology, course offerings continued to diversify. In December 1912, the department added Diseases of Field Crops. By the 1914–15 academic year the department was offering seven basic courses, in addition to a thesis course, seminar, and research. The fact that many of these courses were the responsibility of students who were working toward M.A. or Ph.D. degrees under Jones demonstrated the need to appoint new professors as quickly as possible.

Recognizing the need and the opportunity for graduate education in the future of plant pathology, Jones set out to build a strong graduate program. The first Master of Science degree in plant pathology at the University of Wisconsin was awarded to George W. Keitt in 1911, and the first Ph.D. degree in 1912 to I. E. Melhus, who came to Wisconsin from Iowa. Charlotte Elliott received her Ph.D. degree in 1918 for research in phytobacteriology and went on to a productive career with the USDA. The first decade of graduate education in plant pathology at Wisconsin produced twenty-eight Ph.D. degrees and forty-two master's degrees. All but six of these graduates had L. R. Jones as their major professor.

Although teaching occupied a large amount of Jones's time, his research program was not allowed to suffer. The successes achieved early in Jones's tenure at Wisconsin firmly established both his reputation and that of the Department of Plant Pathology for solving disease problems and aiding the farmers of the state. Two of those diseases—pea blight and cabbage yellows—along with Jones's continuing interest in bacterial diseases of plants, would occupy the research agenda of the department for its entire first decade.

Pea blight, attributed to *Septoria* and *Ascochyta* spp., was a problem that was driving canning companies to other vegetables or completely out of business. Jones was eager to see the pea blight problem solved, but by the time the call came for research on the problem, all experiment station research funds for 1910–11 had already been allocated. By 1912 twelve canning companies that had been in operation in 1909 were gone, due primarily to pea blight.[85]

Like Whetzel in New York, Jones turned to agriculture-related industries for support. In 1911 the Wisconsin Pea Packers' Association established a fund for research on the diseases of canning peas.[86] With an industry fellowship of $1,000 for two years' salary and expenses, in 1911 Jones hired Richard E. Vaughan as a research assistant. By 1915 Vaughan had begun to establish criteria for seed selection and field sanitation and had been able to select plants resistant to the pea blight.[87]

The impact of Vaughan's success with pea blight paled, however, in the light of the successes the department was to have in solving an even more pressing disease problem, fusarium yellows of cabbage. It was Dean Russell who pointed out to Jones the importance of the cabbage problems in the Racine area. Not one to ignore the advice of the dean who had hired him, Jones set out to survey the area around Racine. He found hundreds of acres affected by the yellows disease.[88] To confirm that the disease was the same one reported by E. F. Smith in the Hudson River valley in 1899, he sent isolates of a *Fusarium* sp. from cabbage plants with yellows to *Fusarium* authority H. W. Wollenweber, who at the time was working at the USDA in Washington, D.C. In 1913 Wollenweber named the fungus *Fusarium conglutinans*.

During his travels, Jones occasionally found healthy plants flourishing in cabbage fields that were otherwise nearly destroyed by the yellows disease. Aware of the recent successes of W. A. Orton of the USDA in selecting and developing wilt-resistant lines of cotton, Jones selected and saved a number of these healthy plants. He planted them the next spring, allowed the flowers to be cross-pollinated by bees and saved the seed from each plant separately. The progeny of these original seed plants were quite promising, with a "fairly high percentage of resistant plants."

In 1914 Jones arranged to have J. C. Walker, a native of Racine and a new M.S. student in the department, stationed in Racine to take charge of the cabbage plots, and Walker planted trial rows of the second generation of resistant lines in several commercial fields. That summer Dean Russell took university president Van Hise and the university regents on a tour of several branch research stations with Racine as the final stop. The cabbage plots at the research station demonstrated clearly that plants resistant to the yellows had been found. Even more impressive, however, were the demonstration trials in the commercial fields. There, the president and the regents observed single rows of healthy-looking plants in the midst of fields in which some or most of the crop had succumbed to the disease and been destroyed.[89] This disease-resistant strain would be called Wisconsin Hollander.

Throughout his career, L. R. Jones remained a passionate promoter of research on the effects of environment on plant disease. One of the research focuses of his department was the effect of soil temperature on the development of disease. Between 1912 and 1914, J. C. Gilman, working with very crude equipment, showed that cabbage yellows was favored by high temperatures. In 1916, when W. H. Tisdale, a research assistant, reported that resistance to flax wilt was inherited as a dominant trait, he also noted that in the greenhouse the disease was always more severe in flats near heat pipes.[90] Tisdale found that he could grow apparently healthy flax in soil heavily infested with the wilt organism, *Fusarium oxysporum* f.sp. *lini*, in a glass jar if he kept the jar cooled in a sink into which tap water was allowed to run continuously and slowly.

To Jones, the need for controlled equipment for soil temperature experiments was clear. He encouraged James Johnson, who was actually a full-time member in the Horticulture Department, but worked closely for many years with a number of the faculty in Plant Pathology, to build the first Wisconsin soil temperature tank. The tanks were subsequently used for studies on many different diseases.[91]

It was clear that the key to success at the agricultural experiment station lay not only in finding ways to solve disease problems of importance to growers, but also in communicating that information to growers throughout the state. This mandate to communicate was certainly not a new idea. In the 1906–7 biennium, Russell had actually initiated a diverse and loosely organized precursor to an extension service. In 1907 he requested unsuccessfully that the Wisconsin legislature allocate funds to establish an agricultural extension organization.[92] Several of the university's departments had already undertaken extension teaching through farmers' institutes, lectures, and correspondence. In addition, on-farm demonstration projects covered topics such as spraying potato fields and orchards. By July 1909, Russell had established a combined Agricultural Education and Extension Department within the college. Nevertheless, he expected each department within the College of Agriculture to maintain and expand its responsibilities in extension as well as in research and teaching.[93] When L. R. Jones arrived in 1910, he knew that his responsibilities would cover all three of the areas associated with land-grant universities: teaching, research, and extension.

EXTENSION

The increasing complexity of the science involved in plant pathology began to create problems for a discipline that took pride in its applied side. As the twentieth century progressed, plant pathologists had to spend more and more of their time in the laboratory or in the fields and greenhouses that they set up and controlled, rather than in the fields of cooperative farmers. Since funding for agricultural sciences was generally on the rise, the need to encourage agriculturists to write letters to legislators asking for more money for plant disease research was on the decline. This resulted in less direct need for, or time for, the direct responsiveness of the late nineteenth century, when plant pathologists had often based their research agendas on the mail they received from growers. Nevertheless, the field of plant pathology had arisen in response to the tradition of agricultural improvement. To continue justifying its existence as a discipline separate from botany or mycology or genetics, it had to continue to perform its service mission, bringing relief from plant disease to American agriculturists.

The answer to this growing dichotomy in the field was to create extension services, which would take complex research from the laboratory and translate it into practical, useful information for the farmer. Extension services also offered a response to critics of the agricultural science establishment, who claimed that the scientific achievements of the late nineteenth and early twentieth centuries were not reaching the farmer at the level that they should. Extension exemplified Progressive Era thinking, which held that the benefits of the nation's chief institutions, such as government and universities, needed to be extended to more of the nation's people.

Among the earliest organized extension work involving plant pests was undertaken at Cornell University in 1895. The previous year, grape growers in Chautauqua County requested help from the experiment station. When the station did not have funding to meet their request, Liberty Hyde Bailey cooperated with S. F. Nixon, an influential member of the state assembly and owner of the Chautauqua and Erie Grape Company, to acquire state funding for the grape work.[94]

Although similar to the scientist/agriculturist/politician alliances formed to combat specific problems that had long characterized the practice of plant pathology, the new alliance soon became more. Bailey demonstrated various elements of grape horticulture in western New York, including spraying techniques and information on plant disease, but he added a new feature, staging itinerant horticultural schools that lasted several days. Over the next several years, these activities continued to expand along with the funding. Soon Bailey was pushing the scope of activities beyond grape problems and horticulture. The most unusual element that he introduced was a correspondence course for rural, primary schools on basic agriculture and "nature study."

Bailey believed that agricultural extension would ultimately be wasted if the rural population was unfamiliar with fundamental concepts of science, particularly biology.[95] Plant pathology played a role in the work. As early as 1903, he appointed H. H. Whetzel as an "Assistant in Plant Pathology in the Extension Dept."[96] Under Bailey's guidance, extension work at Cornell developed rapidly, becoming a model for others.

Specialized work in plant pathology extension continued to expand at land-grant colleges across the United States. One of the earliest plant pathologists specifically hired for extension purposes was Mortier F. Barrus, who was appointed as assistant professor of extension work in Cornell University's Department of Plant Pathology in 1911. Barrus subsequently traveled around New York giving lectures at farmers' institutes and holding short, itinerant extension schools. In 1911, when the University of Wisconsin hired R. E. Vaughan as Jones's assistant to study pea blight, part of his mission was to demonstrate spraying techniques to pea farmers. In 1915 Vaughan received an official appointment as an extension plant pathologist.[97]

Around the same time, the federal government began to work toward expanding extension work on a national scale. Beverly Galloway had always understood the value and necessity of his scientists in the Bureau of Plant Industry going into the field and demonstrating disease-fighting tactics to farmers, but such activity was not a substantial and separate part of his bureau. In 1904 Seaman A. Knapp of the BPI was sent to Texas as head of the newly created Farmer's Cooperative Demonstration Work to combat the boll weevil, which was then threatening the Southern cotton industry with extinction. In 1906 Knapp appointed the first "county agent," W. C. Stallings of Smith County, Texas.[98] In 1909 W. J. Spillman's Office of Farm Management, another subsection of the BPI, began demonstration work in the northern states. In 1910 Spillman appointed his first county agent.

The agitation and funding by local agricultural organizations were very important in preparing the way for the rapid expansion of the county-agent system. In 1909 the BPI had taken the first steps to institutionalize cooperative demonstration work by making an agreement with the Alabama Polytechnic Institute to employ jointly "a demonstration expert" to be located at the college. By 1912, the first comprehensive cooperative demonstration agreement was made with Clemson College. Agreements with numerous other states quickly followed.[99]

In 1914 the Smith-Lever Extension Act provided each state with $10,000 annually, with additional appropriations depending on the state's percentage of rural population.[100] Smith-Lever began the rapid institutionalization of extension work on a national level, on a scale barely dreamed of a decade before. County agents began to appear across the nation, reaching areas and farmers who had not been reached by farmers' institutes or bulletins. With this new stratum of experts between the field and the laboratory, plant pathologists, as well as other agricultural scientists, could concentrate on research, leaving the touring, demonstrations, and speechmaking to the professional extension agents.

The Smith-Lever Act was evidence of a vigorous, growing nation, and plant pathology was caught up in that growth. The general feeling in the United States, and particularly in agriculture, favored expansion and growth, and the federal government seemed to embody the Progressive Era mantra that centralized, rational management was the wave of the future. The system of support for agriculture, including the science of plant pathology, grew and developed with unprecedented speed and complexity.

During the late nineteenth century, when the central concepts of plant pathology did not go very far beyond basic biology, scientists like Bessey and Scribner had confidence that an educated person could read their scientific bulletins and grasp them in large part. As the science of plant pathology became increasingly more complicated, however, the ability to communicate research goals and results to a general audience was becoming increasingly less likely. On the other hand, it was also becoming less necessary. The science was distancing itself from the general agriculturist. For example, the important breakthroughs in the discovery of specific races in pathogenic fungi held little immediate interest for farmers, who just wanted to know how to combat smut or anthracnose in the next growing season. Such breakthroughs did, however, demonstrate to excited scientists that the concepts of pathogenicity were far more complex than the previous generation had believed, and that there was still much to be discovered.

The growing nation demanded more competent, professional scientists who could bring rational, organized thought to bear on problems and solve them, thereby raising the standard of living. Plant pathology, unlike some "pure" sciences that had little connection to the public at large, had been founded on the premise of interaction with agriculturists. As the science developed and removed itself from the everyday world of farmers, it looked to extension agents for a bridge back to its roots.

$\mathcal{\smallsetminus}$ CHAPTER 12 $\mathcal{\smallsetminus}$

The American Phytopathological Society

Under the leadership of President L. R. Jones and Secretary-Treasurer Cornelius Lott Shear, and with fifty members in attendance, the first annual meeting of the American Phytopathological Society was held at the Harvard Medical School, Boston, Massachusetts, on December 30 and 31, 1909. The decision to form the new scientific society had been neither easy nor unanimous. Many wondered whether it was prudent to divide botanical workers into separate societies. Others wondered whether the new society would be better affiliated with the American Association for the Advancement of Science or the Society for the Promotion of Agricultural Science. But whatever questions remained, it was clear that phytopathology was a discipline of growing prominence and one worthy of recognition among the ranks of the biological and agricultural sciences.

The creation of the American Phytopathological Society was part of the general trend toward scientific organization in the United States and around the world. This movement was defined by societies that were both specialized in discipline and national in scope.[1] As national, specialized societies proliferated, local and regional societies, long the mainstay of scientific intercourse, particularly in the agricultural sciences, decreased in prominence. In order to compete in the new world of science societies, many local organizations began to transform themselves, literally broadening their horizons. Examples from the period included the New York Mathematical Society (1891), which became the American Mathematical Society in 1894; and Harvard College Observatory's organization of the Astronomical and Astrophysical Society of America (1898), which became the American Astronomical Society in 1914.[2]

Scientists located in centers of education and culture quite often took advantage of their situations to organize and dominate their disciplines. For example, in 1876 a group of prominent New York chemists complained that "many chemists of this city and vicinity have felt the want, and deplored the absence of an association, such as exists among other professions, which would lead to a better understanding and a closer acquaintance among its members."[3] This desire for organization expressed by a relative handful of New York City chemists soon developed into the American Chemical Society.

Organizations that already had a nationwide constituency, including the American Association for the Advancement of Science, were affected by the other major change that was occurring in scientific societies, specialization. The senior general science societies found themselves increasingly dividing along disciplinary lines. Many of the resulting subsections formed the basis for individual societies that subsequently broke away from the parent organization. By 1882 the AAAS had already divided itself into nine sections, and subdivision would continue for several decades.[4] It was the AAAS sections dedicated to the biological sciences and then specifically to botany that would play host to the organizational activity that resulted in the formation of the American Phytopathological Society.

The founders of the American Phytopathological Society had to confront the problems associated with the movement toward nationalization and specialization. There was, for example, concern about fairness in geographic representation. The science of plant pathology was distributed across the country at colleges and experiment stations, but it had a tradition of leadership from the eastern universities and from the USDA in Washington, D.C. Furthermore, the founders had to tackle the debate arising in many of the sciences at the time that increasing specialization meant isolation. Strong voices spoke out against fracturing the existing botanical organizations, in which plant disease research had found a professional home for the second half of the nineteenth century. Support for the existing organization generally came from a generation of venerable botanists who had an interest in plant disease. Eventually, however, voices of the next generation, of scientists who were educated and worked specifically as plant pathologists, would win out.

Plant pathology in botany organizations

Botanical and mycological papers appeared on the AAAS program nearly from its beginning. At the second meeting of the association, held in August 1849 in Cambridge, Massachusetts, Asa Gray read a botanical paper for the author

Major Benjamin Alvord in the Section on Natural History and Geology.[5] In 1850, the noted Southern mycologist Henry W. Ravenel offered a paper, "A Catalogue of the Natural Orders of Plants," in which he mentioned the presence of seven orders of "cryptogamous plants" of about 1,338 species.[6] But botanical and mycological papers were never the mainstay of the early meetings of the AAAS; only forty-eight contributions in the broad field of botany were offered in the first thirty-two years of the association.[7]

In 1880, work on plant disease was first reported by T. J. Burrill and B. D. Halsted in Section B on Natural History. Burrill read "Anthrax of Fruit Trees" on his research on the cause of fire blight. The full text of Burrill's paper was published in the proceedings of AAAS for that year.[8] Halsted, then of New York, read "An Investigation of the Peach Yellows," but only the title of the paper was published in the proceedings.

With continued growth in the biological sciences, in 1882 the AAAS established Section F, "excluding such borderline fields as microscopy, histology, anthropology and geology." The top offices of this section were filled in alternate years by botanists and zoologists, respectively. The botanists who served as officers of Section F, and thus as vice presidents of the AAAS, included a number of prominent individuals who played seminal roles in the development of the nascent science of plant pathology: W. J. Beal (1883), T. J. Burrill (1885), W. G. Farlow (1887), G. L. Goodale (1889), and J. M. Coulter (1891), who served as chairman of Section F while C. E. Bessey (1884), J. C. Arthur (1886), B. E. Fernow (1888), J. M. Coulter (1890), and B. D. Halsted (1892) served as secretary.

The 1883 meeting of Section F was "a memorable one for botanists, there being more in attendance than ever before, and for the first time botanical papers were in the majority in the section of biology." J. M. Coulter, with feelings similar to those New York chemists who founded that science's leading society, noted perceptively that "the interest at such a time is not so much in the papers read as in the personal contact of the workers who may long have known of each other but never have met, and for whom the clasp of the hand and the glance of the eye cements a friendship already formed. At such times beginners meet with the leaders whose names are household words, and find them genial, hearty, whole-souled men, with a cheering word for all, and they return with fresh zeal to their work."[9]

Botanical zeal at the meeting ran high, and the result was the formation of an

American Botanical Club. This club, founded with thirty members from thirteen states, was not a formal organization or society—it had no constitution or by-laws. It was rather "simply to be an association of botanists who are members of the A.A.A.S., for the purpose of general botanical conferences and excursions during meetings of the American Association." The club was to serve as a general forum for the exchange of ideas and to "provide opportunity for informal discussions that could find no place in the regular sectional meetings of biology." [10] The formation of the American Botanical Club reflected the growing numbers of botanical researchers and the growing need for opportunities for communication among them, including the regular presentation of papers on mycology and aspects of plant disease. The Botanical Club would be active for more than twenty years, with the last meeting having been held in Philadelphia in 1904.

Although plant disease was the subject of many of the papers read in Section F, such as W. G. Farlow's 1887 vice presidential address "Vegetable Parasites and Evolution," no move was made to form a special section or subsection for such papers until 1893. [11] That year marked several important events in the march toward a separate professional organization for plant disease workers. At the Madison, Wisconsin, meeting of the AAAS in August, Section F was redesignated for zoology and a Section G was established formally for botany, with C. E. Bessey of Nebraska as chairman and AAAS vice president and F. V. Coville of Washington, D.C., as secretary. Among the thirty-four papers read in Section G, six dealt specifically with plant disease. [12] E. F. Smith presented synopses of two papers on new and destructive diseases of cucurbits; P. H. Rolfs described "A Sclerotium Disease of Plants" from Florida; W. T. Swingle reported on "The Principal Diseases of Citrus Fruits now Being Studied at Eustis, Florida;" and B. T. Galloway presented two papers on rusts—one on the Jersey or scrub pine and one on wheat.

In conjunction with the 1893 meeting, a botanical congress was convened. One of the general topics suggested for consideration by the congress was the issue of nomenclature within various botanical disciplines. A committee was appointed to consider the nomenclature of plant diseases. [13] The committee report noted that it "finds itself with a somewhat difficult task upon its hands. Names of fungous diseases that are now in general use will be difficult, if not impossible, to uproot and set aside." [14] The formation of the nomenclature committee demonstrated the need for a more standardized approach to nomenclature and communication in plant disease research. Notables such as C. E. Bessey, B. T. Galloway, G. F. Atkinson, B. D. Halsted, W. T. Swingle, and L. R. Jones undertook the study. Their report presented six principal points:

(1) When a good name has become firmly established no effort should now be made to change it, except when it is manifestly inappropriate, as the so-called "strawberry rust" and "celery rust."

(2) There should be terms to distinguish between the parasite and the disease it produces. . . .

(3) The diseases need to have, for popular use, English or at least Anglicised names, . . .

(4) Names should be as far as possible descriptive, and indicate the plant attacked; . . .

(5) After the pathology of plant diseases is understood much better than now a scientific classification of them can be made, and appropriate names given to each; at present only an artificial system can be hoped for. . . .

(6) From what has been presented above it is clear that the results obtained by the Committee are far from final and therefore it is suggested that the work be continued and that it be along the following among other lines:

[a] Collate and tabulate the common names of plant diseases now in use.

[b] Construct a working scheme in which every plant disease is assigned a place with a distinctive (scientific) name followed by an English name, the last to be, when possible, the one already in use.

[c] It is recommended that the parasite should be distinguished from the disease in all cases.

[d] It goes without saying that mycologists are urged to apply names to plant diseases instead of leaving the matter of a choice to a popular verdict.

In addition to organizing plant pathology information, the Madison congress also saw the birth of the first formal society of botanical scientists in the United States. For some years botanists in America had desired their own society, but opinions had differed on the type of organization that would be most desirable.[15] The previous year, at a meeting of the Botanical Club of the AAAS, Liberty Hyde Bailey of Cornell University had proposed the organization of a "permanent American Botanical Society."[16] There was some discussion "showing a general belief that such a society was desirable, but with some doubts as to the advisability of establishing it at the present time."[17] Despite differences of opinion over who should be allowed as members and whether the new society would just duplicate the old Botanical Club of the AAAS, the Botanical Club took formal action to create the Botanical Society of America. As Bailey wrote Farlow on October 12, "I think the prospect good for a *pure science* society."[18]

Organization of the Botanical Society of America was completed at the Brooklyn meeting of the AAAS in August 1894, with the selection of William Trelease as president. The membership of the new society was restricted to older and better-known botanists who had published outstanding work and were actively engaged in botanical research. Candidates for membership had to be recommended by three active members of the society. At the end of five years the membership had increased to thirty-three.[19] Throughout the 1890s, formal, more or less competing programs were held at the AAAS meetings by Section G, the Botanical Club, and the Botanical Society of America.

In late December 1897, with about thirty botanists present, still another botanical society, the Society for Plant Morphology and Physiology, was established at a meeting of the American Society of Naturalists in Ithaca, New York.[20] The new society was formed with the goal of including "all who are actively engaged in botanical studies not purely floristic," a membership concept more democratic than that of the more restrictive Botanical Society. The primary criterion for membership was that "no one shall be admitted to the society who has not published valuable papers or given satisfactory evidence of ability to do original work."[21]

Thirty-one persons were initially elected to membership in the Society for Plant Morphology and Physiology, with the hope that the membership would soon rise to forty. The first president was W. G. Farlow, who had declined charter membership in the Botanical Society of America. A number of prominent botanists were affiliated with both the Botanical Society of America and this new society.[22] However, the new society was limited by geography. Meetings of the society were to be held, at least initially, in conjunction with the American Society of Naturalists but not farther west than Buffalo nor farther south than Washington, D.C.

The three somewhat similar botanical organizations—the Botanical Club of the AAAS, the Botanical Society of America, and the Society for Plant Morphology and Physiology—provided ample opportunity for the presentation of papers on a wide range of topics. But the numbers of individuals engaged in botanical research was growing, and this research was becoming more specialized. One of the areas of specialization that included many plant disease workers was mycology, and the sentiment was soon sufficiently strong among mycologists to explore the possibilities of forming a specific American organization dedicated to mycology.

When the AAAS held its first winter meeting in Washington, D.C., from December 27, 1902, to January 2, 1903, the mycologists present among the 975 meeting attendees held an "informal conference to consider the advisability of forming an organization."[23] The general sentiment of those in attendance was strongly in favor of such an idea, in part due to the situation of the "younger mycologists [being] barred by the rigid requirements from membership in the Botanical Society of America." Thus, Frederic E. Clements, Franklin S. Earle, and Cornelius L. Shear were appointed as an organizing committee to correspond with mycologists around the country and solicit their views. In April the committee sent a letter to twenty-five mycologists who might be interested in such a society. An enthusiastic, affirmative response came from twenty-four of the individuals contacted and an organizational meeting to establish the new American Mycological Society was held in conjunction with the AAAS's St. Louis meeting on December 29, 1903.

Although the currents of division were thriving among the botanical scientists, there were also currents of consolidation. When leading botanists/mycologists C. L. Shear, G. F. Atkinson, and T. J. Burrill conferred with colleagues concerning possible affiliation of the new American Mycological Society with other societies, they found a growing sentiment among their peers in the Botanical Society of America and the Society for Plant Morphology and Physiology that the interests of the botanical sciences as a whole could be best advanced by a single organization. The challenge was to find a way to bring members of all three societies into one common organization. As H. M. Fitzpatrick noted, "The task of reorganization of the botanists and the preparation of a constitution acceptable to all members proved difficult and negotiations were carried on over a period of months before success was attained."[24] At the twelfth annual meeting of the

Botanical Society of America and the third annual meeting of the American Mycological Society, held in conjunction with the AAAS in New Orleans on January 1, 1906, a new constitution for a unified organization, as recommended by the committees of the Botanical Society of America, the Society of Plant Morphology and Physiology, and the American Mycological Society, was adopted.[25] This action resulted in the federation of the three organizations under the name of the Botanical Society of America. G. F. Atkinson, who had been a member in all three of the uniting organizations, was chosen as the first president of the new, 119-member society.

But the movement toward specialization was not yet at an end. This movement reflected changing needs in science that established generalist organizations could not meet. The new Botanical Society of America, for example, retained the distinction between full membership and associate membership and continued to meet in association with the AAAS. It made no effort to broaden its scope, and it continued to emphasize the more basic components of the botanical sciences. This resulted in "the applied branches of botany being essentially ignored."[26]

A PROFESSIONAL SOCIETY

When Donald Reddick, then in his last year of graduate work and an instructor in the recently established Department of Plant Pathology at Cornell University, traveled to Washington, D.C., in mid-December 1908 to examine some of the literature needed for the completion of his doctoral thesis, he probably did not anticipate that he would be invited to a momentous meeting and would become part of a committee to establish a new professional society for plant pathologists. Reddick's name was the final entry on an invitation calling for "all workers in Plant Pathology . . . to meet at 4 P.M. today in room 314 [the laboratory of William M. Scott] to take preliminary steps for the organization of a general society of Plant Pathologists."[27] Printed on stationery from the Bureau of Plant Industry and dated December 15, 1908, the invitation to this meeting in Washington was the brainchild of Cornelius Shear, who had "the necessary vision and initiative" to serve as a central organizing figure for the creation of the American Phytopathological Society.[28]

Cornelius Lott Shear was born in Coeyman's Hollow, near Albany, New York, on March 26, 1865, the son of Henry and Mary (Speenburg) Shear. He grew up in the village and on an adjacent farm, attending country schools as circumstances permitted. He had the good fortune to graduate from Albany State Normal School in 1888, a student of E. A Burt, one of America's noted mycologists, and an acquaintance of Charles Peck, the pioneer New York State botanist and mycologist. His association with Peck would continue through frequent correspondence over the course of many years and he would receive constant assistance and encouragement from him.

After several years as a schoolteacher, C. L. Shear entered the University of Nebraska in 1894, coming under the tutelage of C. E. Bessey. At Nebraska he met other students, such as F. E. Clements, A. F. Woods, and E. Bessey, who would also become noted pathologists and botanists. In 1897 he graduated from Nebraska with a B.S. degree. During the summer months of his undergraduate years he had served as a special agent of the Division of Agrostology of the USDA; from 1898 to 1901 he served as assistant agrostologist in the same division. He received his M.A. degree from his alma mater in 1901, and, still having an interest in fungi, transferred to a position as mycologist and plant pathologist in the new Bureau of Plant Industry. Although his first research was on Texas root rot of cotton, he soon transferred to the small-fruit-disease project and quickly became its head. His studies on cranberry diseases were the subject of his Ph.D. dissertation in 1906 at George Washington University. In 1923 he was named head of the USDA Division of Mycology and Disease Survey.

An avid mycologist, Shear did pioneering work on the life-histories of many economically important fungi, particularly with *Glomerella* and its *Colletotrichum* and *Gleosporium* stages and with the fungus then known as *Endothia parasitica*, which caused chestnut blight, and its American relatives. In collaboration with B. O. Dodge, he published the results of basic studies that launched the fungus *Neurospora* into genetic fame. He was known internationally for his efforts in the areas of nomenclature of organisms, particularly fungi.

Shear excelled as an organizer and leader, and he served well beyond the normal call of duty. He served as the vice president of the Botanical Society of America in 1908, as charter member and secretary-treasurer of the American Mycological Society until it combined with the Botanical Society of America in 1906, and, later, as a charter member and second president of the Mycological Society of America. He would serve the American Phytopathological Society as secretary-treasurer for the first nine years of it existence, as a member of the board of editors of the journal that would be launched, and, finally, as president in 1919.[29] Through all his tireless service, Shear maintained a clear vision of what the new phytopathological society could be.

In the summer of 1908, when Shear first suggested the desirability of an organization of plant pathologists to several of his colleagues in the Bureau of Plant Industry (BPI), he was a senior plant pathologist with the USDA.[30] In November of that year, he corresponded with several experiment station pathologists to seek their opinion concerning "the question of the [advisability] of organizing an association of plant pathologists."[31] As he later recalled, "I first suggested the desirability of organizing the Phytopaths with Orton, Waite, Scott, Smith & Metcalf. Meeting with a favorable reaction in most cases we arranged for an informal meeting of the Bureau pathologists. It was held in Scott's Lab as it was the largest and most convenient."[32]

On December 15, 1908, twenty-three members of the BPI's staff of pathologists (seven were on leave) met to discuss forming a new society for plant pathologists. The only contentious point of this discussion, as raised by E. F. Smith, was the scope and membership of the proposed society. Questioning the appropriateness of the restricted nature of membership in the original Botanical Society of America, those present decided that the new phytopathological society "should be very broad in its scope and should admit to membership all persons engaged in phytopathological work." But the basic plan for an organization of plant pathologists met with unanimous approval.

The scientists who met in Washington on that day differed greatly from those who had spearheaded the Botanical Society of America. The Botanical Society of America's founders were primarily botanists, cryptogamic or otherwise. The scientists at the Washington gathering had varying backgrounds, but each was a member of the first generation that could truly be called "plant pathologists."

The actual formation of the new society was deferred until the next meeting of the AAAS, soon to be held in Baltimore. C. L. Shear, D. Reddick, and W. A. Orton were appointed "to communicate with the plant pathologists of the country and make preliminary arrangements for the organization of such a society." After consultation "with several experiment station pathologists," this organizational committee drafted and sent a general invitation letter to 130 plant pathologists throughout the country. The letter related the outcome of the Washington, D.C., meeting and announced a formal meeting of those interested following one of the AAAS sessions. Each addressee was "cordially invited to be present December 30, 1908, at an hour and place to be announced later and take part in the organization of the proposed society which, it is believed, can exercise great influence in advancing the study of phytopathology in America."[33]

Responses to the invitational letter were mostly, although not unanimously, positive. Florence Hedges wrote, "I should be very glad to join such a society and I should think that the best time and place of meeting would be that of the annual meeting of State Experiment Station workers."[34] H. H. Whetzel wrote: "The scheme meets with my hearty approval and I trust that we may at the coming meeting of the Association see the final completion of such an organization."[35] L. R. Jones wrote, from Burlington, Vermont, "I now hope to attend the meeting on Dec. 30 looking to the organization of a society of plant pathologists. It certainly meets my approval."[36] However, G. F. Atkinson of the Department of Botany at Cornell replied, "I have your circular letter of Dec. 16th in regard to organization of a Pathological Society. . . . I believe there are certain problems of plant pathology which perhaps might warrant the organization of such a Society distinct from the Botanical Society of America. I do not feel ready at present to promise to join such a Society, chiefly because of the time already consumed in connection with other Scientific Societies."[37]

The meeting of the AAAS held in Baltimore, Maryland, from December 28, 1908, to January 2, 1909, was momentous for plant pathologists. The overall attendance of botanists was "unusually large and representative, and so many papers were offered for presentation that it was found necessary to divide into subsections."[38] Frank Lincoln Stevens of the North Carolina College of Agriculture and Mechanical Arts was chosen to preside over the papers of Subsection B, which was formed exclusively for plant pathologists.

The announced meeting to discuss the formation of a phytopathological society was held in the late afternoon of December 30, 1908, and met at the Eastern High School, with an attendance of fifty-four persons. The organizing committee reported, "It is our opinion that an American Phytopathological Society, placed upon a broad and generous foundation, may be of invaluable aid in promoting the future development of this important and rapidly growing subject in America, and that its influence may be made of international importance." The committee's report apparently resulted in an "animated discussion regarding the desirability of starting a new society." [39]

Although no complete written record of this meeting was made, a hint of the nature of the discussion is gained from A. D. Selby's obituary, which noted that Selby "was one of the earliest and most persistent advocates of the organization of the American Phytopathological Society, and took a very active part in the stormy discussions that attended its birth."[40] Several present, including G. F. Atkinson and J. C. Arthur, thought that such a move would be detrimental to the advancement of botanical unity and organization.[41] The relatively recent union of the three previous botanical organizations into the Botanical Society of America was a factor in the minds of some, who saw the creation of a new society as a "retrograde movement." Some feared that those assembled in Baltimore were "organizing an entirely new society which would break up present organizations."[42] Proponents of the new society, however, maintained adamantly that the "aims and purposes of the proposed organization . . . could not be attained without a separate society."[43] After the full discussion, the committee's report was approved by a vote of 32 to 12. The assembled body went on to discuss issues of the time and place of an annual meeting and whether the society should undertake publication of a journal.

The meeting ended with the unanimous election of a council dominated, not surprisingly, by scientists from the eastern United States—L. R. Jones, Vermont Agricultural Experiment Station, as president; Augustus D. Selby, Ohio Agricultural Experiment Station, as vice president; C. L. Shear as secretary-treasurer; and as councilmen, J.B.S. Norton of the Maryland Experiment Station, and Benjamin Minge Duggar of the Cornell University Agricultural Experiment Station. Significantly, all of the officers were relatively young men, and whether they worked at the USDA or at state agricultural experiment stations, they were all genuine plant pathologists. The council was charged with deciding on the time and place of the first official meeting.

FIRST MEETING OF THE NEW SOCIETY

The new president of the American Phytopathological Society, L. R. Jones, left the 1908 meeting of the AAAS believing that the issue of whether to form a new, independent phytopathological society had been resolved. In fact, however, doubts remained. When Jones contacted Shear, Selby, Norton, and Duggar to propose a meeting of the council, he found that the possibility of remaining a subsection of AAAS was still a point open for discussion. Shear wrote that "it was the opinion of Dr. [H. C.] Cowles, the secretary of Section G, that an arrangement could be made which would secure for us practically all the advantages which we had hoped to secure by a separate organization."[44] Shear, Selby, and Norton agreed that "it may be possible to make some arrangement with Section G which would meet practically all our requirements, and at the same time avoid some of the opposition which has been made to our undertaking."

After Jones had left the Baltimore meeting, several individuals had restated their opposition, which was focused on the fact that the organization of an entirely new society would break up the newly formed, unified organizations. They thought that the plant pathologists might be better organized as a subsection of Section G, so long as they were permitted to arrange a program at the annual meeting and were free, if the need arose, to publish a journal. Jones was amenable to this solution, writing from Vermont that he was personally hopeful that the AAAS council might succeed in "some arrangement such as your outline for a subsection of section G which will accomplish the most that we wish with the minimum of trouble."[45]

The meeting of the council of the phytopathological society took place in Washington, D.C., on March 26 and 27, 1909. The members of the council considered carefully the question of the affiliation of the society and the possibilities of establishing a journal.[46] The sentiment at the meeting was to continue to explore the possibilities of remaining a subsection of Section G, but to proceed to hold the first regular meeting of plant pathologists from December 28 to 31 in conjunction with the AAAS convocation week in Boston. The council also recommended that plans be developed for the publication of a phytopathological journal, suggesting that such a publication might be initiated by the BPI.

In May, W. J. Beal of Michigan State Agricultural College and a member of the sectional committee for Section G wrote a letter to L. R. Jones about the desirability of the plant pathologists' keeping their association with the AAAS. "The A.A.A.S. now has all the sections it can take care of. Separate societies have already multiplied to an unprofitable degree," he wrote, and he went on to suggest that the plant pathologists "form a section or subsection of S.P.A.S. or of [the] Botanical Soc. of America."[47] Although the possibility of an association of the plant pathologists with the Society for the Promotion of Agricultural Science had been explored, the option had not proved tenable. Furthermore, Jones did not believe that the Botanical Society of America, as reorganized, could begin to break up once again into subsocieties. He was "decidedly of the opinion that it is better for the American Association to welcome subsectional meetings rather than to force us to organize further distinct societies in order to secure the opportunity we need for technical discussion."[48]

Ultimately, the impetus to continue with the formation of an independent American Phytopathological Society came from outside the ranks of plant pathologists. At the April 1909 meeting of the council of the AAAS, attended by C. L. Shear, a resolution was made directing sectional secretaries of the association to arrange a program for each section that would not occupy more than two sessions. Further, it was recommended that "the papers presented at the [section] sessions should be of a general nature, and that all other papers presented to the sections should be referred to the affiliated societies." Although this plan was to apply initially only to the upcoming Boston meeting, there was a strong expectation that the plan would become permanent. As Shear wrote to Jones, "If this should become a permanent policy of the Association, there would be little advantage, so far as I can see, to us or to other botanical organizations in our carrying out the proposed arrangement to become a sub-section of section G."[49]

The strong desire of most plant pathologists was to affiliate and cooperate as closely as possible with the Botanical Society of America and Section G of the AAAS.[50] But Jones and Shear, in consultation with the other council members, decided to organize the strongest possible program of papers by plant pathologists for the upcoming Boston AAAS meeting. Jones's goal was to ensure a "high grade program that will command respect and attention."[51] After consultation with the council and officers of other botanical organizations, the decision was affirmed that the plant pathologists should indeed form a separate society.

To organize the first meeting of the new society, Shear once again employed his outstanding skills as an expediter and implementer. In October, he sent an invitation and call for papers to 130 pathologists for the first APS meeting in Boston, to be held in conjunction with the AAAS meeting. At Boston, Shear promised, all remaining issues concerning the independence or affiliation of the APS would be resolved. By early December, a sufficient number of titles had been received to plan for at least two half-day sessions for the new society. Jones judged this magnitude of a program to be "sufficient to establish the character

& dignity of our organization & to permit of discussion & acquaintance."[52] In mid-December he wrote Shear, "It is a splendid showing & a complete demonstration of the need of independent sessions for phytopathology."[53]

The first annual meeting of the American Phytopathological Society, as judged by any standard, was a great success. The sessions, held at the Harvard Medical School, were attended by fifty members.[54] They included a joint session for paper presentations with Section G of the AAAS on Thursday afternoon, December 30, 1909, and two separate sessions on Friday, December 31. The program comprised forty-five papers authored by thirty-eight men and three women from fifteen states (California, Delaware, Louisiana, Maine, Michigan, North Carolina, North Dakota, New York, Ohio, Oregon, Pennsylvania, Tennessee, Virginia, Vermont, and Wisconsin), Washington, D.C., and Ottawa, Canada. Major support for the meeting came from the scientists of the USDA's Bureau of Plant Industry and Laboratory of Plant Pathology, who contributed nineteen papers. Topics included diseases of fruit, fiber, forage, field crops, and forest trees caused by fungi and bacteria, malnutrition of vegetables, and efficacy of fungicides. Results were presented of recent research on diseases, including potato canker, late blight and blackleg, apple scab, chestnut bark disease, white pine blister rust, sugar beet curly top, peach yellows, and banana blight in the United States, Canada, Cuba, and Central America.[55] Even those who had harbored doubts about the new organization now agreed "that the large membership list, the numerous papers presented, and the great enthusiasm showed by the men had thoroughly convinced them that the organization of the Society was fully justified."[56]

The business meeting of the new society was also productive and successful.[57] A total of 130 individuals were accepted as charter members. Newly elected officers, demonstrating a wider geographical range than the initial council, were:

President: F. L. Stevens, North Carolina College of Agriculture and
Mechanic Arts
Vice president: A. F. Woods, University of Minnesota
Secretary-treasurer: C. L. Shear, USDA
Councilors: L. R. Jones, University of Wisconsin; A. D. Selby, Ohio Agricultural Experiment Station; and H. H. Whetzel, Cornell University.

F. L. Stevens, H. Von Schrenk, E. M. Freeman, W. A. Orton, and G. P. Clinton, continuing the attempts initiated at the Madison Botanical Congress in 1893, were appointed as a committee of five "to draw up rules and make recommendations concerning the common names of plant diseases." Because of the recent introduction of two serious plant diseases, wart disease of potato (*Synchytrium endobioticum*) and white pine blister rust (*Cronartium ribicola*), members of the society also unanimously adopted a motion directing the president to appoint a committee of five members to "draft appropriate resolutions regarding these diseases and take steps to secure such actions as would prevent their further introduction and spread." The members appointed were H. Metcalf (USDA); H. T. Güssow (Ottawa, Canada); H. L. Bolley (North Dakota); A. D. Selby (Ohio); and W. A. Orton (USDA).

A NEW JOURNAL

A subject of considerable interest to those attending the first meeting of the APS was the possibility of publishing a specialized phytopathological journal. The society empowered the council of officers to initiate the publication of such a journal if the "necessary financial and editorial arrangements could be made."[58] The subject of a journal had been a recurring theme dating back to the very first discussions concerning the benefits of forming a separate association of plant pathologists. In November 1908, Shear had asked A. D. Selby to "give me your idea . . . whether the annual report would be sufficient or whether a periodical journal would be more desirable."[59] Selby responded "that the effectiveness of such an Association will be more or less influenced by a possible avenue of publication."[60]

One such avenue of publication, discussed at the meeting of the first council in March 1909, was the underwriting or publishing of a journal devoted to phytopathological research by the USDA through the Bureau of Plant Industry.[61] This proposal, however, never came to fruition, perhaps because it did not receive the necessary early administrative support, and perhaps because in 1913 the USDA began its own *Journal of Agricultural Research*. This journal would serve initially as an outlet for the publication of research in plant pathology from government researchers, but later would also publish papers from experiment station researchers.[62]

The formation of the APS was accompanied by a continuous, cautious movement toward the creation of a new, society-sponsored journal. After the initial council meeting, Jones and Shear continued a regular correspondence about affairs of the new society, including the journal. Jones wrote in May 1909, "I think all are agreed that there is need of a distinctively phytopathological journal in America. If, however, such is to be published by the newly organized Phytopathological Society, I feel that we must wait at least a year, and perhaps two, until the organization of this Society is completed and its relation to the American Association definitely settled before we may venture upon it."[63]

Shear saw Jones as the prime candidate to become the first editor in chief of the new journal. Jones was a prominent plant pathologist recognized widely as a capable leader, and he had a reputation for excellence in research. He was also a "station man," on excellent terms with both workers at the state experiment stations and the "government men" working in Washington, D.C., or at federal laboratories around the country. At the first formal meeting of the APS in December 1909, Jones was appointed to the editorship. Now plans for the new journal could proceed. Correspondence on plans for the new journal continued among the leaders of the society, and Donald Reddick was recruited to be the business manager.

Although all were united in desiring the publication of such a journal, the purpose and audience remained points of debate. Whetzel favored a journal "which should include illustrated papers and bring together by the advantages offered all American papers on the subject." He envisioned a showcase of American excellence in research in phytopathology that would be aimed primarily at an audience of accomplished scientists. Selby, and, apparently, Jones, favored a "journal which shall appeal to a large number of people who are not investigators in pathology, but who are interested in the results of investigations when presented clearly and concisely from the standpoint of economic results."[64] This latter philosophy reflected the attitude of the successful experiment station scientists, who directed their research publications, by necessity, to a more general audience. The issues raised were not resolved easily, for they reflected the dichotomy between science and service that had characterized the profession since its beginnings.

Jones pressed forward with nominations for associate editors and the remaining members of the editorial board. He nominated C. L. Shear and H. H.

Whetzel as the two associate editors, based on "the professional position and geographic location of these men as well as their scientific attainments."[65] But after only three full months at Wisconsin, Jones was feeling the "pressures of other work" and was having second thoughts about agreeing to act as editor in chief. He wrote confidentially to Shear, "If you and I & Whetzel are to be the chief victims then I must beg we act from now on as an editorial committee so that you two men share equally the responsibilities with me."[66] Shear and Whetzel approved a proposal to serve along with Jones as "editors." Sharing the load, the three coeditors continued the organizational efforts to launch the journal.

By September the editors were ready to request that President Selby and the council of APS appoint "the following persons as associate editors of the proposed journal 'Phytopathology': G. P. Clinton, E. M. Freeman, H. T. Güssow, F. D. Heald, Haven Metcalf, W. A. Orton, W. M. Scott, A. D. Selby, E. F. Smith, R. E. Smith, F. L. Stevens, and R. Thaxter."[67] The new slate of editors was approved by mail, with President Selby commenting that he was "much pleased that you have represented the Pacific Co[a]st in the appointment of Ralph E. Smith, and Canada in the appointment of H. T. Güssow."[68] Secretary-treasurer Shear sent notices to the twelve men selected as associate editors of "the proposed official journal of the Society, which is to be known as *Phytopathology*."[69]

It must have given Jones, Shear, and Whetzel a great deal of pleasure in late October to be able to send the "Plan of Organization Adopted by the Council" and "Plans for Publication Submitted for Consideration of the Associate Editors" to the twelve men who had accepted the new positions on the editorial board.[70] The plans called for volume one to be completed by December 31, 1911. A general announcement concerning plans for the new journal was sent to all members, along with the meeting notice and calls for papers for the second annual meeting of APS in Minneapolis, Minnesota, December 28–30, 1910.

Dated February 1911, the thirty-eight-page inaugural issue of *Phytopathology* included six research papers, C. L. Shear's review of F. L. Stevens's and J. G. Hall's 1910 book *Diseases of Economic Plants*, and an introductory portrait and brief biographical sketch of Anton De Bary prepared by Erwin F. Smith. The new journal was characterized by a broad disciplinary base and an open and international character. Within the first year, two papers came from authors outside the United States—one by James B. Rorer of the Board of Agriculture, Trinidad, British West Indies, "A Bacterial Disease of Bananas and Plantains," and one from associate editor H. T. Güssow of Ottawa, Canada, "Preliminary Note on 'Silver Leaf' Disease of Fruit Trees." From the beginning, the authors of research papers included women: Florence Hedges of the USDA's Laboratory of Plant Pathology contributed "*Sphaeropsis tumifaciens*, Nov. Sp., the Cause of the Lime and Orange Knot" in the April 1911 issue; Caroline A. Black coauthored a paper with Charles Brooks of the New Hampshire Agricultural Experiment Station on "Apple Fruit Spot and Quince" in the April 1912 issue; and Margaret DeMerit, also of the New Hampshire Station, coauthored a paper with Charles Brooks on "Apple Leaf Spot" for the October 1912 issue.

The successful launching of *Phytopathology* accomplished one of the major goals of those who had agitated for the formation of the American Phytopathological Society. The key to the journal's continuing success, however, would be a subscription base "especially from institutional and public libraries and state horticultural and other societies."[71] Targeting this audience meant that the journal must maintain its accessibility and its contact with the practical and useful side of plant pathology. The editorial board continued defining and defending their niche, especially after *Phytopathology* was challenged with competition from the USDA's *Journal of Agricultural Research* (1913) and the *American Journal of Botany* (1914).

In his retrospective address "Our Journal, *Phytopathology*," delivered in December 1918, L. R. Jones emphasized, "*Phytopathology* began as the Society's journal . . . the expression and embodiment of the three prime characteristics of our Society; youthful vigor, zeal for service, and American democracy."[72] *Phytopathology*, Jones acknowledged, had not reached the level of prestige accorded older, more widely recognized journals such as *Science* or *The American Naturalist*. As Jones stated without apology, "No one who has caught the spirit of American botanical literature would think of turning to *Phytopathology* for such articles as Smith's final publications on crown gall, Shear's on Glomerella, or Freeman and Stakman's on disease resistance in wheat."[73]

While openly acknowledging the regret of some members that the journal "was not more largely or exclusively the channel of publication of our maturer members," Jones expressed his belief that "if the journal of the Society is to be primarily of service to the Society it must help chiefly our younger members" with its columns "in the future as in the past to be freely open to their 'maiden manuscripts.'"[74] Although he emphasized that editorial standards could not be compromised, he maintained that the "encouragement of our young men must continue as one of our peculiar opportunities and large responsibilities."[75] As he pointed out, "Few, if any, lengthy American manuscripts representing properly matured research in phytopathology had . . . lacked adequate opportunity for publication even before the appearance of our journal."[76]

Jones's audience had good reason to extol the democratic system on which the American Phytopathological Society had been founded, characterized by national geographic representation and cooperation between pathologists at the various levels of government. As Whetzel wrote that year, "This society and the journal it publishes has done more to stimulate and unify the phytopathologic work and workers of this country than any other one thing."[77] The establishment of *Phytopathology* was the final necessary piece of a mature and fully functioning scientific society. It was a fitting final act to the series of events that created the science of plant pathology in the United States. Plant pathology now had the organizational and publishing mechanisms to unite the scientists in the discipline as well as to establish and maintain the professional standards by which future work in the science would be judged.

An emerging discipline

The American Phytopathological Society emerged from a period of frequently chaotic development in sciences and scientific organization during the late nineteenth and early twentieth centuries. Science societies were growing in scope, seeking to represent their members on a national level. In addition, the sciences were splintering into new disciplines and subdisciplines, each desiring representation for its unique aspects. The creation of the APS was the work of the first generation of true plant pathologists, scientists who had been educated in a wide variety of scientific subdisciplines such as mycology, bacteriology, chemistry, even rudimentary genetics, but with a specific goal of understanding and controlling plant disease.

The previous generation of plant disease researchers, including Farlow, Bessey, and Burrill, who were botanists by education, had generally been satisfied with plant pathology's place inside the botanical societies. This place, however, depended on a conservative argument that was doomed to failure in a fundamentally progressive era. Like other sciences with a strong practical, economic basis, plant pathology began to feel the need and the right for self-definition. By the early twentieth century, phytopathology had a roster of prominent scientists with a sizable corpus of knowledge sufficient to allow them to leave the confines of traditional botany with no fear of isolation. Acting on this feeling of independence, they created their own society and journal, establishing beyond a doubt the weight and importance of the discipline as a separate branch of botanical science significant not only economically, but also scientifically.

The second decade of the twentieth century was marked by the deaths of plant pathology's founding fathers: Charles Bessey in 1915; Thomas J. Burrill in 1916; William Farlow in 1919. Together, these three symbolized the most important elements of the creation of the science of plant pathology. Bessey represented the stage when the practically minded agricultural improvement movement of the early nineteenth century united with the developing biological sciences of mid-century, including the new botany. Burrill, through his discovery that bacteria caused plant diseases, demonstrated that the science could explore completely untrammeled areas of research. And Farlow brought rigorous scientific method to a science that was often, in its formative years, a disjointed combination of botany, mycology, and old-fashioned naturalism. All three of these men taught and wrote and influenced future generations to take the science to new levels. By the time of their deaths, they had passed on the mantle of leadership in plant pathology to their students and the students of their students. But they lived to see plant pathology evolve from its days as a little-known scientific sideshow combining botanical studies and practical agriculture into a mature science.

Several elements describe a mature science—a unique body of knowledge; an interconnected corps of professional scientists who share similar research goals and methods; institutions to support and organize research; and an educational system that guides, trains, and certifies practitioners. In addition, plant pathology's tradition of public service created observable goals that acted as practical measures of the science's success and maturity.

The passing of plant pathology's three great founders—Bessey, Burrill, Farlow—happened to coincide with the period of World War I, a watershed period for U.S. history, sparking a reexamination and redefinition of the American national identity. Most of American society, including sciences like plant pathology, was affected by this time of change. "Every American science is mobilizing for the grand assault on the enemy," wrote H. H. Whetzel in 1918. "One of the youngest of American Sciences, Plant Pathology, is preparing to do its full share. . . . No agricultural scientist is now more important . . . than the plant pathologist, be he in research, teaching or extension work."[1] Whetzel's claim for the importance of plant pathology in the national war effort lay in the role it might play in improving food production and distribution, both within the United States and in Europe. National attention was focused on the quantities of food consumed by the Allied armies, on the quantities lost as cargoes sunk at sea, and on the poor wheat crop of 1916. Under the emergency conditions of war, plant pathologists came together to examine their science, to study their institutions, and to find ways to improve both. To this end they created the War Emergency Board, the Plant Disease Survey, and the barberry eradication program.

The national scope of plant pathology was clearly evident from the fact that the War Emergency Board and the Plant Disease Survey were organized and administered on a national level. Furthermore, although the program for barberry eradication was generally confined to the great wheat states of the Northern Plains, it was created partially through the influence and funding of the USDA as part of that organization's nationwide outreach. All three bodies demonstrated the spirit of national and local communication and cooperation that existed in plant pathology by the second decade of the twentieth century. The scientists who administered these bodies represented institutions like the USDA and universities that were capable of supporting research—institutions that were increasingly interlocked in pursuit of shared goals.

The War Emergency Board was an adjunct of the American Phytopathological Society. It was born at the APS annual meeting in January 1918, at the height of American involvement in World War I. The board's purpose was to encourage cooperation with and among existing organizations at all levels of government and in private industry, and among all the workers of the phytopathological profession. It was "charged with the responsibility of stimulating and accelerating phytopathological work to the end that, in this present world crisis, the reduction of crop losses from diseases should be made most effective as a factor in the increase of our food supply."[2] The board had eight commissioners: one from each of the six APS districts, one from the federal government, and one from Canada. The board had no real authority. Its mission was to "make suggestions which may increase the efficiency of plant disease control."[3] But the members of the board had status in their field, and their suggestions held great weight among their colleagues, leading to changes in plant pathology.

The activities of the board provide an excellent snapshot of the structure of phytopathological education and research at this time, including the thoughts of the leadership concerning the future of the science. It was first necessary to determine the size of the pool of professionals in plant pathology and to estimate the need for future growth. Commissioner F. D. Kern, head of the Department of Botany at the Pennsylvania State College, ran a manpower census on the availability and expertise of plant pathologists and botanists who could work on plant diseases. Then, H. H. Whetzel of Cornell, chairman of the War Emergency Board, was charged with the task of coordinating innovations and opportunities in education to fill the "inadequate and depleted ranks" of plant pathologists.[4] The result was that the board recommended recruiting and training an increasing number of students, which, along with developments in the science, led to the rapid expansion of university graduate programs in plant pathology.

Commissioner G. H. Coons of the Department of Botany and Plant Pathology at Michigan Agricultural College was charged with gathering data on the tools of applied plant pathology, the "supply of dependable fungicides, disinfectants and machinery for their application."[5] Shortages of necessary fungicides and sprayers were reported and expected across the country due to uncertainties of price and demand as well as war emergency railroad embargoes. Coons, with Whetzel's help, secured additional allocations of supplies associated with the manufacture of fungicides from the War Industries Board for the duration of the war. More important, he held meetings with sprayer manufacturers, which led to a more rapid standardization of spray machinery than would ever have been possible under normal conditions.[6]

E. C. Stakman, head of the Section of Plant Pathology at the University of Minnesota, assembled important information on the availability and qualifications of workers who might redirect their work to include "problems in seed disinfection; in promising new field fungicides; in disease resistant strains and varieties; and in many other directions [that] call imperatively for solution."[7] Such work in promoting cooperative research was thought by the board to have the most "far-reaching possibilities affecting American plant pathology."[8] Similar efforts in the areas of communication and cooperation helped complete the transformation of American pathologists from a group of relative isolationists into a community of interacting scientists who shared a common goal and vision: to increase the production of food crops through the practical application of knowledge on the control of plant diseases. "The free and open exchange of views and data" allowed pathologists to push "forward the solution of many problems far beyond what could have been expected under the former conditions of suspicion and personal isolation in which [they] worked."[9] The result was an emphasis on studies to solve the most pressing and important plant disease problems and on the rapid dissemination of practical information.

Choosing wisely among disease-control projects depended on the availability of reliable information on disease prevalence, severity, and geographical distribution. These essential data were provided by Commissioner G. R. Lyman of the USDA, who was attached to the War Emergency Board by virtue of his role as head of the U.S. Plant Disease Survey. Created by the federal government prior to World War I, the Plant Disease Survey became an official part of the Bureau of Plant Industry in July 1917. It was perhaps the largest public service project undertaken during the early twentieth century by the BPI. The purpose of the survey was to "gather information on plant diseases, not to investigate the diseases themselves nor to demonstrate their control."[10] Thus, the survey was intended not as competition with plant disease research, but rather as a central repository for information. The intent was that the researcher could use the information on the geographic location and severity of diseases to help prioritize research projects by identifying the most pressing problems for study. To have such information at hand "would be important in times of peace, but is imperative now in times of war, when, with our numbers reduced by military service, we must make sure that we are wasting no time on problems of lesser importance."[11]

Associated with the growing self-awareness in plant pathology, aided by the War Emergency Board and the Plant Disease Survey, was the creation of a plan for the eradication of the barberry plant. The barberry eradication program was a direct result of the staggering losses, estimated at up to 300 million bushels in North America, of wheat due to the epidemic of stem rust in 1916. Spurred by the vital need for a grain surplus to supply the Allies in their prosecution of the war in Europe, barberry eradication was a massive applied program that sought to exploit the increasing ability of plant pathologists to perceive and prioritize problems, seek a scientific understanding of the problem, formulate solutions, and enact a far-reaching control program.

The United States had suffered a serious epidemic of wheat rust in 1904 and, although professional pathologists knew that the common barberry was the source of the inoculum for the stem rust phase on wheat, the institutional infrastructure was not sufficiently developed at that time to take positive action to eliminate the common barberry and reduce the threat of future epidemics. However, in 1903, Denmark embarked on a barberry eradication program that virtually eliminated wheat stem rust in that nation over the next decade.

Prior to World War I, many cereal pathologists in the United States had pushed for a national barberry eradication program. Several influential individuals, including H. L. Bolley of North Dakota, E. C. Stakman of Minnesota, and M. A. Carleton of the USDA, realized the magnitude of the barberry problem and had a vision of how to solve it. Bolley had previously been successful in implementing such a program at the state level in North Dakota. Stakman decided to use his War Emergency Board pulpit to push for barberry eradication as a central goal of his Great Plains district. Officials of the USDA brought the influence and funding of the federal government to bear, pressuring states to pass barberry eradication laws, even though a federal law was never passed.

Because of its appeal to patriotism, the barberry eradication program was a unique effort in the history of plant pathology. For example, a press release from the South Dakota State College of Agriculture proclaimed the barberry bush "pro-German" and told citizens that "it is decidedly disloyal to allow the common barberry bush to live—it must be treated as a dangerous enemy alien."[12] A widespread educational program presented this dire situation to farmers and town dwellers alike, harnessing the manpower of agricultural and civic organizations. In the Upper Midwest, more than 1.6 million barberry bushes were dug out and destroyed in 1918 alone.

The barberry eradication program outlived the frenzy of the war years. In 1920 it provided the reason for a farm-to-farm survey in each of the thirteen states participating in the program. This was the most extensive survey of American farmland undertaken to that time. By 1922, with 5.1 million barberry bushes located and destroyed, the Second Annual Conference for the Prevention of Grain Rust in Minneapolis found itself hosting several state governors, commissioners of state departments of agriculture, presidents of Farm Bureau Federations, representatives of the state agricultural colleges and agricultural experiment stations, and federal plant pathologists. The conference demonstrated clearly the degree of cooperation that had developed within and among states and the federal government on pathological matters.

The coordinated efforts of the War Emergency Board, the Plant Disease Survey, and the barberry eradication program did much to solidify plant pathology as a field. Even before World War I had ended, W. W. Robbins, a botanist at the Colorado Agricultural College and Experiment Station, wrote to Whetzel that "in years to come when we look back upon the development of the science, the board will have much to its credit in placing plant pathology on a firm basis, in establishing a program and viewpoint of work, and in creating an esprit de corps among pathologists, which does not have its equal in any other field."[13]

In addition to bolstering camaraderie, the activities of plant pathologists during World War I had an effect on the perception of the importance of the field. The recognition by the public and policy makers that plant diseases caused significant "leaks in crop production" was heightened by the vital need for food in a world at war. During a time of national crisis, it was vitally important that "plant doctors" had within their arsenal effective weapons against many important plant diseases. Both federal and state governments increased funding for programs such as the Plant Disease Survey and the barberry eradication campaign. Barriers to communication and cooperation among plant pathologists within the regions of the United States were brought down, and farmers benefited from the rapid and open dissemination of the results of studies on controlling plant disease. American plant pathologists also began to reach out to their international colleagues. In time, American plant pathology emerged, as did the nation, from the relative isolation of the previous decades.

The successes achieved by the War Emergency Board resulted from a combination of interlinking factors. The commissioners were enthusiastic, dedicated individuals who knew their science and were excited about the contributions plant pathologists could make to the welfare of the nation and the world. Major epidemics of devastating plant diseases such as stem rust of wheat, late blight of potato, and chestnut blight attracted public attention because they had a great impact on the production of food and the appearance of the landscape. Plant pathology was a mature science with demonstrated successes in teaching, research, and extension. It had at its disposal the technology necessary to solve practical problems based upon an ever-growing understanding of biological processes. Finally, the World War supplied an identified, critical need to reduce waste in the production of food, fiber, and forest materials. The time was right for plant pathologists to demonstrate their worth and the real value of their science. Patriotism, national need, accumulated knowledge equal to the task, a spirit of cooperation, and the enthusiastic leadership of dedicated scientific visionaries converged.[14] Near the end of the second decade of the twentieth century, plant pathology graduated from a science in its formative years to one ready to assume its rightful place as one of the mature, natural sciences in the United States.

Name used in text	Current name
FUNGI	
Aschersonia aleyrodis	entomogenous fungus
Botrytis acinorum	No current name known
Botrytis infestans	*Phytophthora infestans*
Botrytis viticola	*Plasmopara viticola*
Botrytis vitis viticola	*Plasmopara viticola*
Ceratostomella fimbriata	*Ceratocystic fimbriata*
Colletotrichum phomoides	*Colletotrichum gloeosporiodes*
Cylindrosporium padi	*Blumeriella jaapii*
Diaporthe parasitica	*Cryphonectria parasitica*
Dibotryon morbosum	*Apiosporina morbosa*
Endothia parasitica	*Cryphonectria parasitica*
Entomosporium maculatum	*Entomosporium mespili*
Fusarium arcuatum	*Fusarium avenaceum*
Fusarium conglutinans	*Fusarium oxysporum* f. sp. *conglutinans*
Guignardia vaccinii	*Phyllosticta vaccinii*
Macrosporium solani	*Alternaria solani*
Noemaspora ampelicida	*Guignardia bidwellii*
Oidium tuckeri	*Uncinula necator*
Ozonium auricomun	*Phymatotrichum omnivorum*
Peronospora infestans	*Phytophthora infestans*
Peronospora viticola	*Plasmopara viticola*
Phoma uvicola	*Guignardia bidwellii*
Phyllosticta labruscae	*Guignardia bidwellii*
Phyllosticta viticola	*Guignardia bidwellii*
Plowrightia morbosum	*Apiosporina morbosa*
Puccinia dispersa	*Puccinia triticina*

Puccinia pruni-spinosae — *Tranzschelia pruni-spinosae*

Ramularia areola — *Ramularia gossypii*

Rhizopus nigricans — *Rhizopus stolonifer*

Sclerotium clavus — *Claviceps purpurea*

Sphaerotheca mali — *Sphaerotheca fuliginea* or *macularis*

Sphoerella gossypina — *Mycosphaerella gossypina*

Tilletia foetida — *Tilletia laevis*

Urocystis cepulae — *Urocystis magica*

Ustilago kolleri — *Ustilago segetum*

Ustilago nuda — *Ustilago tritici*

BACTERIA

Bacillus carotovorus — *Erwinia carotovora*

Bacillus oleae-tuberculosis — *Pseudomonas savastanoi*

Bacillus solanacearum — *Ralstonia solanacearum*

Bacillus tracheiphilus — *Erwinia tracheiphila*

Bacterium tumefaciens — *Agrobacterium tumefaciens*

Bacterium teutlium — Possibly *Erwinia carotovora* subsp. *betavasculorum*

Micrococcus amylovorus — *Erwinia amylovora*

Pseudomonas campestris — *Xanthomonas campestris* pv. *campestris*

Pseudomonas solanacearum — *Ralstonia solanacearum*

Streptococcus diphtheriticus — *Corynebacterium diptheriae*

vibrion butyrique (of Pasteur) — *Clostridium butyricum*

Xanthomonas malvacearum — *Xanthomonas campestris* pv. *malvacearum*

NEMATODES

Dorylaimus bulbiferus — *Discolaimoides bulbiferus*

Heterodera radicicola — Now over 100 species of *Meloidogyne*

Source: Booth 1971; Farr et al. 1989; Holt et al. 1994; Stevens 1919; Thomson et al. 1981; Young et al. 1996. We appreciate the assistance of Kenneth R. Barker, David F. Ritchie, and Amy Y. Rossman in locating information for this list of current names.

❧ NOTES ❧

Part I

1. Stevens 1933, 435; Stevenson 1959, 16.
2. Rossiter 1976, 279–83.
3. Dupree 1957, 40–41; Bates 1965, 27–33.
4. Peterson, Campbell, and Griffith 1992.
5. Gates 1960; Hurt 1994, 117–18.
6. Rossiter 1976, 279.

Chapter 1

1. Carrier 1923, 239; Crosby 1986, 164.
2. Josselyn 1674, 190.
3. Hedrick 1950, 39–40; Fisher and Upshall 1976, 101–2.
4. Antill 1771, 126–27; Eliot 1934, 216.
5. Antill 1771, 163.
6. Josselyn 1674, 188–89.
7. Mayo 1936, 197–98.
8. Stevenson 1959, 15.
9. Russell 1976, 133.
10. General Assembly, Connecticut 1873; General Court of the Commonwealth 1878; General Assembly, Rhode Island and Providence 1861.
11. Daniels 1971, 69–70.
12. Bates 1965, 4–11; Stearns 1970, 670–74; McClellan 1985, 22.
13. Dupree 1957, 6–9; Bates 1965, 26–27.
14. Bates 1965, 6; Daniels 1971, 101–7.
15. American Philosophical Society 1843, 15.
16. Rossiter 1976, 284.
17. Bates 1965, 21–24.
18. Daniels 1971, 153–54.
19. True 1929, 7–17.
20. Rossiter 1976, 285.
21. McClellan 1985, 38–40.
22. Rossiter 1976, 284–88; Thornton 1989, 105.
23. Philadelphia Society for the Promotion of Agriculture 1854, 2.
24. Rossiter 1976, 285; Thornton 1989, 107–8.
25. Colman 1963, 19–20.
26. True 1929, 2.
27. Colman 1963, 20.
28. Lee 1904, 314; Hall 1934, 7–8; "Sketch of Samuel Latham Mitchill" 1890–1891, 691–92.
29. Guralnick 1975, 12–13.
30. Humphrey 1976, 294.
31. Lee 1904, 328–29; Hall 1934, 3.
32. Mitchill 1792, 27.
33. Ibid. 1794, 11–12.
34. True 1929, 9–10; Hall 1934, 94–101.
35. Rossiter 1976, 286–87; Marti 1979, 11–12.
36. Marti 1980, 30.
37. Fletcher 1931, 14–20; Thornton 1989, 147–72.

38. Van Der Zwet and Keil 1979; Campbell 1979.

39. Fisher and Upshall 1976, 306.

40. Campbell 1979, 64–65.

41. Denning 1794.

42. True 1937, 4–5; Hedrick 1950, 210–11.

43. Agricola [a North Carolina Farmer] 1819, 153.

44. Coxe 1817, 1.

45. Ibid., 174–75.

46. Ibid., 174–76.

47. Beecher 1844, 56.

48. For lightning, see "Blight in pear trees, etc." 1829. For insects, see Lowell 1826; Buel 1827; Belden 1837; Kenrick 1841, 173–74. For overheating, see Castor 1839, 8; [C.W.E.] 1844. For cultivation, see Kenrick 1841, 174; Beecher 1844, 444.

49. Lowell 1826, 17.

50. Lazell 1840, 415; Beecher 1844, 447–55; Downing 1872, 646–50.

51. Beecher 1844, 452.

52. McMahon 1806, 40; Thacher 1822, 181–82.

53. Fletcher 1932, 45.

54. Forsyth 1802, 229.

55. Buel 1832.

56. "Cure for blight of pear trees" 1836–1837.

57. Lodeman 1916, 17–18.

58. C. D. 1844.

59. Waters 1839.

60. Lazell 1840, 415; Beecher 1844, 455.

61. Lowell 1826; Kenrick 1841, 173–74; Belden 1837; Elliot 1844, 255.

62. Lazell 1840, 415.

63. Smith 1888, 11.

64. Fletcher 1931, 6–14.

65. Smith 1888, 18.

66. American Philosophical Society 1799.

67. Ellis 1802, 326.

68. Coulter 1802, 327.

69. Peters 1808a, 21.

70. Ibid.

71. Ibid., 22–24.

72. Ibid., 24.

73. Ibid. 1808b, 184–89.

74. Coxe 1817, 215–17.

75. Darling 1845, 60; Downing 1872, 587.

76. Coxe 1817, 215–16.

77. Fisher and Upshall 1976, 307.

78. D. T. [David Thomas] 1831, 46.

79. Mitchill 1826, 21.

80. Prince 1828, 14–15.

81. For weather, see Lindley and Floy 1846, 365. For insects, see [Pascalis] 1814–1815, 394; Schaeffer 1816–1817, 199. For cultivation, see Thacher 1822, 180–84.

82. "Peach Prospects" 1838.

83. Hedrick 1950, 234–35; Campbell 1979, 65.

84. Biggs 1993, 2.

Chapter 2

1. Farley 1977, 48–50.

2. McNew 1963, 167; Baker 1969, 82.

3. Duhamel du Monceau 1728; Micheli 1729; Fontana [1767] 1932, 10–11; Targioni-Tozzetti [1767] 1952, 54–55.

4. Tillet [1755] 1970; Prévost [1807] 1939; Ainsworth 1981, 28–31.

5. Daniels 1971, 63.

6. Deane 1797, 301.

7. Mease 1818, 79.

8. Pickering 1811, 164–65.

9. Ibid., 165–66; Mease 1814, 423.

10. Deane 1797, 301–2; Mease 1818, 78.

11. "Fothergill, Anthony" 1937–1938.

12. Fothergill 1808, 65–66.

13. Ibid., 81–83.

14. Prescott 1813, 4–6.

15. Rogers 1977, 8.

16. De Schweinitz 1834.

17. Rogers 1977, 8–11.

18. Thomas 1844, 202.

19. Ibid., 209–11.

20. Ibid., 203.

21. Ibid., 205.

22. Ibid., 213–14.

23. Niederhauser and Cobb 1959.

24. Stevens 1933, 441–45; Bourke 1993, 31–32.

25. Hawkes 1967, 217–21.

26. Carefoot and Sprott 1967, 71–72.

27. Bourke 1993, 25–31.

28. Niederhauser and Cobb 1959.

29. Stevens 1933, 442–45; Bourke 1993, 141.

30. True 1937, 25.

31. Ibid., 20–21; Dupree 1957, 15–16.

32. Hedrick 1950, 253 and 429.

33. Gates 1960, 322.

34. True 1937, 23–26; Harding 1947, 9–12; Dupree 1957, 47.

35. U.S. Patent Office 1844, 62.

36. Ibid., 62.

37. U.S. Patent Office 1845, 8.

38. Bush 1846, 342.

39. U.S. Patent Office 1845, 79.

40. Ibid., 215.

41. U.S. Patent Office 1846, 221.

42. U.S. Patent Office 1845, 216.

43. Bush 1846, 344.

44. Ibid., 346–47.

45. Rossiter 1975, 91–148.

46. U.S. Patent Office 1845, 80–81; U.S. Patent Office 1846, 212–15.

47. U.S. Patent Office 1846, 503.

48. Ibid., 505.

49. Teschemacher 1844a.

50. Warren 1853; "Obituary [J. E. Teschemacher]" 1854, 150; Bouvé 1880, 59–60.

51. Teschemacher 1844b.

52. Ibid.

53. "Agricultural meeting at the State House" 1845, 237.

54. Teschemacher 1845c.

55. U.S. Patent Office 1846, 487–514.

56. Ibid., 517.

57. Teschemacher 1844a.

58. Teschemacher 1845b.

59. U.S. Patent Office 1846, 517–18.

60. Teschemacher 1844a.

61. Bourke 1993, 133–39.

62. U.S. Patent Office 1846, 206–7.

63. "The Patent Office report and the potato rot" 1847, 11.

64. U.S. Patent Office 1848, 140.

65. Ibid., 139–40.

66. Ibid., 141.

67. "The Patent Office report and the potato rot" 1847, 11.

68. U.S. Patent Office 1848, 150.

69. "The Patent Office report and the potato rot" 1847, 15.

70. Berkeley [1846] 1948, 15; De Bary [1853] 1969.

71. Jones 1914, 99.

72. Bidwell and Falconer 1925, 377.

73. Day 1954, 164–65.

74. Bruce 1987, 7–13.

75. Miller 1970, 6–7.

CHAPTER 3

1. Hedrick 1950.

2. Dodge 1850, 276.

3. Hedrick 1950, 230–32; Fisher and Upshall 1976, 303.

4. S. S. 1844.

5. Prince 1847, 318.

6. Boyd 1854.

7. Mason 1857.

8. "The blight in the pear tree" 1847, 59.

9. Thomas 1848, 20.

10. Overman 1855, 447.

11. For bitter rot, see Goodsell 1849. For rust, see Gillett 1855. For leaf blight, see "Blight of the Pryor's red apple tree" 1857; "Apple tree blight" 1858; Elliott 1858, 62. For scab, see Nichols 1858.

12. "Fruit culture in North Carolina" 1853, 371; Van Buren 1854; Hillsides 1861.

13. Goodsell 1849.

14. Webster 1819, 86; Arboreum 1820; "Diseases and enemies of fruit trees" 1835–1836, 341–42.

15. Wyman 1847; Tomlinson 1852; Pierce 1859, 387.

16. "Remedy for black knot in plum and cherry trees" 1858, 57.

17. Clarke 1830, 528; Howe 1856; Read 1858.

18. "On the vine and rot in grapes" 1833.

19. Richards 1851.

20. Clay 1859.

21. "New remedy for mildew of grapes" 1852, 225.

22. For pear leaf blight, see Hooker 1851, 117–19; Field 1858, 176–77; Warder 1862, 250–51. For gooseberry mildew, see "To prevent mildew on gooseberries" 1848, 107; "Gooseberries— cause of failure" 1856; Wilson 1857. For cherry black wart, see J.J.T. 1844; M. W. 1857. For peach mildew, see Dagge 1848. For peach leaf curl, see Goodrich 1854; Harry 1856; Elliott 1858, 267–68.

23. McAllister et al. 1854, 258.

24. Gates 1960, 160–61.

25. McNall 1976, 148.

26. U.S. Patent Office 1844, 18.

27. "Wheat culture-uredo (or smut), etc." 1848; Mason 1848; Younger 1850.

28. Goodwin 1856, 186.

29. Smith, R. A. 1961b, 712.

30. "Digging potatoes—corn smut— plaster" 1844; U.S. Patent Office 1848, 129; Brown 1858, 354–55.

31. "Rust on oats" 1858, 329.

32. Capell 1858.

33. P.W.J. 1846.

34. Harden 1837–1838; Farmington 1843; Glover 1858.

35. A Georgia Planter 1830, 23.

36. Spalding 1835.

37. Miner 1856, 336.

38. "Diseases of Plants" 1852.

39. "Gardening-cabbage (*Brassica oleracea*)" 1855, 343–44.

40. Proctor 1857.

41. McAllister et al. 1854, 258.

42. Overman 1855, 450.

43. Danhof 1969, 288–89; Rossiter 1976, 291–93.

44. Darling 1846.

45. Darling 1845.

46. G.S.B. 1847, 254–55.

47. "Smut in wheat" 1845.

48. "Wheat" 1848, 108.

49. U.S. Patent Office 1845, 40–41.

50. Johnston 1850, 192–93.

51. For wheat, see "A Farmer of Tompkins County" 1846; D.J.B. 1855, 184; Wadsworth 1858, 126. For oat, see "Rust on oats" 1858, 329. For corn, see "Digging potatoes—corn smut—plaster" 1844; Brown 1858, 354–55. For rye, see Ohio State Board of Agriculture 1858, 387–88.

52. For pear and apple, see Townley 1851, 354; "Blight of the Pryor's red apple tree" 1857, 27; "Apple tree blight" 1858. For cedar-apple, see Crosier 1858, 299. For peach, see "Curl in the peach" 1853. For plum, see Kirtland 1855, 35. For grape, see Allen 1853, 118–20; Saunders 1855, 129; Elliott 1858.

53. Hildreth 1850, 88.

54. Burkett 1900, 82–83; Jones 1983, 217.

55. Kirtland 1855, 34–35.

56. Ibid.

57. Ibid., 35–36.

58. "Klippart" 1933.

59. Klippart 1858, 776–78.

60. Danhof 1969, 69.

61. Kirtland 1855, 35.

62. R.R.S. 1858.

63. Van Buren 1859.

64. Massachusetts Board of Agriculture, Committee on Diseases of Vegetation 1860, 49–50; Agricola 1860, 55.

65. Massachusetts Board of Agriculture, Committee on Diseases of Vegetation 1859, 33–34.

66. Agricola 1860, 56.

67. Gates 1962, 292–93; Paarlberg 1988, 60–66.

68. Gates 1962, 297–302; Cochrane 1979, 235–45.

69. Johnston 1850, 192.

70. Gates 1962, 293–94; Rossiter 1975, 59–60.

71. "The editor to the reader" 1855; "Agricultural schools and colleges" 1856.

72. Rossiter 1975, 149–71.

73. Allen 1844; "Memorial from the agriculturalists" 1846; "Agricultural education" 1847.

74. Lee 1850, 18.

75. Turner 1852b, 38.

76. True 1929, 99–106.

77. "Agricultural Department at Washington" 1859, 103.

78. U.S. Congress, Committee on Agriculture 1856, 8.

79. Agricola. 1860.

80. True 1937, 34–40; Harding 1947, 13–22.

PART 2

1. Rodgers 1944, 145.

2. Coleman 1977, 22–23.

3. Dupree 1957, 384–409; Marcus 1985, 217–21.

4. Bruce 1987, 146–47, 317–19.

5. Griffith, Peterson, and Campbell 1994.

CHAPTER 4

1. Rossiter 1976.

2. Berkeley [1846] 1948.

3. Berkeley 1854–1857.

4. Coleman 1977, 23–34.

5. De Bary [1853] 1969.

6. Morton 1981, 429, 433–34.

7. De Bary 1861.

8. De Bary 1865–1866.

9. Kühn 1858.

10. Bessey 1955, 241–42.

11. Sander 1878–1886; Spaulding 1909; Bornstein 1963.

12. Hedrick 1950, 311–13, 380–82.

13. Engelmann 1861, 165.

14. Engelmann 1883; "September 16, 1861."

15. Dupree 1959; Shaw 1986.

16. Dupree 1959, 355–83; Kastner 1977, 302–4.

17. Rodgers 1944, 2

18. Klein 1930, 716–17; Bruce 1987, 145–48.

19. Ross 1942, 9–10.

20. "Agricultural education" 1863.

21. Colman 1963, 57; Fitzharris 1974, 203.

22. J. C. Arthur 1925. Botany and botanists in the eighties, 4, in Biographical Files, National Fungus Collections.

23. Colman 1963, 54.

24. Kirkendall 1986, 9.

25. Klein 1930, 21.

26. Busch and Lacy 1983, 9.

27. Marcus 1985, 64–65; Bruce 1987, 336.

28. Ross 1942, 113–35.

29. Rodgers 1944, 198.

30. Bessey 1935, 227.

31. Walsh 1971, 389.

32. Arthur, supra note 22, 5.

33. Thaxter 1920, 89.

34. Ibid.

35. Ibid.

36. Harris 1945, 15.

37. Setchell 1927, 3.

38. Thaxter 1920, 90–91; Harris 1945.

39. Harris 1945, 13.

40. Ibid., 17.

41. Farlow 1913, 83.

42. Harris 1945, 26–27; Peterson 1998, 717–19.

43. Rodgers 1952, 92–94.

44. Harris 1945, 13.

45. Ibid., 25.

46. Farlow 1913, 84.

47. Rossiter 1975, 138–39.

48. Elliot and Rossiter, eds. 1992, 343–44.

49. Thaxter 1920, 91; Setchell 1927, 5–6.

50. Smith 1962, 119.

51. Farlow 1875; ibid. 1876a; ibid. 1876b; ibid. 1876c.

52. Ibid. 1877.

53. Ibid. 1878.

54. Ibid. 1879.

55. Ibid. 1875, 319.

56. Peck 1872; Farlow 1876c.

57. Farlow 1875, 319.

58. Rodgers 1944, 129.

59. Thaxter 1920, 91.

60. Osborne 1913, 217.

61. Clinton 1920, 7.

62. True 1937, 110; Sidar 1976, 206–7.

63. Rodgers 1952, 96.

64. Reed 1943, 158; Rodgers 1952, 98.

65. Setchell 1927, 18; Peterson 1998, 717–19.

66. Pool 1915, 506.

67. C. L. Shear [n.d.] Charles Edwin Bessey, 1, in Bessey Papers.

68. Pool 1915, 506.

69. Overfield 1993, 7.

70. Shear, supra note 67, 1; Pool 1915, 506–7; Overfield 1993, 7.

71. Gillispie 1970, 102–3.

72. Shear, supra note 67, 1.

73. H. S. McNabb 1969. An historical outline of the first century of botanical instructions at the Iowa State University of Science and Technology, 9, unpublished paper in Bessey Papers.

74. Overfield 1993, 12–14.

75. Bessey 1935, 227–30; Rodgers 1944, 230; Overfield 1993, 13–14.

76. Pool 1915.

77. Gillispie 1970, 103.

78. Overfield 1993, 37.

79. Bessey 1874.

80. H. B. Ward 1915. Charles Edwin Bessey: An appreciation, in Papers read in appreciation of the services of Charles Edwin Bessey as a part of a memorial program of the Nebraska School Masters Club at Lincoln, 21, in Bessey Biographical/bibliographical File; Gillispie 1970, 102.

81. Pool 1915, 511; Overfield 1993, 25.

82. Bessey 1935, 232.

83. Tobey 1981, 10

84. Bessey 1935, 232.

85. L. H. Pammel 1928. Prominent men I have met, 19, in Bessey Papers.

86. C. E. Bessey [n.d.] A summary history of the Department of Botany in the Iowa State College from 1870 to 1884, 11, in Bessey Papers.

87. Bessey 1935, 232; Rodgers 1944, 227–28.

88. Arthur, supra note 22, 7; Bessey, supra note 86, 11.

89. Bessey, supra note 86, 11; Walsh 1971, 390.

90. McNabb, supra note 73, 10.

91. Arthur 1885c, 396–403.

92. Overfield 1993, 24–25.

93. Ibid., 27–28.

94. Rodgers 1944, 230.

95. G. A. 1880, 337.

96. Overfield 1993, 29.

97. "Review of *The Essentials of Botany*" 1884.

98. Rothrock 1884, 1248.

99. Pool 1915, 511.

100. Ibid., 508.

101. Walsh 1971, 391; ibid. 1972, 32.

102. Ibid. 1972, 47–48.

103. Ibid. 1971, 396.

104. Pool 1915, 512.

105. Ibid., 508; Rodgers 1944, 228.

106. Bessey 1875; ibid. 1877; ibid. 1882; ibid. 1883b; ibid. 1886.

107. Ibid. 1886, 35.

108. Ibid. 1882, 96.

109. Tobey 1981, 11.

110. Walsh 1971, 391–93.

111. Ibid., 397.

112. Pool 1915, 511.

113. Ibid., 508.

114. Tobey 1981, 12–13.

115. Pool 1915, 509.

116. Bessey 1885, 803.

117. Tobey 1981, 12.

118. Illegible letter to L. H. Pammel, April 24, 1906, in Pammel Papers.

Chapter 5

1. Barrett 1918, 1; Glawe 1992, 17–18.

2. Barrett 1918, 1; Thornberry, H. H. 1964. Thomas Jonathan Burrill's contribution to the history of microbiology and plant pathology, 24–25, in Thornberry Papers; Solberg 1968, 152.

3. Thornberry, supra note 2, 25–26.

4. Burrill 1869, 327.

5. Ibid., 328.

6. Solberg 1968, 153.

7. Rogers 1977, 15.

8. Thornberry, supra note 2, 28.

9. Ibid.

10. Story told by Hottes 1940, 6, in Burrill Exhibit File.

11. Rogers 1977, 14–15.

12. Baker 1971, 603–5.

13. Turner 1852a, 164.

14. Burrill's first report on fire blight investigation, September 13, 1876, 1, in Thornberry Papers; Thornberry, supra note 2, 34.

15. Burrill 1878, 114.

16. Ibid., 116.

17. Burrill 1879, 80.

18. Ibid. 1881b, 527–30.

19. Ibid. 1881a, 588–96.

20. Ibid. 1881d, 69.

21. Ibid. 1881e.

22. Ibid., 161.

23. Ibid. 1881a, 588–96.

24. Ibid. 1881d.

25. Ibid. 1881e.

26. Bessey 1881, 950.

27. Burrill 1883b.

28. Ibid. 1881a, 588.

29. Ibid. 1881d, 80.

30. Peffer 1882; Burrill 1883c, 49.

31. Peffer 1882, 196–98.

32. Briggs 1873.

33. Baker 1971, 610–11.

34. Salisbury and Salisbury 1864, 452.

35. Meehan 1868, 49.

36. G.S.B. 1847, 254–55; Wendell 1850, 447.

37. Hull 1869, 35; Baker 1971, 612.

38. Burrill 1878; ibid. 1881a, 585.

39. Ibid. 1881d, 73.

40. Ibid. 1878.

41. Brock 1988, 28–33.

42. Burrill 1879, 80; ibid. 1883a.

43. Reed 1943, 162–63; Coleman 1977, 165–66; Brock 1988, 101–3.

44. Burrill 1881e, 163; Kennedy, Widkin, and Baker 1979, 2.

45. Smith 1911, 7–9.

46. Burrill 1879, 80.

47. Baker 1971, 612.

48. Ibid., 617–18; Kennedy, Widkin, and Baker 1979, 3.

49. Thornberry, supra note 2, 38.

50. Baker 1971, 621.

51. Burrill 1881e, 164.

52. Ibid., 165.

53. Ibid. 1883d, 46.

54. Ibid. 1881c, 272; ibid. 1883a, 134.

55. Ibid. 1889; Moore 1894, 368–69.

56. Burrill 1887.

57. Baker 1971, 621.

58. Burrill 1883d, 48.

59. Ibid. 1883b.

60. Thornberry, supra note 2, 38

61. Ibid., 26–27; Solberg 1968, 251, 331.

62. T. J. Burrill to E. J. Townsend, February 23, 1907, in Thornberry Papers.

63. Thornberry, supra note 2, 30.

64. Trelease 1916.

65. Burrill 1882a; Burrill 1882b.

66. Smith 1916, 270.

67. Horsfall 1992, 9–15.

68. Knoblauch, Law, and Meyer 1962, 14–17; Rossiter 1975, 149–71; Hadwiger 1982, 15–17.

69. Colman 1963, 71–73; ibid. 1965, 41.

70. Fitzharris 1974, 207.

71. Rosenberg 1971; Busch and Lacy 1983, 10–12.

72. Rossiter 1975, 149–71.

73. J. C. Arthur 1925. Botany and botanists in the eighties, 13, in Biographical Files, National Fungus Collections.

74. Galloway 1900, 193–94.

75. Kern 1942, 617.

76. Cummins 1978, 21.

77. Ibid., 20.

78. Arthur [1878] 1981.

79. Arthur, supra note 73; Trelease 1925, 61; Cummins 1978, 20.

80. Kern 1942, 617; Rodgers 1944, 243.

81. Arthur, supra note 73, 8–12.

82. Thwaites, ed. 1900, 772; Arthur, supra note 73, 12 and 16.

83. Ibid., 14.

84. Ibid.

85. Rodgers 1952, 111.

86. Arthur, supra note 73, 13–14.

87. Ibid. 1885b, 343.

88. Ibid. 1885b.

89. Brock 1988, 95.

90. Arthur 1885b, 344; Baker 1971, 618–19.

91. Arthur 1885b, 345.

92. Ibid., 343–45.

93. Ibid. 1885a, 1180.

94. Ibid.

95. Ibid., 1180–81.

96. Ibid., 1183–84.

97. Arthur 1887, 126.

98. Ibid., 127.

99. Arthur 1885a, 1184.

100. Arthur 1887, 127.

101. Ibid.

102. Ibid., 128.

103. Ibid., 127–28.

104. Rodgers 1952, 111; Kennedy, Widkin, and Baker 1979, 3.

105. Osborne, ed. 1913, 145.

CHAPTER 6

1. "Department of Agriculture" 1863, 26.

2. Beal 1883, 328.

3. Dupree 1957, 152–57.

4. Ross 1946, 138; White 1958, 233; Gates 1965, 309–23.

5. Harding 1942, 34–36.

6. Newton 1863, 20–21.

7. Ross 1946, 138.

8. Dupree 1957, 155–56.

9. Harding 1942, 42; Ross 1946, 138.

10. True 1937, 54.

11. Ibid., 55; Marcus 1985, 152–53.

12. Ibid., 153.

13. Ross 1946, 138–39.

14. True 1937, 58.

15. Marcus 1985, 128–29.

16. Ibid., 150–55.

17. Ibid., 164.

18. Saunders 1862, 543.

19. Bartlett 1863, 103–4.

20. Bollman 1863, 77–80.

21. Ibid. 1865, 97–110.

22. Ibid., 102.

23. Saunders, 1866, 16.

24. Erni, 1866.

25. Ibid., 331–32.

26. Lodge 1866, 206–7.

27. Unpublished biographical paper [n.d.] in Taylor Topical Files, National Fungus Collections, 1–4; Grace 1988, 25–26.

28. Taylor 1872.

29. Ibid., 111–12.

30. Ibid., 120.

31. Ibid., 118.

32. Ibid., 116; ibid. 1875, 176–77.

33. Ibid. 1872, 116–18; ibid. 1875, 174–76.

34. "Science the Loser" 1910.

35. Taylor 1873; ibid. 1874, 186–96.

36. Ibid. 1876, 206.

37. Ibid. 1874, 210.

38. Ibid., 200.

39. Ibid., 207.

40. Watts 1876, 12.

41. Taylor 1875, 161–73; ibid. 1876, 193–206.

42. Watts 1876, 12.

43. Russell to T. Taylor, February 27, 1871, in Taylor Topical Files, National Fungus Collections.

44. M. C. Cooke to T. Taylor, June 7, 1875, in Taylor Topical Files, National Fungus Collections; Burrill 1881a, 586.

45. Resolutions of the Medical Society of

the District of Columbia [n.d.] in Taylor Topical Files, National Fungus Collections.

46. Grace 1988, 26; Unpublished biographical paper [n.d.] in Taylor Topical Files, National Fungus Collections, 8; T. Taylor to C. H. Peck, May 8, 1884, in Taylor Topical Files, National Fungus Collections.

47. Taylor 1877.

48. Loring 1882.

49. Taylor 1876, 206.

50. B. T. Galloway [n.d.] Fifty years in the Department of Agriculture, in Taylor Topical Files, National Fungus Collections, 2.

51. E. F. Smith [n.d.] A ward politician in the United States Department of Agriculture, unpublished paper, in Taylor Topical Files, National Fungus Collections, 1.

52. T. Taylor to J. S. Morton, June 29, 1895, in Taylor Topical Files, National Fungus Collections.

53. Miller 1970, 182–87.

54. Beal 1883, 332.

55. Ibid., 328.

56. Bessey 1883a, 544.

57. Dupree 1957, 163–67.

58. True 1937, 59–60; Harding 1942, 47.

59. AAAS 1885, xvii.

60. AAAS 1886, 38.

61. N. J. Colman to W. G. Farlow, October 21, 1885, in USDA Letters Sent (General) 1882–1897.

62. AAAS 1886, 38.

63. Ibid.

64. Harding 1942, 48–49; Marcus 1985, 191–92.

65. Knoblauch, Law, and Meyer 1962, 49.

66. F. L. Scribner to C. E. Bessey, May 29, 1885, in Bessey Papers 1865–1915.

67. Gray 1863; A. E. Jenkins, Conversation with F. L. Scribner, April 12, 1937, in Scribner Topical Files, National Fungus Collections.

68. Ibid.

69. F. L. Scribner to C. E. Bessey, June 28, 1885, in Bessey Papers 1865–1915.

70. Scribner 1885.

71. Saunders 1885.

72. N. J. Colman to W. G. Farlow, October 21, 1885, in USDA Letters Sent (General) 1882–1897.

73. AAAS 1886, 39–40; N. J. Colman to C. E. Bessey, January 8, 1887, in USDA Letters Sent (General) 1882–1897.

74. Botanical Club, American Association for the Advancement of Science to the Honorable Commissioner of Agriculture, [1886], in Bessey Papers 1865–1915.

75. G. Vasey to C. E. Bessey, [1886], in Bessey Papers 1865–1915.

76. C. E. Bessey to A. J. Weaver, January 27, 1886, in Committee on Agriculture, 49th Congress.

77. T. J. Burrill to W. H. Hatch, March 23, 1886, in Committee on Agriculture, 49th Congress.

PART 3

1. Hurt 1994, 165–218.
2. Dupree 1957, 158–63.
3. Peterson, Griffith, and Campbell 1996.

CHAPTER 7

1. Scribner 1887, 95.
2. Ibid. 1885, 76–77.
3. Pinney 1989, 156–202.
4. LeDuc 1881, 26.
5. Pinney 1989, 172.
6. Bush 1880, 43.
7. Woods 1905, 127.
8. Lodeman 1916, 303–4.
9. Riley 1886.
10. Scribner 1885, 80.
11. R. E. Smith 1961a, 701.
12. Large 1940, 225–28.
13. Riley 1886, 52.
14. Scribner 1885, 84–85.
15. USDA Division of Botany, Section of Mycology 1886.
16. USDA Division of Botany, Section of Vegetable Pathology 1886, 1.
17. Scribner 1887, 100–1.
18. Ibid. 1886, 7, 52.
19. Colman 1888, 27–28.
20. Scribner 1888b, 144; ibid. 1888c, 323–31; ibid. 1888d.
21. Ibid. 1886, 28–34, 40–42.
22. Ibid. 1888b, 134; Scribner and Viala 1888, 5.
23. Scribner 1888a, 70–72; Scribner and Viala 1888, 18–24.
24. F. L. Scribner to A. W. Pearson, May 5, 1888, in USDA Letters Sent 1885–1888; Scribner 1888e.
25. F. L. Scribner to H. Jaeger, August 8, 1888, in USDA Letters Sent 1885–1888.
26. Scribner and Viala 1888, 29.
27. Galloway [n.d.] Fifty years ago in the Department of Agriculture, 2, in Galloway Topical Files, National Fungus Collections.
28. Assistant Secretary of Agriculture, a Former Pharmacist, May 1913, 56, in Galloway Topical Files, National Fungus Collections, 56; Beverly Thomas Galloway, unpublished autobiography, 1, in Galloway Topical Files, National Fungus Collections; Woods 1938.
29. G. Vasey to S. M. Tracy, September 21, 1886, in Galloway Topical Files, National Fungus Collections.
30. S. M. Tracy to Speer, March 28, 188[8], in Galloway Topical Files, National Fungus Collections; Beverly Thomas Galloway, unpublished autobiography, 1–2, in Galloway Topical Files, National Fungus Collections.
31. Stevenson, 1938, 597.
32. Beverly Thomas Galloway, unpublished autobiography, 2, in Galloway Topical Files, National Fungus Collections.
33. Stevenson 1958, 217–18.
34. Colman 1889.
35. Galloway 1890c.
36. Ibid. 1891, 26.

37. Colman 1889, 34.

38. Galloway 1889a.

39. Galloway 1889b, 420.

40. B. T. Galloway to A. W. Howell, December 11, 1888, USDA Letters Sent 1885–1888.

41. Galloway 1889b, 398; ibid. 1890a, 393.

42. Colman 1889, 34.

43. Ibid., 33–34; Galloway 1889a; ibid. 1889b.

44. B. T. Galloway to L. M. Underwood, July 6, 1889, USDA Letters Sent 1885–1888; Galloway 1889b, 420–21.

45. Stevenson 1938, 597.

46. B. T. Galloway to V. M. Spalding, November 10, 1888, USDA Letters Sent 1885–1888.

47. G. Vasey to S. M. Tracy, September 21, 1886, in Galloway Topical Files, National Fungus Collections.

48. "Notes and news" 1887; Cattell and Cattell 1927, 923.

49. Smith, E. F. 1929, 45.

50. Southworth 1890.

51. Galloway and Fairchild 1891a; Galloway 1890a, 394–98.

52. Galloway 1891.

53. Galloway and Fairchild 1891b.

54. Jenkins interview, in Scribner Topical Files, National Fungus Collections.

55. F. L. Scribner to Lyman, May 10, 1888, in USDA Letters Sent 1885–1888.

56. Lodeman 1916, 188–89.

57. Ibid., 207–24.

58. Galloway 1890b, plates V and VI.

59. Rusk 1890, 35.

60. Galloway 1890b, 37–38, 59–60; ibid. 1890a, 400–1.

61. Ibid. 1889b, 397–98.

62. Ibid., 397.

63. Rodgers 1952, 118.

64. M. B. Waite to O. P. Bennett, July 17, 1889, in USDA Letters Sent 1885–1888.

65. M. B. Waite to O. P. Bennett, June 28, 1889, in USDA Letters Sent 1885–1888; Galloway 1890a, 393.

66. B. T. Galloway to F. S. Earle, July 22, 1889, in USDA Letters Sent 1885–1888.

67. E. F. Smith 1929, 22.

68. Galloway 1892b, 373.

69. Ibid. 1917; Rodgers 1952, 177.

70. B. T. Galloway to J. B. Ellis, December 8, 1888, in USDA Letters Sent 1885–1888.

71. Rodgers 1952, 170.

72. Galloway 1889b, 423–24.

73. Gardner and Hewitt 1974, 63–65.

74. Galloway 1889b, 429.

75. Gardner and Hewitt 1974, 67–117.

76. Galloway 1889b, 423–29.

77. Gardner and Hewitt 1974, 106–7.

78. Pierce 1890, 176.

79. Gardner and Hewitt 1974, 111–13.

80. E. F. Smith 1888, 170–71; Rodgers 1952, 240.

81. E. F. Smith 1888, 173.

82. Campbell 1983, 21.

83. Jones 1939, 4.

84. Smith Papers 1885–1927.

85. Jones 1939, 5.

86. Ibid., 17–18.

87. Ibid., 47.

88. Ibid., 18.

89. Rodgers 1952, 139.

90. Commission Note of E. F. Smith by N. J. Colman, July 1, 1887, in Smith Papers 1865–1940.

91. Rodgers 1952, 154.

92. Galloway [n.d.] Fifty years ago in the Department of Agriculture, 2, in Galloway Topical Files, National Fungus Collections.

93. True 1937, 92–93.

94. T. T. Lyon to E. F. Smith, February 17, 1888, in Smith Papers 1865–1940.

95. Jones 1939, 18.

96. E. F. Smith 1891, 25.

97. Darling 1845; E. F. Smith 1888, 149.

98. Ibid., 179.

99. F. L. Scribner to E. F. Smith, September 14, 1887, in Smith Papers 1865–1940.

100. F. L. Scribner to V. M. Spalding, July 10, 1888, in Scribner Papers 1885–1938.

101. W. S. Maxwell to E. F. Smith, February 22, 1888, in Smith Papers 1865–1940.

102. W. Webb to E. F. Smith, April 14, 1888; L. H. Bailey to E. F. Smith, May 16, 1888. Both in Smith Papers 1865–1940.

103. C. W. Garfield to E. F. Smith, March 26, 1888, in Smith Papers 1865–1940.

104. F. L. Scribner to V. M. Spalding, July 10, 1888, in Smith Papers 1865–1940.

105. Galloway 1889b, 397.

106. Rodgers 1952, 177.

107. Smith, E. F. 1894, 10.

108. Ibid.

CHAPTER 8

1. Waggoner and Gough 1975, 2.

2. Rodgers 1952, 129.

3. True 1937, 92–93.

4. Ibid., 88.

5. Allen 1922, 87–91; Fitzharris 1974, 204.

6. Rodgers 1944, 241.

7. True 1937, 130–31; Knoblauch, Law, and Meyer 1962, 55–80.

8. Chester 1892, 60–62.

9. Ibid. 1889, 69–109; ibid. 1891, 44–91.

10. Galloway and Southworth 1889, 210; E. S. Goff to B. T. Galloway, March 15, 1892, in USDA Letters Received (General) 1891–1900.

11. Goff 1890.

12. Osborne 1913, 205–6; Horsfall 1992, 81.

13. Weston 1933.

14. Thaxter 1890, 129–53; ibid. 1891, 81–95.

15. Ibid., 130–31; Rodgers 1952, 200.

16. Thaxter 1890, 131–32.

17. Ibid., 137.

18. Ibid., 146.

19. Ibid., 146–53.

20. Sturgis 1894, 100; ibid. 1896, 176–82.

21. Walker 1957, 398–99.

22. Horsfall 1963.

23. Ibid. 1992, 82.

24. Ibid. 1979, 33.

25. Stevens, Pammel, and Cook 1920, 305; Woodward and Waller 1932, 251–52.

26. Halsted 1884, 88; B. D. Halsted to B. T. Galloway, April 5, 1893, in USDA Letters Received (General) 1891–1900.

27. Pammel 1919, 31; Stevens, Pammel, and Cook 1920.

28. Fairchild 1919, 1; Rodgers 1952, 172.

29. Fairchild 1919, 5; Woodward and Waller 1932, 254–55.

30. Fairchild 1919, 5.

31. Halsted and Fairchild 1891.

32. Halsted 1889, 332–39.

33. Ibid. 1891.

34. Ibid., 333.

35. Woodward and Waller 1932, 258.

36. Ibid., 259; S. M. Tracy to B. T. Galloway, April 15, 1891, in USDA Letters Received (General) 1891–1900.

37. Halsted 1892.

38. Woronin 1934.

39. Halsted 1893, 3.

40. Ibid., 5.

41. Ibid., 13–15; Woronin 1934, 28.

42. Keitt and Rand 1946, 2.

43. Jones 1904. The diseases of the potato in relation to its development, a lecture delivered before the New Jersey State Board of Agriculture, Trenton, N.J., January 13, 1904, 2, in Jones Papers 1892–1943.

44. Jones [n.d.] Potato blight and its remedies, unpublished paper, 2, in Jones Papers 1892–1943.

45. B. T. Galloway to L. R. Jones, January 9, 1893, in USDA Letters Received (General) 1891–1900.

46. Jones 1891.

47. Ibid. 1893.

48. B. T. Galloway to L. R. Jones, June 15, 1891; L. R. Jones to B. T. Galloway, August 12, 1892. Both in USDA Letters Received (General) 1891–1900.

49. Jones 1896b.

50. H. W. Collingwood to L. R. Jones, January 8, 1896, in Jones Papers 1896–1909; Jones 1896a; Jones 1899, 25–32.

51. Jones 1899, 27.

52. Jones 1901a; Jones 1910, 281–360.

53. Jones 1901b.

54. Jones [n.d.] Lecture, 1 in Jones Papers 1892–1943.

55. Hurt 1994, 173–81.

56. Large 1940, 240–47; Rodgers 1952, 141–42.

57. Arthur 1896, 25.

58. Arthur and Stuart 1900.

59. Kellerman 1891a, 101–4.

60. Arthur 1898, 29.

61. Arthur and Stuart 1900, 86–89.

62. Danbom 1990, 13–14.

63. Bolley 1891a, 24–28.

64. Ibid. 1889; ibid. 1891c.

65. Ibid. 1890; ibid. 1891b, 11–12.

66. Kellerman and Swingle 1890a, 113–15; Kellerman and Swingle 1890b, 41.

67. Bolley 1891b, 25.

68. Ibid., 14.

69. Arthur 1897, 20.

70. Bolley 1897, 128, 154–55.

71. Ibid., 149.

72. Ibid. 1900.

73. Ibid. 1901.

74. Kommedahl, Christensen, and Frederiksen 1970, 8.

75. Bolley 1906.

76. Pohl 1985; L. H. Pammel to Crocker, 1928, in Pammel Papers; H. S. McNabb, An historical outline of the first century of botanical instruction at the Iowa State University of Science and Technology, unpublished paper, May 9–10, 1969, 21–22, in Pammel Papers.

77. Pammel 1889.

78. F. L. Scribner to L. H. Pammel, August 8, 1888, in Pammel Papers.

79. Pammel 1889, 61–62.

80. Ibid. 1894b, 987.

81. Ibid. 1891; ibid. 1892c; ibid. 1894b; Pammel and Carver 1895.

82. Pammel 1894a, 470; ibid. 1894d, 92; ibid. 1900, 182.

83. Kellerman 1891b, 90–93; Pammel 1892a, 324–29.

84. Pammel 1894b, 985–87.

85. Ibid. 1892b, 499–502.

86. Ibid. 1894c.

87. [H. Garman] 1892; ibid. 1894.

88. Whetzel 1919.

89. Atkinson 1889, 4.

90. Farlow, Thaxter, and Bailey 1919, 302.

91. Fitzpatrick 1919.

92. Atkinson 1889, 5–6.

93. Hilty and Peterson 1997, 24.

94. L. H. Pammel to G. F. Atkinson, April 18, 1890, in Atkinson Papers.

95. Atkinson 1890a. 111.

96. B. T. Galloway to G. F. Atkinson, January 20, 1890, in Atkinson Papers.

97. Atkinson 1890b.

98. Ibid. 1891b.

99. Ibid. 1892c, 4.

100. Ibid. 1892b, 27–29, 55–61.

101. Walker 1957, 112.

102. Atkinson 1891a.

103. E. A. Southworth to G. F. Atkinson, November 28, 1890, in Atkinson Papers.

104. B. T. Galloway to G. F. Atkinson, November 19, 1890, in Atkinson Papers.

105. Atkinson 1892a; ibid. 1893a; ibid. 1894.

106. Ibid. 1895.

107. Atkinson 1893b.

108. Ibid. 1889, 49.

109. True 1937, 130–31; Carstensen 1960, 13.

110. Marcus 1985, 215–18.

111. White 1958, 250–52; Kerr 1987, 40–41.

Chapter 9

1. Galloway [n.d.] Vegetable pathological and physiological investigations, 2, unpublished paper, in Galloway Topical Files, National Fungus Collections.

2. Act of March 3, 1887.

3. Act of July 14, 1890.

4. Galloway [n.d.] Vegetable pathological

and physiological investigations, 2, unpublished paper, in Galloway Topical Files, National Fungus Collections.

5. Hurt 1994, 165.

6. Pierce 1892, 208.

7. Galloway 1892b, 371.

8. Wetmore 1891, 10.

9. N. B. Pierce to B. T. Galloway, August 24, 1891. Unless otherwise designated, all letters cited in this chapter are in USDA Letters Received (General) 1891–1900.

10. N. B. Pierce to B. T. Galloway, August 27, 1891.

11. N. B. Pierce to B. T. Galloway, November 5, 1891.

12. Galloway 1893, 231–33.

13. Morton 1894, 28.

14. Galloway 1894b, 271–72.

15. N. B. Pierce to B. T. Galloway, March 6, 1892.

16. N. B. Pierce to E. F. Smith, January 30, 1893, in Smith Papers 1865–1940.

17. Galloway 1894b, 270–74; ibid. 1895a, 145–46; ibid. 1895b, 172.

18. Gardner and Hewitt 1974, 141.

19. Ibid., 203.

20. N. B. Pierce to E. F. Smith, January 30, 1893, in Smith Papers 1865–1940.

21. Gardner and Hewitt 1974, 68.

22. [Rusk] 1892, 49.

23. Secretary to W. T. Swingle, June 15, 1891; B. T. Galloway to J. S. Morton, April 1, 1893.

24. Venning 1977, 7–9.

25. Galloway 1892b, 373–74; Cooper 1995, 2.

26. Fisher and Upshall 1976, 37.

27. W. T. Swingle, Sketch of diseases, July 23, 1891, in USDA Letters Received (General) 1891–1900, 3–9.

28. B. T. Galloway to J. S. Morton, April 1, 1893.

29. J. C. Neal to B. T. Galloway, May 25, 1891; B. T. Galloway to J. C. Neal, June 2, 1891.

30. Galloway 1892b, 374.

31. Ibid. 1893, 243–44.

32. Swingle 1892, 97.

33. B. T. Galloway to J. S. Morton, April 1, 1893.

34. Galloway 1893, 244; B. T. Galloway to J. S. Morton, April 1, 1893.

35. "Webber, Herbert John" 1927, 387.

36. Elliott, Lodgson, and Yowler 1992, 59–60.

37. Galloway 1893, 244.

38. Ibid. 1894b, 265–70; ibid. 1895a, 148; Cooper 1995, 4–13.

39. Swingle 1894, 73.

40. Galloway 1894b, 269; Swingle and Webber 1896, 31–33.

41. Galloway 1895a, 145.

42. Swingle and Webber 1896, 23.

43. W. J. Webber to B. T. Galloway, October 8, 1894; Webber 1897.

44. Swingle and Webber, 1896, 25–28; Webber 1897.

45. Fairchild 1893, 240.

46. [Rusk] 1892, 48.

47. D. G. Fairchild to B. T. Galloway, March 23, 1891.

48. D. G. Fairchild to B. T. Galloway, April 20, 1891; Galloway 1892b, 57–60, 62–68.

49. Galloway 1892a, 179.

50. D. G. Fairchild to B. T. Galloway, April 20, 1891.

51. Fairchild 1893.

52. Ibid., 241–42.

53. Galloway 1894a, 30–31.

54. [B. T. Galloway] to J. S. Morton, November 8, 1894.

55. J. S. Morton to D. G. Fairchild, October 5, 1893.

56. Fairchild 1938, 494.

57. Galloway 1894b, 276.

58. Ibid., 258–62.

59. Ibid. 1895a, 146.

60. Ibid. 1893, 216–24.

61. Ibid., 245–46.

62. W. T. Swingle to B. T. Galloway, December 6, 1892.

63. B. T. Galloway to W. T. Swingle, March 27, 1893.

64. W. T. Swingle to B. T. Galloway, March 30, 1893.

65. Rand 1949, 314.

66. Galloway 1895b, 169.

67. Galloway [n.d.] Plant diseases, unpublished paper, in USDA Letters Received (General) 1891–1900, 1–2.

68. [B. T. Galloway] to J. S. Morton, November 8, 1894.

69. Galloway 1893, 216–17.

70. Ibid. 1895b, 171.

71. Ibid., 169–74; ibid. 1896; ibid. 1897.

72. Rodgers 1952, 184.

73. M. B. Waite [n.d.] When Dr. Erwin F. Smith became interested in bacteriology, unpublished paper, in Smith Papers 1865–1940.

74. Jones 1939, 4.

75. Campbell 1983, 24.

76. Smith 1896, 631; Jones 1939, 8–9.

77. Rodgers 1952, 146.

78. Smith 1896, 626–27.

79. Ibid., 635–36.

80. Ibid., 626–43; Jones 1939, 24–26.

81. Jones 1939, 22.

82. Rodgers 1952, 267–70.

83. E. F. Smith to B. T. Galloway, July 1, 1894; Jones 1939, 20–22.

84. Smith 1899.

85. Burrill 1890.

86. Halsted 1892, 11–12.

87. Rodgers 1952, 294.

88. L. R. Jones to E. F. Smith, February 20, 1897, in Smith Papers 1865–1940.

89. [Garman] 1894; Pammel 1895, 130–34.

90. E. F. Smith to A. F. Woods, September 1, 1897.

91. Beardsley 1969, 22–24.

92. Smith 1899; ibid. 1911, 300–44.

93. Fischer 1897, 131–32.

94. Beardsley 1969, 10.

95. Campbell 1981, 4–8.

96. Ibid., 4–8.

97. Ibid., 9–16.

98. Ibid., 17–23.

99. Rodgers 1952, 348.

100. W. H. Welch to E. F. Smith, November 10, 1897, in Smith Papers 1865–1940.

101. Rodgers 1952, 345.

102. Ibid., 351.

103. Campbell 1981, xiii

PART 4

1. Wiebe 1967.

2. Hurt 1994, 221–22.

3. Ibid., 266–68.

4. Peterson and Campbell 1997.

CHAPTER 10

1. W. T. Swingle to B. T. Galloway, June 9, 1893; H. J. Webber to B. T. Galloway, May 1893, in USDA Letters Received (General) 1891–1900.

2. Baker, Rasmussen, Wiser, and Porter 1963, 39–43; Dupree 1957, 175.

3. Galloway 1926, The genesis of the Bureau of Plant Industry, 5–6, unpublished paper, in Galloway Topical Files, National Fungus Collections.

4. Stevenson 1954, 161.

5. Bureau of Plant Industry, December 1, 1912, 33, in USDA Correspondence of the Office of the Chief (General) 1908–1939.

6. Von Schrenk and Spaulding 1903; Scott 1908.

7. Galloway 1907, 187.

8. Hurt 1994, 256.

9. Galloway 1907, 187.

10. Smith 1902, 612.

11. Jones 1914, 108.

12. Bureau of Plant Industry, December 1, 1912, 33, in USDA Correspondence of the Office of the Chief (General) 1908–1939.

13. Rodgers 1952, 428–29.

14. Toumey 1900.

15. Galloway 1902, 66.

16. Hedgcock 1905; Von Schrenk and Hedgcock 1907, 13–24.

17. Smith and Townsend 1907, 672.

18. Smith, Brown, and McCulloch 1912, 11.

19. Rodgers 1952, 451–66.

20. Ibid., 532–33.

21. Smith 1902, 609.

22. Webber and Bessey 1900.

23. Swingle and Webber 1896, 32–33; Rodgers 1952, 365.

24. Gardner and Hewitt 1974, 141.

25. Bowler 1989; Kimmelman 1983.

26. Gilbert 1930; Jones 1931, 1–2.

27. Galloway, supra note 3, 8.

28. Orton 1900, 5.

29. W. A. Orton to B. T. Galloway, July 18, 1899, in USDA Letters Received (General) 1891–1900.

30. Webber and Bessey 1900, 490.

31. W. A. Orton to B. T. Galloway, July 18, 1899, in USDA Letters Received (General) 1891–1900.

32. Jones 1931, 3.

33. Ibid., 2–3.

34. Rodgers 1952, 405.

35. Orton 1900, 11–15.

36. Ibid., 14.

37. Jones 1931, 3.

38. W. A. Orton to A. F. Woods, July 27, 1902, in USDA Letters Received (General) 1901–1902.

39. D. Morris to W. A. Orton, March 24, 1905, in USDA Letters Received (General) 1903–1908.

40. Orton 1902, 20.

41. H. J. Webber to A. F. Woods, September 4, 1901, in USDA Letters Received (General) 1901–1902.

42. Smith 1902, 609.

43. W. A. Orton to B. T. Galloway, March 24, 1908, in USDA Letters Received (General) 1903–1908.

44. Bureau of Plant Industry, December 1, 1912, 28–30, in USDA Correspondence of the Office of the Chief (General) 1908–1939.

45. Orton and Gilbert 1912, 9.

46. Ward 1902; Ward 1905.

47. Arthur 1906, 14.

48. Orton 1909.

49. Thorne 1961, 3.

50. Ibid., 4.

51. Bessey 1911, 61–63.

52. Ibid., 8–9.

53. Neal 1889, 5–6.

54. Atkinson 1889.

55. Stone and Smith 1898, 5–6.

56. Webber and Orton 1902, 30.

57. Ibid., 23–36.

58. H. J. Webber to A. F. Woods, September 4, 1901, in USDA Letters Received (General) 1901–1902.

59. Shamel and Cobey 1907, 61.

60. Bessey 1911, 71–72.

61. Ibid., 8–10.

62. Huettal and Golden 1991, 15.

63. Blanchard 1957, 205–13.

64. Huettal and Golden 1991, 16–17; Sayre 1994, 7.

65. Blanchard 1957; Sayre 1994, 27–77.

66. Cobb 1894.

67. Bessey 1911, 8–9.

68. Huettal and Golden 1991, 22.

69. Cobb 1891; ibid. 1896; ibid. 1897.

70. Ibid. 1890a; ibid. 1890b.

71. Ibid. 1906, 177.

72. Thorne 1961, 14.

73. Blanchard 1957, 248.

74. Cobb 1917; ibid. 1918; Sayre 1994, 100–9.

75. Sayre 1994, 95–97.

76. Cobb 1915, 490.

77. Ibid., 459–61.

78. Ibid., 462.

79. Ibid., 465.

80. Ibid., 484, 486, 488.

81. Murrill 1906a; ibid. 1906b; Griffin 1986, 293.

82. Metcalf and Collins 1909, 45.

83. Griffin 1986, 291–93; Cochrane 1990.

84. Hartley 1950, 446.

85. Metcalf 1908, 3.

86. Acree 1940; Meinecke 1941, 289.

87. Metcalf 1904.

88. Ibid. 1906.

89. Woods 1898, 30–31.

90. Buchanan 1976, 5–8; Hartley 1950, 445–46.

91. Metcalf and Collins 1909, 49–52.

92. Ibid. 1911, 10–23.

93. Hepting 1974, 62.

94. Metcalf 1908.

95. Metcalf and Collins 1911, 11.

96. Hepting 1974, 62–63.

97. Ibid., 64.

98. Ibid.

99. Ibid., 65.

100. Gravatt 1949, 4–5.

101. Anderson 1914, 5–36.

102. Heald and Studhalter 1914.

103. Heald, Gardner, and Studhalter 1915.

104. Rumbold 1920.

105. Spaulding and Field 1912, 5.

106. Ibid., 29.

107. Weber 1930, 4–8.

108. Rosenberg 1989, 61–67.

109. Weber 1930, 2.

110. McCubbin 1954, 176–79; True 1937, 191.

111. Weber 1930, 3.

112. Metcalf, H. 1910. Statement, Inspection of Nursery Stock, Committee on Agriculture, in House Committee on Agriculture, Hearings on Miscellaneous Bills, 519.

113. Orton, W. A. 1910. Statement, Inspection of Nursery Stock, Committee on Agriculture, in House Committee on Agriculture, Hearings on Miscellaneous Bills, 516–19.

114. Weber 1930, 11.

115. Ibid., 159.

116. Allen 1978.

CHAPTER 11

1. Cremin 1961, 41–50; Veysey 1965, 125–33.

2. Veysey 1965, 121–79.

3. Glover 1952, 138–42.

4. Dudley 1889, 172; ibid. 1891, 31.

5. Kent and Newhall 1983, 1.

6. Korf 1991, 108–13.

7. Colman 1963, 128; Korf 1991, 113.

8. Colman 1963, 99–100, 140–41.

9. Ibid., 142; Lodeman 1916.

10. Korf 1991, 111.

11. Fitzpatrick 1945.

12. G. Atkinson to T. F. Crane, September 30, 1904, in Atkinson Papers.

13. Whetzel 1945, 99.

14. T. F. Crane to G. Atkinson, November 9, 1906, in Atkinson Papers.

15. Fitzpatrick 1945.

16. Whetzel 1945, 99.

17. Korf 1991, 113; Whetzel 1909, 39; ibid. 1910, lxiii–lxxi.

18. Korf 1991, 115.

19. Barrus and Stakman 1945; Whetzel 1918.

20. Newhall 1980; Whetzel and McCallan 1930.

21. Whetzel 1910, lxiii.

22. Ibid. 1911, 64.

23. Reddick 1915, liv; Whetzel 1916, xxxviii.

24. Fitzpatrick 1945, 402; Korf 1991, 114.

25. Whetzel 1945, 99–100.

26. Ibid., 100.

27. Ibid. 1910, lxviii.

28. Ibid. 1945, 100–101.

29. Korf 1991, 115.

30. Barrus 1911; ibid. 1915; Burkholder 1918; ibid. 1923.

31. Kent and Newhall 1983, 8, 18; Whetzel 1945, 102; ibid. 1914, lv.

32. "The Cornell University Agricultural Experiment Station" 1937, 144–45; Whetzel 1914, liv; ibid. 1911, 62.

33. Korf 1991, 113.

34. Jones 1922.

35. Gray 1951, 94–104.

36. Lugger [n.d.] Flax wilt disease, 5, unpublished paper, Institute of Agriculture.

37. Ibid. 1890, 21.

38. Carleton 1905, 5.

39. Stakman and Christensen 1954, 285; Freeman 1947, 3; Edward Monroe Freeman [n.d.] unpublished paper, 1, in Freeman Biographical File.

40. Freeman 1905, viii.

41. Ibid., vii.

42. Gray 1951, 120.

43. Freeman 1947, 3–4.

44. Beardsley 1969, 72.

45. Freeman 1947, 3–4.

46. Ibid.

47. Stakman and Christensen 1954, 286.

48. *University of Minnesota Bulletin, The College of Agriculture, 1908–1909*, 43; *University of Minnesota Bulletin, The College of Agriculture, 1909–1910*, 23.

49. Christensen 1984, 20–21; ibid. 1992, 331–34; Wilcoxson and Kommedahl 1992, 223–25.

50. Christensen 1992, 334; Wilcoxson and Kommedahl 1992, 228.

51. Rowell 1983, 79.

52. Wood 1975, 34.

53. Gray 1951, 227.

54. Ibid., 232; Edward Monroe Freeman [n.d.] supra note 39, 3.

55. *University of Minnesota Bulletin, The College of Agriculture, 1913–1914*, 57–58; Freeman 1947, 11.

56. Freeman 1947, 6–7.

57. Ibid., 6.

58. Ibid., 6–9.

59. Stakman and Christensen 1954, 286.

60. Rodgers 1952, 103.

61. Scribner 1886, 82.

62. Rodgers 1952, 119.

63. A. B. Seymour to Classmates, February 18, 1931, in Biographical Files, National Fungus Collections.

64. Ibid., 3.

65. Jones 1934.

66. Cranefield 1902, 411.

67. Goff 1975, 57.

68. Beardsley 1969.

69. H. L. Russell 1942. Getting started in bacteriology, 4–5, unpublished paper in University of Wisconsin College of Agriculture, Administration Records, in H. L. Russell (Miscellaneous Files).

70. Beardsley 1969, 18–19.

71. Ibid., 21.

72. Smith 1897a.

73. Ibid. 1897b.

74. Russell 1898.

75. Beardsley 1969, 68–69.

76. Ibid., 194.

77. Ibid., 71–73.

78. Glover 1952, 274.

79. L. R. Jones February 17, 1910, Course syllabus, in University of Wisconsin College of Agriculture, Administration Records, Committees.

80. Andrews and Marosy 1986, 248.

81. *University of Wisconsin Catalogue, 1908–1909*, 232.

82. Beardsley 1969, 72.

83. Ibid., 194.

84. *University of Wisconsin Catalogue, 1910–1911*, 401–2.

85. Russell 1913, 5.

86. Glover 1952, 292; Jenkins 1991, 66.

87. Russell 1916, 8.

88. Walker 1986, 7–9.

89. Ibid., 8.

90. Arny 1986.

91. Jones, Johnson, and Dickson 1926.

92. Glover 1952, 213.

93. Jenkins 1991, 67.

94. Scott 1970, 145–46.

95. Colman 1963, 153; Scott 1970, 145–47.

96. Williams to H. H. Whetzel, June 23, 1903, in Whetzel Papers.

97. Haskell 1953, 570.

98. Rasmussen 1989, 35.

99. True 1928, 70–72.

100. Rasmussen 1989, 49.

CHAPTER 12

1. Bates 1965, 85.

2. Ibid., 97–98.

3. Browne and Weeks 1952, 14.

4. Bates 1965, 125–31.

5. Alvord 1850.

6. Ravenel 1850.

7. Fitzpatrick 1937, 5.

8. Burrill 1881a.

9. Coulter 1883, 291.

10. Ibid. 1883.

11. Farlow 1888.

12. "Papers read. Section G at the 42nd meeting in Madison, Wisconsin in August 1893" 1894, 258–59, 260, 262.

13. "Proceedings of the Madison Botanical Congress" 1893, 351.

14. "Proceedings of Section G, AAAS, Brooklyn Meeting" 1894.

15. "Proceedings of the Botanical Club of the AAAS" 1892, 289; [Fairchild] 1892, 294.

16. Rodgers 1949, 178.

17. "Proceedings of the Botanical Club of the AAAS" 1892, 289

18. Rodgers 1949, 180.

19. Fitzpatrick 1937, 13.

20. Smith 1898, 96.

21. Ibid., 97.

22. Fitzpatrick 1937, 14.

23. [Clements] 1904, 46.

24. Fitzpatrick 1937, 15.

25. Shear 1906a, 85; ibid. 1906b, 186.

26. Fitzpatrick 1937, 16.

27. [Memo from] Bureau of Plant

Industry to all workers in Plant Pathology, December 15, 1908, in American Phytopathological Society Records (hereafter APS Records).

28. McCallan 1959, 24.

29. Stevenson 1957a; Stevenson 1957b.

30. McCallan 1959, 24; Shear 1919, 165.

31. C. L. Shear to A. D. Selby, November 25, 1908, in Selby Files.

32. Fitzpatrick 1983, 26.

33. C. L. Shear, D. Reddick, and W. A. Orton to F. Hedges, December 16, 1908, in APS Records.

34. Ibid.

35. H. H. Whetzel to C. L. Shear, December 19, 1908, in APS Records.

36. L. R. Jones to C. L. Shear, December 22, 1908, in APS Records.

37. G. F. Atkinson to C. L. Shear, December 17, 1908, in APS Records.

38. Cowles 1909, 903.

39. Shear 1919, 167.

40. Stover 1925, 3.

41. Fitzpatrick 1983, 25.

42. [C. L. Shear] to L. R. Jones, February 23, 1909, in APS Records.

43. Shear 1919, 168.

44. C. L. Shear to L. R. Jones, February 23, 1909, in APS Records.

45. L. R. Jones to C. L. Shear, February 26, 1909, in APS Records.

46. Shear 1919, 168.

47. W. J. Beal to L. R. Jones, May 31, 1909, in APS Records.

48. L. R. Jones to W. J. Beal, June 8, 1909, in APS Records.

49. C. L. Shear to L. R. Jones, April 27, 1909, in APS Records.

50. Shear 1919, 169.

51. L. R. Jones, handwritten note on C. L. Shear, June 19, 1909, in APS Records.

52. L. R. Jones to C. L. Shear, December 3, 1909, in APS Records.

53. L. R. Jones to C. L. Shear, December 15, 1909, in APS Records.

54. Shear 1910a, 746.

55. Shear 1910a; ibid. 1910b.

56. McCallan 1959, 25.

57. Shear 1910a, 746–47.

58. Ibid., 746.

59. C. L. Shear to A. D. Selby, November 25, 1908, in Selby Files.

60. A. D. Selby to C. L. Shear [n.d.] in Selby Files.

61. C. L. Shear to L. R. Jones, April 27, 1909, in APS Records.

62. Galloway 1913, i.

63. L. R. Jones to C. L. Shear, May 10, 1909, in APS Records.

64. A. D. Selby to L. R. Jones, March 23, 1910, in Selby Files.

65. L. R. Jones to F. L. Stevens, April 11, 1910, in APS Records.

66. L. R. Jones to C. L. Shear, April 21, 1910, in APS Records.

67. L. R. Jones to A. D. Selby, September 19, 1910, in APS Records.

68. A. D. Selby to L. R. Jones, September 21, 1910, in Selby Files.

69. C. L. Shear to A. D. Selby, November 3, 1910, in Selby Files.

70. L. R. Jones, C. L. Shear and H. H. Whetzel to A. D. Selby, October 25, 1910, in Selby Files.

71. Board of Editors to the Members of the American Phytopathological Society, March 16, 1912, in APS Records.

72. Jones 1919, 163.

73. Ibid., 160.

74. Ibid., 163.

75. Ibid., 163.

76. Ibid., 160.

77. Whetzel 1918, 111.

Epilogue

1. Whetzel 1918, 1.

2. War Emergency Board 1918b, 1.

3. War Emergency Board 1918c, 1.

4. Whetzel 1918, 3.

5. Ibid., 2.

6. War Emergency Board 1918a, 19–20.

7. Whetzel 1918, 2.

8. Lyman 1918a, 212.

9. War Emergency Board 1918a, 2.

10. Lyman 1918b, 220.

11. Ibid., 221.

12. C. Larsen press release, April 22, 1918, in Johnson Topical Files, National Fungus Collections.

13. W. W. Robbins to H. H. Whetzel, October 26, 1918, in Whetzel Papers.

14. Campbell, Peterson, and Griffith 1998.

❧ BIBLIOGRAPHY ❧

AAAS (American Association for the Advancement of Science). 1885. *Proceedings*, ed. by F. Putnam. Salem, Mass.: Permanent Secretary, AAAS. [Ch. 6]

———. 1886. *Proceedings*, ed. by F. Putnam. Salem, Mass.: Permanent Secretary, AAAS. [Ch. 6]

Acree, S. F. 1940. Obituary: Haven Metcalf, 1875–1940. *Science* 92:98–99. [Ch. 10]

Act of July 14, 1890, 26 Stat. L. 283, 285, Statutes at large of the United States of America, 1798–1915. [Ch. 9]

Act of March 3, 1887, 24 Stat. L. 495, Statutes at large of the United States of America, 1798–1915. [Ch. 9]

Agricola [A North Carolina farmer]. 1819. *A series of essays on agriculture and rural affairs in forty-seven numbers.* Raleigh, N.C.: Joseph Sales. [Ch. 1]

———. 1860. Diseases of vegetation. *The Magazine of Horticulture, Botany, and All Useful Discoveries and Improvements in Rural Affairs* 26:55–57. [Ch. 3]

"Agricultural Department at Washington: What it might be, ought to be, but is not." 1859. *American Agriculturist* 18:103–4. [Ch. 3]

"Agricultural education." 1863. *Country Gentleman* 22:187. [Ch. 4]

"Agricultural education: Accumulating signs in favor of a general system of appropriate education for agriculturists—voice of 'the cultivator.'" 1847. *Monthly Journal of Agriculture* 3:88–89. [Ch. 3]

"Agricultural meeting at the State House—Tuesday evening, January 14." 1845. *New England Farmer and Horticultural Register* 23:237–38. [Ch. 2]

"Agricultural schools and colleges." 1856. *The Valley Farmer* 8:173–74. [Ch. 3]

Ainsworth, G. C. 1981. *Introduction to the history of plant pathology.* Cambridge, England: Cambridge University Press. [Ch. 2]

Allen, G. E. 1978. *Life science in the twentieth century.* Cambridge, Mass.: Cambridge University Press. [Ch. 10]

Allen, J. F. 1853. *A practical treatise on the culture and treatment of the grape vine: Embracing its history, with directions for its treatment, in the United States of America, in the open air, and under glass structures, with and without artificial heat.* New York: C. M. Saxton. [Ch. 3]

Allen, J. M. 1922. *United States Department of Agriculture Library bibliographic contributions: Check list of publications of the State Agricultural Experiment Stations on the subject of plant pathology, 1876–1920.* Washington, D.C.: GPO. [Ch. 8]

Allen, R. L. 1844. Agricultural colleges. *The American Agriculturist* 3:52–54. [Ch. 3]

Alvord, B. 1850. The polar plant, or *Silphium laciniatum.* In *Proceedings of the American Association for the Advancement of Science, second meeting held at Cambridge, August 1849,* 2:12–16. Boston: Henry Flanders and Co. [Ch. 12]

American Philosophical Society. 1799. Premiums. *Transactions of the American Philosophical Society, held at Philadelphia, for promoting useful knowledge* 4:5. [Ch. 1]

———. 1843. Discourse. *Proceedings of the American Philosophical Society, held at Philadelphia, for promoting useful knowledge* 3:3–36. [Ch. 1]

American Phytopathological Society Records. Department of Special Collections. The Parks Library. Iowa State University. Ames, Iowa. [Ch. 12]

Anderson, P. J. 1914. *The morphology and life history of the chestnut blight fungus.* The Commission for the Investigation and Control of the Chestnut Tree Blight Disease in Pennsylvania Bulletin 7. Harrisburg, Pa. [Ch. 10]

Andrews, J. H., and M. Marosy. 1986. The academic experience. In *With one foot in the furrow: A history of the first seventy-five years of the Department of Plant Pathology at the University of Wisconsin—Madison,* ed. by P. H. Williams and M. Marosy, 248–64. Dubuque, Iowa: Kendall/Hunt Publishing Co. [Ch. 11]

Antill, E. 1771. An essay on the cultivation of the vine, and the making and preserving of wine, suited to the different climates in North America. In Section 2, *Essays on agriculture,* of *Transactions of the American Philosophical Society* 1:117–97. [Ch. 1]

"Apple tree blight." 1858. *The Valley Farmer* 10:281. [Ch. 3]

Arboreum. 1820. Canker on plum trees. *Rural Magazine and Literary Evening Fire-side* 1:174. [Ch. 3]

Arny, D. C. 1986. Field crops. In *With one foot in the furrow: A history of the first seventy-five years of the Department of Plant Pathology at the University of Wisconsin—Madison,* ed. by P. H. Williams and M. Marosy, 164–74. Dubuque, Iowa: Kendall/Hunt Publishing Co. [Ch. 11]

Arthur, J. C. [1878] 1981. On the structure of the *Echinocystis lobata.* Reprint in *Iowa State Journal of Research* 55:219–34. [Ch. 5]

————. 1885a. Pear blight and its cause. *American Naturalist* 19:1177–85. [Ch. 5]

————. 1885b. Proof that bacteria are the direct cause of the disease in trees known as pear blight. *Botanical Gazette* 10:343–45. [Ch. 5]

————. 1885c. Some botanical laboratories of the United States. *Botanical Gazette* 10:395–406. [Ch. 4]

————. 1887. Pear blight *Micrococcus amylovorus* (Bur.). In *Report of the Commissioner of Agriculture 1886*, 125–29. Washington, D.C.: GPO. [Ch. 5]

————. 1896. Report of the Botanical Department. In *Purdue University, eighth annual report of the Agricultural Experiment Station, Lafayette, Ind., 1895*, 21–28. Indianapolis, Ind.: Wm. B. Burford. [Ch. 8]

————. 1897. *Formalin for prevention of potato scab.* Purdue University Agricultural Experiment Station Bulletin 65. Lafayette, Ind. [Ch. 8]

————. 1898. Report of the Botanical Department. In *Purdue University, tenth annual report of the Agricultural Experiment Station, Lafayette, Ind., 1897*, 25–34. Indianapolis, Ind.: Wm. B. Burford. [Ch. 8]

————. 1906. An address on the history and scope of plant pathology. In *Congress of Arts and Science, Universal Exposition, St. Louis, 1904*, vol. 5, *Biology, Anthropology, Psychology, Sociology*, 149–64. Boston: Houghton Mifflin. [Ch. 10]

Arthur, J. C., and W. Stuart. 1900. Corn smut. In *Purdue University, twelfth annual report of the Indiana Agricultural Experiment Station, Lafayette, Ind., for the year ending June 30, 1899*, 84–135. Lafayette, Ind.: Home Journal Co. [Ch. 8]

Atkinson, G. F. 1889. Nematode root-galls. In *Report of Agricultural Experiment Station, Agricultural and Mechanical College, Auburn, Alabama, December 1889*, 4–6. Bulletin 9. Auburn, Ala. [Ch. 8, 10]

————. 1890a. Experiment station abstracts: Nematode root galls. *Agricultural Science* 4:111–12. [Ch. 8]

————. 1890b. A *new root rot disease of cotton.* Agricultural Experiment Station, Agricultural and Mechanical College Bulletin 21. Auburn, Ala. [Ch. 8]

————. 1891a. Anthracnose of cotton. *Journal of Mycology* 6:173–78. [Ch. 8]

————. 1891b. *Black "rust" of cotton.* Agricultural Experiment Station, Agricultural and Mechanical College Bulletin 27. Auburn, Ala. [Ch. 8]

————. 1892a. *Cryptogamic botany and plant pathology.* Cornell University Agricultural Experiment Station, Bulletin 49, 404–17. Ithaca, N.Y. [Ch. 8]

————. 1892b. *Some diseases of cotton.* Agricultural Experiment Station, Agricultural and Mechanical College Bulletin 41. Auburn, Ala. [Ch. 8]

————. 1892c. *Some leaf blights of cotton.* Agricultural Experiment Station, Agricultural and Mechanical College Bulletin 36. Auburn, Ala. [Ch. 8]

————. 1893a. *Cryptogamic botany and plant pathology.* Cornell University Agricultural Experiment Station, Bulletin 61, 305–10. Ithaca, N.Y. [Ch. 8]

————. 1893b. *Edema of the tomato.* Cornell University Agricultural Experiment Station, Botanical Division, Bulletin 53. Ithaca, N.Y. [Ch. 8]

————. 1894. *Leaf curl and plum pockets. Contribution to the knowledge of the Prunicolous Exoasceae of the United States.* Cornell University Agricultural Experiment Station, Botanical Division, Bulletin 73. Ithaca, N.Y. [Ch. 8]

————. 1895. *Damping off.* Cornell University—Agricultural Experiment Station, Botanical Division, Bulletin 94. Ithaca, N.Y. [Ch. 8]

Atkinson, G. F. Papers. 1880–1918. Rare and Manuscript Collections, Carl A. Kroch Library, Cornell University, Ithaca, N.Y. [Ch. 8, 11]

Baker, G. L., W. D. Rasmussen, V. Wiser, and J. M. Porter. 1963. *Century of service: The first 100 years of the United States Department of Agriculture.* Washington, D.C.: GPO. [Ch. 10]

Baker, K. F. 1969. Plant pathology and mycology. In *A short history of botany in the United States,* ed. by J. Ewan, 82–88. New York: Hafner Publishing Company. [Ch. 2]

————. 1971. Fire blight of pome fruits: The genesis of the concept that bacteria can be pathogenic to plants. *Hilgardia* 40:603–33. [Ch. 5]

Barrett, J. T. 1918. Thomas J. Burrill (1839–1916). *Phytopathology* 8:1–4. [Ch. 5]

Barrus, M. F. 1911. Variation of varieties of beans in their susceptibility to anthracnose. *Phytopathology* 1:190–95. [Ch. 11]

————. 1915. An anthracnose-resistant red kidney bean. *Phytopathology* 5:303–11. [Ch. 11]

————, and E. C. Stakman, 1945. Herbert Hice Whetzel, 1877–1944. *Phytopathology* 35:661. [Ch. 11]

Bartlett, L. 1863. Wheat-growing in New Hampshire. In *Report of the Commissioner of Agriculture for the year 1862,* 96–104. Washington, D.C.: GPO. [Ch. 6]

Bates, R. S. 1965. *Scientific societies in the United States.* 3d ed. Cambridge, Mass.: MIT Press. [Ch. 1, 12; Pt. 1 intro.]

Beal, W. J. 1883. Agriculture: Its need and opportunities. *Science* 2:328–33. [Ch. 6]

Beardsley, E. H. 1969. *Harry L. Russell and agricultural science in Wisconsin.* Madison: University of Wisconsin Press. [Ch. 9, 11]

Beecher, H. W. 1844. The blight in the pear tree, its cause and a remedy for it. *The Magazine of Horticulture, Botany, and All Useful Discoveries and Improvements in Rural Affairs* 10:441–56. [Ch. 1]

Belden, H. 1837. Blight in pear trees. *Farmers' Register* 5:461. [Ch. 1]

Berkeley, M. J. [1846] 1948. *Observations, botanical and physiological on the potato murrain, together with selections from Berkeley's Vegetable Pathology.* Reprint, Phytopathological Classics 8. East Lansing, Mich.: American Phytopathological Society. [Ch. 2, 4]

———. 1854–57. Vegetable pathology [173 articles]. *Gardener's Chronicle,* January 7, 1854; October 3, 1857. [Ch. 4]

Bessey, C. E. 1874. Choosing a microscope. *Aurora,* April 1874, 7. [Ch. 4]

———. 1875. On injurious fungi. In *Biennial report of the Board of Trustees of the Iowa State Agricultural College and Farm to the Governor of Iowa,* 128–33. Des Moines: G. W. Edwards. [Ch. 4]

———. 1877. On injurious fungi. *Biennial report of the Board of Trustees of the Iowa State Agricultural College and Farm to the Governor of Iowa,* 185–204. Des Moines: G. W. Edwards. [Ch. 4]

———. 1880. *Botany for high schools and colleges.* New York: Henry Holt. [Ch. 4]

———. 1881. A sketch of the progress of botany in the United States in 1880. *American Naturalist* 15:947–55. [Ch. 5]

———. 1882. The diseases of plants. *Transactions of the Iowa State Horticultural Society for 1881* 16:85–98. [Ch. 4]

———. 1883a. A government duty. *American Naturalist* 17:543–44. [Ch. 6]

———. 1883b. On parasitic and other fungi. *Transactions of the Iowa State Horticultural Society for 1882* 17:280–84. [Ch. 4]

———. 1884. *Essentials of botany.* New York: Henry Holt. [Ch. 4]

———. 1885. The Botanical Club of the A.A.A.S. *The American Naturalist* 19:802–3. [Ch. 4]

———. 1886. Injurious fungi, in their relation to the diseases of plants. In *Proceedings of the twentieth session of the American Pomological Society held in Grand Rapids, Mich.,* 35–43. Cleveland, Ohio: American Pomological Society. [Ch. 4]

Bessey, C. E. Biographical/bibliographical file. Archives, University of Nebraska, Lincoln. [Ch. 4]

Bessey, C. E. Papers, 1865–1915. Microfilm Edition, ed. by G. Svoboda and P. Churray, University of Nebraska Studies: New Series 67. Lincoln: University of Nebraska Press. [Ch. 6]

Bessey, C. E. Papers. Archives, Iowa State University, Ames. [Ch. 4]

Bessey, E. A. 1911. *Root-knot and its control.* USDA Bureau of Plant Industry Bulletin 217. Washington, D.C.: GPO. [Ch. 10]

————. 1935. The teaching of botany sixty-five years ago. *Iowa State College Journal of Science* 9:227–33. [Ch. 4]

————. 1955. Mycology. In *A Century of Progress in the Natural Sciences: 1853–1953*, 225–65. San Francisco: California Academy of Sciences. [Ch. 4]

Bidwell, P. W., and J. I. Falconer. 1925. *History of agriculture in the northern United States, 1620–1860*. Washington, D.C.: Carnegie Institution. [Ch. 2]

Biggs, A. R., ed. 1993. *Handbook of cytology, histology, and histochemistry of fruit tree diseases*. Boca Raton: CRC Press. [Ch. 1]

Biographical files, National Fungus Collections. U.S. Department of Agriculture Systematic Botany and Mycological Laboratory, Beltsville, Md. [Ch. 4, 5, 11]

Blanchard, F. C. 1957. Nathan A. Cobb, botanist and zoologist, a pioneer scientist in Australia. *Asa Gray Bulletin* 3: 205–13. [Ch. 10]

"Blight in pear trees, etc." 1829. *New England Farmer* 8:57. [Ch. 1]

"The Blight in the pear tree." 1847. *The Horticulturist and Journal of Rural Art and Rural Taste* 1:58–62. [Ch. 3]

"Blight of the Pryor's red apple tree." 1857. *The Valley Farmer* 9:27–28. [Ch. 3]

Bolley, H. L. 1889. The wintering of rust in winter wheat. *Agricultural Science* 3:105–7. [Ch. 8]

————. 1890. Potato scab: A bacterial disease. *Agricultural Science* 4:243–56. [Ch. 8]

————. 1891a. *Grain smuts.* Government Agricultural Experiment Station for North Dakota Bulletin 1. Fargo, N.D. [Ch. 8]

————. 1891b. *Potato scab and possibilities of prevention. A disease of beets identical with "deep scab" of potatoes. Hastening the maturity of potatoes.* Government Agricultural Experiment Station for North Dakota Bulletin 4. Fargo, N.D. [Ch. 8]

————. 1891c. Wheat rust: Is the infection local or general in origin? *Agricultural Science* 5:259–64. [Ch. 8]

————. 1897. *New work upon the smuts of wheat, oats and barley, and a resume of treatment experiments for the last three years.* Government Agricultural Experiment Station for North Dakota Bulletin 27. Fargo, N.D. [Ch. 8]

————. 1900. Department of Botany. In *Tenth annual report of the North Dakota Agricultural Experiment Station, Agricultural College, N.D. to the Governor of North Dakota*, 22–25. Grand Forks, N.D.: Herald, State Printers and Binders. [Ch. 8]

————. 1901. *Flax wilt and flax sick soil.* North Dakota Agricultural College Government Agricultural Experiment Station for North Dakota Bulletin 50. Fargo, N.D. [Ch. 8]

————. 1906. *Flax culture.* North Dakota Agricultural College Government Agricultural Experiment Station for North Dakota Bulletin 71. Fargo, N.D. [Ch. 8]

Bollman, L. 1863. The wheat plant. In *Report of the Commissioner of Agriculture for the year 1862*, 65–95. Washington, D.C.: GPO. [Ch. 6]

———. 1865. The hop plant. In *Report of the Commissioner of Agriculture for the year 1864*, 97–110. Washington, D.C.: GPO. [Ch. 6]

Booth, C. M. 1971. *The genus Fusarium*. Kew, England: Commonwealth Mycological Institute. [Appendix]

Bornstein, F. P. 1963. Doctors afield: George Engelmann (1809–1884), botanist of the West. *The New England Journal of Medicine* 268:1293–95. [Ch. 4]

Bourke, A. 1993. *"The visitation of God"? The potato and the Great Irish Famine*. Irish Historical Studies, ed. by Jacqueline Hill and Cormac Ó Gráda. Dublin: Lilliput Press. [Ch. 2]

Bouvé, T. 1880. *Historical sketch of the Boston Society of Natural History; with a notice of the Linnaean Society, which preceded it*. Boston: Boston Society of Natural History. [Ch. 2]

Bowler, P. 1989. *The Mendelian revolution*. Baltimore, Md.: The Johns Hopkins University Press. [Ch. 10]

Boyd, S. S. 1854. Peaches, apricots, and nectarines: condensed correspondence. In *Report of the Commissioner of Patents, for the year 1853*, 284. Washington, D.C.: Beverly Tucker. [Ch. 3]

Briggs, G. W. 1873. Notes on blight of the pear. In *Proceedings of the fourteenth session and quarter centennial celebration of the American Pomological Society*, ed. by H. T. Williams, 71–72. Cleveland, Ohio: American Pomological Society. [Ch. 5]

Brock, T. D. 1988. *Robert Koch: A life in medicine and bacteriology*. Madison, Wis.: Science Tech Publishers. [Ch. 5]

Brown, I. 1858. Indian corn. In *Fifth report of the Indiana State Board of Agriculture, containing the transactions of the board, and reports of county societies, for the year 1856*, 331–55. Indianapolis, Ind.: Joseph J. Bingham. [Ch. 3]

Browne, C. A., and M. E. Weeks. 1952. *A history of the American Chemical Society: Seventy-five eventful years*. Washington, D.C.: American Chemical Society. [Ch. 12]

Bruce, R. V. 1987. *The launching of modern American science, 1846–1876*. New York: Alfred A. Knopf. [Ch. 2, 4; Pt. 2 intro.]

Buchanan, T. S. 1976. *Forest disease research in the western United States and British Columbia, Canada*. Asheville, N.C.: USDA Forest Service, Southeastern Forest Experiment Station. [Ch. 10]

Buel, J. 1827. The disease in pear and apple trees. *New England Farmer* 6:108. [Ch. 1]

———. 1832. Blight in pear trees. *Southern Agriculturist* 5:54. [Ch. 1]

Burkett, C. W. 1900. *History of Ohio agriculture: A treatise on the development of the various lines and phases of farm life in Ohio*. Concord, N.H.: Rumford Press. [Ch. 3]

Burkholder, W. H. 1918. The production of an anthracnose-resistant white marrow bean. *Phytopathology* 8:353–59. [Ch. 11]

———. 1923. The gamma strain of *Colletotrichum lindemuthianum* (Sacc. and Magn.) etc. *Phytopathology* 13:316–23. [Ch. 11]

Burrill, T. J. 1869. Agricultural botany. Illinois Industrial University, minutes of the Board of Trustees. 2d Report. 319–30. [Ch. 5]

———. 1878. Report on botany and vegetable physiology: Pear blight. *Transactions of the Illinois State Horticultural Society for 1877* 11:114–16. [Ch. 5]

———. 1879. Fire blight. *Transactions of the Illinois State Horticultural Society for 1878* 12:79–82. [Ch. 5]

———. 1881a. Anthrax of fruit trees; or the so-called fire blight of pear, and twig blight of apple, trees. In *Proceedings of the American Association for the Advancement of Science, twenty-ninth meeting held at Boston, Mass., August, 1880*, 29:583–97. Salem, Mass.: AAAS. [Ch. 5, 6, 12]

———. 1881b. Bacteria as a cause of disease in plants. *American Naturalist* 15:527–31. [Ch. 5]

———. 1881c. Blight. *Botanical Gazette* 6:271–73. [Ch. 5]

———. 1881d. Blight of pear and apple trees. In *Illinois Industrial University Board of Trustees tenth report for two years, ending August, 1880*, 62–84. Springfield, Ill.: H. W. Rokker. [Ch. 5]

———. 1881e. Report on botany and vegetable physiology: pear and apple-tree blight. *Transactions of the Illinois State Horticultural Society* 14:157–67. [Ch. 5]

———. 1882a. Bacteria and their effects. *Transactions of the Illinois State Horticultural Society* 15:165–84. [Ch. 5]

———. 1882b. The bacteria: An account of their nature and effects, together with a systematic description of the species. In *Illinois Industrial University Board of Trustees eleventh report*, 93–157. Urbana, Ill.: Illinois Industrial University. [Ch. 5]

———. 1883a. New species of *Micrococcus* (bacteria). *American Naturalist* 17:319–20. [Ch. 5]

———. 1883b. Notes on parasitic fungi. In *Proceedings of the first, second, and third meeting of the Society for the Promotion of Agricultural Science*, ed. by W. S. Beal, and L. B. Arnold, 103–10. Syracuse, N.Y.: Farmer and Dairyman Print. [Ch. 5]

———. 1883c. Pear blight and peach yellows. *Transactions of the Illinois State Horticultural Society* 17:46–50. [Ch. 5]

———. 1887. A disease of broom-corn and sorghum. In *Proceedings of the eighth annual meeting of the Society for the Promotion of Agricultural Science*, ed. by W. R. Lazenby, 30–36. Columbus, Ohio: Gazette Printing House. [Ch. 5]

————. 1889. A bacterial disease of Indian corn. In *Proceedings of the tenth annual meeting of the Society for the Promotion of Agricultural Science*, ed. by W. R. Lazenby, 19–25. Columbus, Ohio: Gazette Printing House. [Ch. 5]

————. 1890. Preliminary notes upon the rotting of potatoes. In *Proceedings of the eleventh annual meeting of the Society for the Promotion of Agricultural Science, held at Indianapolis, Ind. 1890*, 21–22. Columbus, Ohio: Gazette Printing House. [Ch. 9]

Burrill, T. J. Exhibit File. University of Illinois Archives, Urbana. [Ch. 5]

Busch, L., and W. B. Lacy. 1983. *Science, agriculture and the politics of research*. Boulder, Colo.: Westview Press. [Ch. 4, 5]

Bush, A. 1846. Rot in potatoes. *Transactions of the New York State Agricultural Society* 5:342–79. [Ch. 2]

Bush, I. 1880. Grape-rot. In *Proceedings of the seventeenth session of the American Pomological Society, held in Rochester, New York, September 17th, 18th, and 19th, 1879*, 41–44. Cleveland, Ohio: American Pomological Society. [Ch. 7]

C. D. 1844. On the blight of pear trees. *Western Farmer* 5:94. [Ch. 1]

C.W.E. 1844. The blight. *Western Farmer* 4:275. [Ch. 1]

Campbell, C. L. 1983. Erwin Frink Smith: Pioneer plant pathologist. *Annual Review of Phytopathology* 21:21–27. [Ch. 7, 9]

————, ed. and trans. 1981. *The Fischer-Smith controversy: Are there bacterial diseases of plants?* Phytopathological Classics 13. St. Paul, Minn.: APS Press. [Ch. 9]

Campbell, C. L., P. D. Peterson, and C. S. Griffith. 1998. The War Emergency Board. *Plant Disease* 82:121–25. [Epilogue]

Campbell, R. N. 1979. Fire blight. *Natural History* 88:62–69. [Ch. 1]

Capell, E. J. 1858. Oats—rust, etc. *Southern Cultivator* 16:238. [Ch. 3]

Carefoot, G. L., and E. R. Sprott. 1967. *Famine on the wind: Man's battle against plant disease*. Chicago: Rand McNally. [Ch. 2]

Carleton, M. A. 1905. *Lessons for the grain-rust epidemic of 1904*. USDA Farmers' Bulletin 219. Washington, D.C.: GPO. [Ch. 11]

Carrier, L. 1923. *The beginnings of agriculture in America*. New York: McGraw-Hill. [Ch. 1]

Carstensen, V. 1960. The genesis of an agricultural experiment station. *Agricultural History* 34:13–20. [Ch. 8]

Castor, J. Y. 1839. Blast in pear trees. *Western Farmer* 1:7–8. [Ch. 1]

Cattell, J. M., and J. Cattell. 1927. Spalding, Prof. Effie S(outhworth). In *American men of science*, 4th ed., s.v. New York: The Science Press. [Ch. 7]

Chester, F. D. 1889. Report of the mycologist. In *Second annual report of the Delaware College*

Agricultural Experiment Station 1889, 69–109. Wilmington, Del.: Delaware Printing Company. [Ch. 8]

———. 1891. Report of the mycologist. In *Third annual report of the Delaware College Agricultural Experiment Station, 1890*, 44–91. Wilmington, Del.: Mercantile Printing Company. [Ch. 8]

———. 1892. *Report of the mycologist.* In *Fourth annual report of the Delaware College Agricultural Experiment Station, 1891*, 40–74. Wilmington, Del.: Mercantile Printing Company. [Ch. 8]

Christensen, C. M. 1984. *E. C. Stakman, statesman of science.* St. Paul, Minn.: American Phytopathological Society. [Ch. 11]

———. 1992. Elvin Charles Stakman, May 17, 1885–January 22, 1979. In *National Academy of Sciences of the United States of America Biographical Memoirs,* s.v. 61:331–49. [Ch. 11]

Clarke, G. 1830. On the rot or mildew of grapes. *Southern Agriculturist* 3:528–33. [Ch. 3]

Clay, C. M. 1859. The grape rot. *The Ohio Farmer* 8:241. [Ch. 3]

[Clements, J.] 1904. The American Mycological Society. *Journal of Mycology* 10:46–47. [Ch. 12]

Clinton, G. P. 1920. William Gilson Farlow. *Phytopathology* 10:1–8. [Ch. 4]

Cobb, N. A. 1890a. *Oxyuris* larvae hatched in the human stomach under normal conditions. *Proceedings of the Linnean Society of New South Wales* 5:168–85. [Ch. 10]

———. 1890b. Two new instruments for biologists. *Proceedings of the Linnean Society of New South Wales* 5:157–67. [Ch. 10]

———. 1891. Parasites in the stomach of a cow. *Agricultural Gazette of New South Wales* 2:524–34. [Ch. 10]

———. 1894. *Tricoma* and other new nematode genera. *Proceedings of the Linnean Society of New South Wales* 8:389–421. [Ch. 10]

———. 1896. Wormy fowls. *Agricultural Gazette of New South Wales* 7:746–53. [Ch. 10]

———. 1897. The sheep-fluke. *Agricultural Gazette of New South Wales* 8:453–81. [Ch. 10]

———. 1906. *Fungus maladies of the sugar cane with notes on associated insects and nematodes.* Report of the work of the Experiment Station of the Hawaiian Sugar Planters' Association Bulletin 5. Honolulu: Hawaiian Gazette Co. [Ch. 10]

———. 1915. Nematodes and their relationships. In *Yearbook of the United States Department of Agriculture, 1914*, 457–90. Washington, D.C.: GPO. [Ch. 10]

———. 1917. Notes on nemas. *Contributions to a Science of Nematology* 5:117–28. [Ch. 10]

———. 1918. *Estimating the nema population of the soil, with special reference to the sugar-beet and root-gall nemas, Heterodera schachtii Schmidt and Heterodera radicicola (Greef) Müller, and with a description of Tylencholaimus aequalis n. sp.* USDA Bureau of Plant Industry, Agricultural Technology Circular. Washington, D.C.: GPO. [Ch. 10]

Cochrane, M. F. 1990. Back from the brink: Chestnuts. *National Geographic* 177:128–40. [Ch. 10]

Cochrane, W. W. 1979. *The development of American agriculture: A historical analysis.* Minneapolis: University of Minnesota Press. [Ch. 3]

Coleman, W. R. 1977. *Biology in the nineteenth century: Problems of form, function, and transformation.* Cambridge, Mass.: Cambridge University Press. [Ch. 4, 5; Pt. 2 intro.]

Colman, G. P. 1963. *Education and agriculture: A history of the New York State College of Agriculture at Cornell University.* Ithaca, N.Y.: Cornell University. [Ch. 1, 4, 5, 11]

————. 1965. Government and agriculture in New York State. *Agricultural History* 39:41–50. [Ch. 5]

Colman, N. J. 1888. Section of Vegetable Pathology. In *Report of the Commissioner of Agriculture, 1887,* 27–30. Washington, D.C.: GPO. [Ch. 7]

————. 1889. Section of Vegetable Pathology. In *Report of the Commissioner of Agriculture, 1888,* 33–34. Washington, D.C.: GPO. [Ch. 7]

Committee on Agriculture (HR 49A-H2.7). 49th Congress. Records of the U.S. House of Representatives, Record Group 233. National Archives, Washington, D.C. [Ch. 6]

Cooper, W. C. 1995. *The United States Research Laboratory: A century of USDA subtropical-horticultural research: 1892–1992.* Florida Citrus Research Foundation. [Ch. 9]

"Cornell University Agricultural Experiment Station: Fifty Years of Research, 1887–1937." 1937. In *New York State College of Agriculture at Cornell University, Ithaca, N.Y., Cornell University Agricultural Experiment Station, fiftieth annual report,* 84–172. Ithaca, N.Y.: Cornell University. [Ch. 11]

Coulter, J. M. 1883. Botany at the Minneapolis meeting of the AAAS. *The Botanical Gazette* 8:291–95. [Ch. 12]

Coulter, T. 1802. Description of a method of cultivating peach trees, with a view to prevent their premature decay; confirmed by the experience of forty-five years, in Delaware state and the western parts of Pennsylvania. In Appendix 2 of *Transactions of the American Philosophical Society, held at Philadelphia, for promoting useful knowledge* 5:327–28. [Ch. 1]

Cowles, H. C. 1909. The American Association for the Advancement of Science Section G—Botany. *Science* 29:903–16. [Ch. 12]

Coxe, W. 1817. *A view of the cultivation of fruit trees, and the management of orchards and cider; with accurate descriptions of the most estimable varieties of native and foreign apples, pears, peaches, plums, and cherries, cultivated in the middle states of America.* Philadelphia: M. Carey and Son. [Ch. 1]

Cranefield, F. 1902. Emmett Stull Goff. *Wisconsin Alumni Magazine* 3:409–13. [Ch. 11]

Cremin, L. A. 1961. *The transformation of the school: Progressivism in American education, 1876–1957.* New York: Alfred A. Knopf. [Ch. 11]

Crosby, A. W. 1986. *Ecological imperialism: The biological expansion of Europe, 900–1900.* New York: Cambridge University Press. [Ch. 1]

Crosier, E. S. 1858. Rust on apple leaves. *The Ohio Cultivator* 14:299–300. [Ch. 3]

Cummins, G. B. 1978. J. C. Arthur: The man and his work. *Annual Review of Phytopathology* 16:19–30. [Ch. 5]

"Cure for blight of pear trees." 1836–1837. *Farmers' Register* 4:395. [Ch. 1]

"Curl in the peach." 1853. *The Cultivator* 1:251. [Ch. 3]

D.J.B. 1855. Wheat Diseases. In *Report of the Commissioner of Patents, for the year 1854,* 184–86. Washington, D.C.: Beverly Tucker. [Ch. 3]

D. T. [David Thomas]. 1831. Peach trees. *New York Farmer and Horticultural Repository, devoted to practical husbandry and gardening, and embracing the most important information in the sciences, intimately connected with rural pursuits* 4:46–47. [Ch. 1]

Dagge, E. 1848. Mode of preserving peach trees from mildew. *The Southern Cultivator* 6:190. [Ch. 3]

Danbom, D. B. 1990. *Our purpose is to serve: The first century of the North Dakota Agricultural Experiment Station.* Fargo: North Dakota Institute for Regional Studies. [Ch. 8]

Danhof, C. 1969. *Change in agriculture: The northern United States, 1820–1870.* Cambridge, Mass.: Harvard University Press. [Ch. 3]

Daniels, G. H. 1971. *Science in American society: A social history.* New York: Alfred A. Knopf. [Ch. 1, 2]

Darling, N. 1845. The "yellows" in peach trees. *The Cultivator* 2:60–62. [Ch. 1, 3, 7]

———. 1846. The yellows. *The Cultivator* 3:141–42. [Ch. 3]

Day, C. A. 1954. A history of Maine agriculture, 1604–1860. *University of Maine Bulletin,* 2d series, no. 68. Orono, Maine: University of Maine Press. [Ch. 2]

De Bary, A. [1853] 1969. *Investigations of the brand fungi and the diseases of plants caused by them with reference to grain and other useful plants.* Reprint, Phytopathological Classics 11, trans. by R.M.S. Heffner, D. C. Arny, and J. D. Moore. Madison, Wis.: American Phytopathological Society. [Ch. 2, 4]

————. 1861. *Die gegenwärtig herrschende Kartoffelkrankeit, ihre Ursache und ihre Verütung.* Leipzig: Arthur Felix. [Ch. 4]

————. 1865–66. *Neue Untersuchungen über Uredineen, inbesondere die Entwicklung der Puccinia graminis und den Zusammenhang derselbe mit Aecidium berberidis.* In two parts, 15–49, 205–15. Berlin: Monatsberichte der Berlinder Academie. [Ch. 4]

De Schweinitz, L. D. 1834. *Dothidea pomigena.* Transactions of the American Philosophical Society, n.s. 4:232. Philadelphia: James Kay, Jun. and Co. [Ch. 2]

Deane, S. 1797. *The New England farmer; or georgical dictionary, containing a compendious account of the ways and methods in which the important art of husbandry, in all its various branches, is, or may be practised, to the greatest advantage, in this country.* 2d ed. Worcester, Mass.: Isaiah Thomas. [Ch. 2]

Denning, W. 1794. On the decay of apple trees. *Transactions of the New York Society for the Promotion of Agriculture, Arts and Manufacturers* 2:219–22. [Ch. 1]

"Department of Agriculture." 1863. *Country Gentleman* 21:25–26. [Ch. 6]

"Digging potatoes—corn smut—plaster." 1844. *The Cultivator* 1:15. [Ch. 3]

"Diseases and enemies of fruit trees." 1835–1836. *Farmers' Register* 3:340–42. [Ch. 3]

"Diseases of plants." 1852. *The Cultivator* 9:386. [Ch. 3]

Dodge, A. W. 1850. Orchards—Their cultivation and management. In *Report of the Commissioner of Patents, for the year 1849,* 276–81. Washington, D.C.: Office of Printers to the House of Representatives. [Ch. 3]

Downing, A. J. 1872. *The fruits and fruit-trees of America; or, the culture, propagation, and management, in the garden and orchard, of fruit-trees generally; with descriptions of all the finest varieties of fruit, native and foreign, cultivated in this country.* New York: John Wiley and Son. [Ch. 1]

Dudley, W. R. 1889. *The strawberry leaf-blight.* Cornell University College of Agriculture Bulletin of the Agricultural Experiment Station Bulletin 14. Ithaca, N.Y. [Ch. 11]

————. 1891. Report of the cryptogamic botanist. In *Third annual report of the Agricultural Experiment Station. Cornell University, College of Agriculture, Ithaca, N.Y.,* 29–34. Ithaca, N.Y.: Cornell University. [Ch. 11]

Duhamel du Monceau, H. L. 1728. Explication physique d'une maladie qui fait périr plusieurs plantes dans le Gastinois et particulièrement le safran. *Memoirs of the Royal Academy of Science,* 100–12. [Ch. 2]

Dupree, A. H. 1957. *Science in the federal government: a history of policies and activities to 1940.* Cambridge, Mass.: Belknap Press. [Ch. 1, 2, 6, 10; Pt. 1, 3 intro.]

————. 1959. *Asa Gray.* Cambridge, Mass: Harvard University Press. [Ch. 4; pt. 2 intro]

"The editor to the reader." 1855. *The Valley Farmer* 7:440–41. [Ch. 3]

Eliot, J. 1934. *Essays upon field husbandry in New England and other papers, 1748–1762*. New York: Columbia University Press. [Ch. 1]

Elliot, C. A., and M. W. Rossiter, eds. 1992. *Science at Harvard University: Historical perspectives*. Bethlehem, Pa.: Lehigh University Press. [Ch. 4]

Elliot, C. W. 1844. The blight. *Western Farmer* 4:254–55. [Ch. 1]

Elliott, B. J., D. G. Lodgson, and P. Yowler. 1992. *Eustis Site Survey*. Orlando, Fla.: The Historic Works. [Ch. 9]

Elliott, F. R. 1858. *Elliott's fruit book; or, the American fruit-grower's guide in orchard and garden. Being a compendium of the history, modes of propagation, culture, etc., of fruit trees and shrubs, with descriptions of nearly all the varieties of fruits cultivated in this country: Notes of their adaptation to localities and soils, and also a complete list of fruits worthy of cultivation*. New York: A. O. Moore, Agricultural Book Publisher. [Ch. 3]

Ellis, J. 1802. Account of a method of preventing the premature decay of peach trees. In Appendix 1 of *Transactions of the American Philosophical Society, held at Philadelphia, for promoting useful knowledge* 5:325–26. [Ch. 1]

Engelmann, G. 1861. The president, Dr. Engelmann, in the chair. *The Transactions of the Academy of Sciences of St. Louis* 2 (1861–1868): 165–66. [Ch. 4]

———. 1883. *The true grape vines of the United States, and the diseases of the grape vines*. St. Louis, Mo.: R. P. Studley & Co. [Ch. 4]

Erni, H. 1866. The grape disease in Europe; its origin, history, phenomena and cure. In *Report of the Commissioner of Agriculture for the year 1865*, 324–38. Washington, D.C.: GPO. [Ch. 6]

[Fairchild, D. G.] 1892. Proceedings of the Botanical Club of the forty-first meeting of the AAAS, Rochester, N.Y., August 18–24, 1892. *Torrey Botanical Club Bulletin* 19:281–97. [Ch. 12]

Fairchild, D. G. 1893. Experiments in preventing leaf diseases of nursery stock in western New York. *Journal of Mycology* 7:240–64. [Ch. 9]

———. 1919. Byron David Halsted, botanist (1852–1918). *Phytopathology* 9:1–6. [Ch. 8]

———. 1938. *The world was my garden: Travels of a plant explorer*. New York: Charles Scribner's Sons. [Ch. 9]

Farley, J. 1977. *The spontaneous generation controversy from Descartes to Oparin*. Baltimore: Johns Hopkins University Press. [Ch. 2]

Farlow, W. G. 1875. The potato rot. *Bulletin of the Bussey Institution* 1:319–38. [Ch. 4]

———. 1876a. On a disease of olive and orange trees, occurring in California in the spring and summer of 1875. *Bulletin of the Bussey Institution* 1:404–14. [Ch. 4]

———. 1876b. On the American grape-vine mildew. *Bulletin of the Bussey Institution* 1:415–25. [Ch. 4]

———. 1876c. The black knot. *Bulletin of the Bussey Institution* 1:440–53. [Ch. 4]

———. 1877. The onion smut. In *Twenty-fourth annual report of the Secretary of the Massachusetts Board of Agriculture*, 164–76. Boston: Albert J. Wright, State Printer. [Ch. 4]

———. 1878. On the synonymy of some species of *Uredineae*. *Proceedings of the American Academy of Arts and Sciences* 13:251–52. [Ch. 4]

———. 1879. Fungi on forest trees. *Botanical Gazette* 4:244–46. [Ch. 4]

———. 1888. Vegetable parasites and evolution. In *Proceedings of the American Association for the Advancement of Science, thirty-sixth meeting held at New York, August, 1887*, 36:233–49. Salem, Mass.: Published by the Permanent Secretary. [Ch. 12]

———. 1913. The change from the old to the new botany in the United States. *Science* 37:79–86. [Ch. 4]

Farlow, W. G., R. Thaxter, and L. H. Bailey. 1919. George Francis Atkinson. *American Journal of Botany* 6:301–8. [Ch. 8]

A farmer of Tompkins County. 1846. Rust on wheat. *The Cultivator* 3:23. [Ch. 3]

Farmington. 1843. Rust in Cotton. *The Southern Cultivator* 1:162. [Ch. 3]

Farr, D. F., G. F. Bills, G. P. Chamuris, and A. Y. Rossman. 1989. *Fungi on plants and plant products in the United States.* St. Paul, Minn.: APS Press. [Appendix]

Field, T. W. 1858. *Pear Culture: A manual for the propagation, planting, cultivation, and management of the pear tree. With descriptions and illustrations of the most productive of the finer varieties, and selections of kinds most profitably grown for market.* New York: Orange Judd and Co. [Ch. 3]

Fischer, A. 1897. *Vorlesungen über Bakterien.* Jena, Germany: Gustav Fischer. [Ch. 9]

Fisher, D. V., and W. H. Upshall, eds. 1976. *History of fruit growing and handling in United States of America and Canada, 1860–1972.* Kelowna, British Columbia: Regatta City Press. [Ch. 1, 3, 9]

Fitzharris, J. C. 1974. Science for the farmer: The development of the Minnesota agricultural experiment station, 1868–1910. *Agricultural History* 48:202–20. [Ch. 4, 5, 8]

Fitzpatrick, H. M. 1919. George Francis Atkinson. *Science* 49:371–72. [Ch. 8]

———. 1937. Historical background of the Mycological Society of America. *Mycologia* 29:1–25. [Ch. 12]

———. 1945. Herbert Hice Whetzel. *Mycologia* 37:395–400. [Ch. 11]

———. 1983. Records and reminiscences in this anniversary year. *Cornell Plant Path. Newsletter* 28:24–26. [Ch. 12]

Fletcher, S. W. 1931. A history of fruit growing in Pennsylvania, Part 1: The colonial period (1623–1827). *Pennsylvania State Horticultural Association News* 8:1–26. [Ch. 1]

———. 1932. A history of fruit growing in Pennsylvania, Part 2: The transition period (1827–1887). *Pennsylvania State Horticultural Association News* 9:27–52. [Ch. 1]

———. 1976. *The Philadelphia Society for Promoting Agriculture, 1785–1955.* Revised edition. Philadelphia: Philadelphia Society for Promoting Agriculture. [Ch. 1]

Fontana, F. [1767] 1932. *Observations on the rust of grain.* Reprint, Phytopathological Classics 2, trans. by P. P. Pirone. Washington, D.C.: American Phytopathological Society. [Ch. 2]

Forsyth, W. 1802. *A treatise on the culture and management of fruit trees; in which a new method of pruning and training is fully described, together with observations on the diseases, defects, and injuries, in all kinds of fruit and forest trees; as also, an account of a particular method of cure, made public by order of the British government, to which are added, an introduction and notes, adapting the rules of the treatise to the climates and seasons of the United States of America by William Cobbett.* Philadelphia: J. Morgan. [Ch. 1]

Fothergill, A. 1808. Remarks on the smut and mildew of wheat; with hints on the most probable means of prevention. *Memoirs of the Philadelphia Society for Promoting Agriculture* 1:65–84. [Ch. 2]

"Fothergill, Anthony." 1937–1938. *The Dictionary of National Biography*, vol. 7, s.v. London: Oxford University Press. [Ch. 2]

Freeman, E. M. 1905. *Minnesota Plant Diseases.* St. Paul: University of Minnesota. [Ch. 11]

———. 1947. The department by decades. *Aurora* 23:3–12. [Ch. 11]

Freeman, E. M. Biographical File, University Archives, University of Minnesota, St. Paul, Minn. [Ch. 11]

"Fruit culture in North Carolina." 1853. *Arator* 1:370–71. [Ch. 3]

G. A. [Gray, A.] 1880. Review of *Botany for high schools and colleges. The American Journal of Science* 20:337. [Ch. 4]

G.S.B. [Gookins, S. B.] 1847. Remarks on the pear blight of the west. *The Horticulturist and Journal of Rural Art and Rural Taste* 1:253–56. [Ch. 3, 5]

Galloway, B. T. 1889a. Report of the Section of Vegetable Pathology. In *Report of the Commissioner of Agriculture*, 325–404. Washington, D.C.: GPO. [Ch. 7]

———. 1889b. Report of the Section of Vegetable Pathology. In *First report of the Secretary of Agriculture, 1889*, 397–432. Washington, D.C.: GPO. [Ch. 7]

———. 1890a. Report of the chief of the Division of Vegetable Pathology. In *Report of the Secretary of Agriculture, 1890*, 393–408. Washington, D.C.: GPO. [Ch. 7]

———. 1890b. *Report of the experiments made in 1889 in the treatment of the fungous diseases of plants.* USDA Botanical Division, Section of Vegetable Pathology Bulletin 11. Washington, D.C.: GPO. [Ch. 7]

————. 1890c. Some recent observations on black-rot of the grape. In *Proceedings of the eleventh annual meeting of the Society for the Promotion of Agricultural Science, held at Indianapolis, Indiana, 1890,* 60–63. Columbus, Ohio: Gazette Printing House. [Ch. 7]

————. 1891. Experiments with fungicides. In *Transactions of the Peninsula Horticultural Society, fourth annual session, held at Easton, Maryland, January 20, 21, and 22, 1891,* 26–27. Wilmington, Del.: A. P. Whitaker. [Ch. 7]

————. 1892a. Outline of the work by Mr. Fairchild in 1891. In *Tenth annual report of the Board of Control of the New York Agricultural Experiment Station, for the year, 1891,* 179–81. Albany, N.Y.: James B. Lyon. [Ch. 9]

————. 1892b. Report of the chief of the Division of Vegetable Pathology. In *Report of the Secretary of Agriculture, 1891,* 359–78. Washington, D.C.: GPO. [Ch. 7, 9]

————. 1893. Report of the chief of the Division of Vegetable Pathology for 1892. In *Report of the Secretary of Agriculture, 1892,* 215–46. Washington, D.C.: GPO. [Ch. 9]

————. 1894a. Effect of spraying with fungicides on the growth of nursery stock. USDA Division of Vegetable Pathology Bulletin 7. Washington, D.C.: GPO. [Ch. 9]

————. 1894b. Report of the chief of the Division of Vegetable Pathology. In *Report of the Secretary of Agriculture, 1893,* 245–76. Washington, D.C.: GPO. [Ch. 9]

————. 1895a. Report of the chief of the Division of Vegetable Pathology. In *Report of the Secretary of Agriculture, 1894,* 143–50. Washington, D.C.: GPO. [Ch. 9]

————. 1895b. Report of the chief of the Division of Vegetable Physiology and Pathology. In *Report of the Secretary of Agriculture, 1895,* 169–74. Washington, D.C.: GPO. [Ch. 9]

————. 1896. Report of the chief of the Division of Vegetable Physiology and Pathology. In *Report of the Secretary of Agriculture,* 15–22. Washington, D.C.: GPO. [Ch. 9]

————. 1897. Report of the chief of the Division of Vegetable Physiology and Pathology. In *Report of the Secretary of Agriculture, 1897,* 7–13. Washington, D.C.: GPO. [Ch. 9]

————. 1900. Progress in the treatment of plant diseases in the United States. In *U.S. Department of Agriculture Yearbook 1899,* 191–99. Washington, D.C.: GPO. [Ch. 5]

————. 1902. Report of the chief of the Bureau of Plant Industry. In *Annual Reports of the Department of Agriculture, 1902,* 47–108. Washington, D.C.: GPO. [Ch. 10]

————. 1907. Report of the Chief of the Bureau of Plant Industry. In *Annual Reports of the Department of Agriculture, 1906,* 175–266. Washington, D.C.: GPO. [Ch. 10]

————. 1913. Foreword. *Journal of Agricultural Research* 1:i. [Ch. 12]

————. 1917. Newton B. Pierce. *Phytopathology* 7:143–44. [Ch. 7]

Galloway, B. T. Topical Files, National Fungus Collections. USDA Systematic Botany and Mycological Laboratory, Beltsville, Md. [Ch. 7, 9, 10]

Galloway, B. T., and D. G. Fairchild. 1891a. Experiments in the treatment of plant diseases. Part 1: Treatment of black rot of grapes. *Journal of Mycology* 6:89–99. [Ch. 7]

Galloway, B. T., and D. G. Fairchild. 1891b. Experiments in the treatment of plant diseases. Part 2: Treatment of pear leaf-blight and scab in the orchard. *Journal of Mycology* 6:137–42. [Ch. 7]

Galloway, B. T., and E. A. Southworth. 1889. Treatment of Apple Scab. *The Journal of Mycology* 5:210–14. [Ch. 8]

"Gardening-cabbage (*Brassica oleracea*)." 1855. In *Report of Commissioner of Patents, for the year 1854*, 341–44. Washington, D.C.: A.O.P. Nicholson. [Ch. 3]

Gardner, M. W., and W. B. Hewitt. 1974. *Pierce's disease of the grapevine: The Anaheim disease and the California vine disease.* Berkeley: University of California. [Ch. 7, 9, 10]

[Garman, H.] 1892. The brown rot fungus of plums, peaches, apples, and cherries. In *Second annual report of the Kentucky Agricultural Experiment Station of the State College of Kentucky 1889*, 31–42. Frankfort, Ky.: Capital Printing Co. [Ch. 8]

———. 1894. A bacterial disease of cabbage. In *Third annual report of the Kentucky Agricultural Experiment Station of the State College of Kentucky for the Year 1890*, 43–46. Frankfort, Ky.: Capital Printing Co. [Ch. 8, 9]

Gates, P. W. 1960. *The farmer's age: Agriculture, 1815–1860.* New York: Holt, Rinehart and Winston. [Ch. 1, 2, 3; Pt. 1 intro.]

———. 1962. The Morrill Act and early agricultural science. *Michigan History* 46:289–302. [Ch. 3]

———. 1965. *Agriculture and the Civil War.* New York: Alfred A. Knopf. [Ch. 6]

General Assembly, Connecticut. 1873. An act concerning barberry bushes. In *The public records of the Colony of Connecticut, from May 1726 to May 1735, inclusive*, ed. by Charles J. Hoadly, 10–11. Hartford, Conn.: Case, Lockwood and Brainard. [Ch. 1]

General Assembly, Rhode Island and Providence. 1861. An act for destroying barberry bushes in Middletown. In *Records of the colony of Rhode Island and Providence Plantations, in New England*, ed. by J. R. Bartlett, 6 (1757–69): 509–10. Providence, R.I.: Knowles, Anthony and Co. [Ch. 1]

General Court of the Commonwealth. 1878. An act to prevent damage to English grain, arising from barberry-bushes. In *The acts and resolves, public and private, of the province of the Massachusetts Bay: To which are prefixed the charters of the province*, 3 (1754–1755): 797–98. Boston: Albert J. Wright. [Ch. 1]

A Georgia planter. 1830. On the rust in cotton. *Southern Agriculturist* 3:23–24. [Ch. 3]

Gilbert, W. W. 1930. William A. Orton. *Science* 71:89–90. [Ch. 10]

Gillett, H. N. 1855. Destructive rust on apple tree leaves. *The Ohio Cultivator* 11:211. [Ch. 3]

Gillispie, C. C. 1970. Charles Edwin Bessey. In *Dictionary of scientific biography*, s.v. New York: Charles Scribner's Sons. [Ch. 4]

Glawe, D. A. 1992. Thomas J. Burrill, pioneer in plant pathology. *Annual Review of Phytopathology* 30:17–24. [Ch. 5]

Glover, T. 1858. Investigations on the insects and diseases affecting the cotton plant. In *Report of the Commissioner of Patents, for the year 1857*, 121–29. Washington, D.C.: William A. Harris. [Ch. 3]

Glover, W. H. 1952. *Farm and college: The College of Agriculture of the University of Wisconsin, a history*. Madison: University of Wisconsin Press. [Ch. 11]

Goff, E. S. 1890. *Prevention of Apple Scab*. University of Wisconsin Agricultural Experiment Station Bulletin 23. Madison, Wis.: Democrat Printing Co. [Ch. 8]

Goff, M. B. 1975. *A biography of Emmett Stull Goff (1852–1902): A nineteenth-century professor of horticulture at the University of Wisconsin*. Sun City, Calif: Published privately. [Ch. 11]

Goodrich, C. E. 1854. Peaches, apricots, and nectarines: Condensed correspondence. In *Report of the Commissioner of Patents, for the year 1853*, 285–86. Washington, D.C.: Beverly Tucker, Senate Printer. [Ch. 3]

Goodsell, N. 1849. Disease in apples. *The Magazine of Horticulture, Botany, and All Useful Discoveries and Improvements in Rural Affairs* 15:23–24. [Ch. 3]

Goodwin, J. R. 1856. *Premium essays read before the Indiana State Board of Agriculture: On the cultivation of wheat*. In *Fourth report of the Indiana State Board of Agriculture, containing the transactions of the board, for the years 1854–1855*, 185–87. Indianapolis: William J. Brown. [Ch. 3]

"Gooseberries—cause of failure." 1856. *Ohio Farmer* 5:185. [Ch. 3]

Grace, J. K. 1988. The role of Thomas Taylor in the history of American phytopathology. *Annual Review of Phytopathology* 26:25–28. [Ch. 6]

Gravatt, G. F. 1949. The chestnut blight in Asia and North America. *Unasylva* 3:3–7. [Ch. 10]

Gray, A. 1863. *Manual of botany and of the northern United States, including Virginia, Kentucky, and all east of the Mississippi (the mosses and liverworts by William S. Sullivant), to which is added garden botany, and introduction to a knowledge of cultivated plants*. 4th revised ed. New York: Ivison, Phinney & Co. [Scribner's notes and an index prepared by him while a student at Orono were found in a copy of this edition of Gray's book at the University of Maine.] [Ch. 6]

Gray, J. 1951. *The University of Minnesota, 1851–1951*. Minneapolis: The University of Minnesota. [Ch. 11]

Griffith, C. S., P. D. Peterson, and C. L. Campbell. 1994. The origins of plant disease research in the United States Department of Agriculture. *Plant Disease* 78:318–21. [Pt. 2 intro.]

Griffin, G. J. 1986. Chestnut blight and its control. *Horticultural Reviews* 8:291–336. [Ch. 10]

Guralnick, S. M. 1975. *Science and the ante-bellum American college*. Philadelphia: American Philosophical Society. [Ch. 1]

Hadwiger, D. F. 1982. *The politics of agricultural research*. Lincoln: University of Nebraska. [Ch. 5]

Hall, C. R. 1934. *A scientist in the early republic: Samuel Latham Mitchill, 1764–1831*. New York: Columbia University Press. [Ch. 1]

Halsted, B. D. 1884. The white mildews. In *Proceedings of the nineteenth session of the American Pomological Society, held in Philadelphia, Penn., September 12th, 13th, and 14th, 1883* 19:88. Cleveland, Ohio: American Pomological Society. [Ch. 8]

———. 1889. *Some fungus diseases of the cranberry*. New Jersey Agricultural College Experiment Station Bulletin 64. New Brunswick, N.J. [Ch. 8]

———. 1891. Report of the Botanical Department. In *Eleventh annual report of the New Jersey State Agricultural Experiment Station and the third annual report of the New Jersey Agricultural College Experiment Station for the Year 1890*, 332–39. Trenton, N.J.: John L. Murphy Publishing Co. [Ch. 8]

———. 1892. *The Southern tomato blight*. Mississippi Agricultural and Mechanical College Experiment Station Bulletin 19. Agricultural College, Miss. [Ch. 8, 9]

———. 1893. *Club-root of cabbage and its allies*. New Jersey Agricultural College Experiment Station Bulletin 98. New Brunswick, N.J. [Ch. 8]

Halsted, B. D., and D. G. Fairchild. 1891. Sweet-potato black rot (*Ceratocystis fimbriata*, Ell. & Hals.). *Journal of Mycology* 7:1–11. [Ch. 8]

Harden, R. 1837–1838. The rot in cotton. *Farmers' Register* 5:22–27. [Ch. 3]

Harding, T. S. 1942. *Some landmarks in the history of the Department of Agriculture*. USDA History Series 2. Washington, D.C.: GPO. [Ch. 6]

———. 1947. *Two blades of grass: A history of scientific development in the U.S. Department of Agriculture*. Norman: University of Oklahoma Press. [Ch. 2, 3]

Harris, H. F. 1945. The correspondence of William G. Farlow during his student days at Strasbourg. *Farlowia* 2:9–37. [Ch. 4]

Harry, A. 1856. Peaches: Condensed correspondence. In *Report of the Commissioner of Patents, for the year 1855*, 298. Washington, D.C.: Cornelius Wendell. [Ch. 3]

Hartley, C. 1950. *The Division of Forest Pathology*. The Plant Disease Reporter Supplement 190, issued by the Plant Disease Survey, Bureau of Plant Industry, Soils, and Agricultural Engineering, Agricultural Research Administration, USDA. Washington, D.C.: GPO. [Ch. 10]

Haskell, R. J. 1953. Extension work in plant pathology. *Plant Disease Reporter* 37:570–74. [Ch. 11]

Hawkes, J. G. 1967. The history of the potato. *Journal of the Royal Horticultural Society* 92:207–24. [Ch. 2]

Heald, F. D., and R. A. Studhalter. 1914. Birds as carriers of the chestnut-blight fungus. *Journal of Agricultural Research* 2:405–22. [Ch. 10]

Heald, F. D., M. M. Gardner, and R. A. Studhalter. 1915. Air and wind dissemination of ascospores of the chestnut-blight fungus. *Journal of Agricultural Research* 3:493–526. [Ch. 10]

Hedgcock, G. G. 1905. *The crown-gall and hairy-root diseases of the apple tree*. Part 2 of USDA Bureau of Plant Industry Bulletin 90. Washington, D.C.: GPO. [Ch. 10]

Hedrick, U. P. 1950. *A history of horticulture in America to 1860*. New York: Oxford University Press. [Ch. 1, 2, 3, 4]

Hepting, G. H. 1974. Death of the American chestnut. *Journal of Forest History* 18:60–67. [Ch. 10]

Hildreth, I. 1850. The fire-blight. *The Cultivator* 7:87–88. [Ch. 3]

Hillsides. 1861. Rot in apples. *Southern Cultivator* 19:61–62. [Ch. 3]

Hilty, J. H., and P. D. Peterson. 1997. Frank Lamson-Scribner: Botanist and pioneer plant pathologist in the United States. *Annual Review of Phytopathology* 35:17–26. [Ch. 8]

"History and properties of ergot." 1858. In *Twelfth annual report of the Ohio State Board of Agriculture with an abstract of the proceedings of the county agricultural societies, to the General Assembly of Ohio, for the year 1857*, 387–91. Columbus, Ohio: Richard Nevins. [Ch. 3]

Holt, J. G., N. R. Krieg, P.H.A. Sneath, J. T. Staley, and S. T. Williams. 1994. *Bergey's manual of determinative bacteriology*, 9th ed. Baltimore: Williams and Wilkins. [Appendix]

Hooker, H. E. 1851. Remarks on leaf blight. *The Horticulturist and Journal of Rural Art and Rural Taste* 6:117–20. [Ch. 3]

Horsfall, J. G. 1963. A vignette on Roland Thaxter. In *Perspectives of biochemical plant pathology*, ed. by S. Rich, 154–62. Connecticut Agricultural Experiment Station Bulletin 663. New Haven, Conn. [Ch. 8]

———. 1979. Roland Thaxter. *Annual Review of Phytopathology* 19:33. [Ch. 8]

———. 1992. *Pioneer experiment station, 1875 to 1975: A history*. Lexington, Ky.: Antoca Press. [Ch. 5, 8]

House Committee on Agriculture. Hearings on Miscellaneous Bills, vol. 3. April 27, 1910. [Ch. 10]

Howe, J. J. 1856. Method of using sulphur for mildew. *The Horticulturist and Journal of Rural Art and Rural Taste* 6:206. [Ch. 3]

Huettal, R. N., and A. M. Golden. 1991. Nathan Augustus Cobb: The father of nematology in the United States. *Annual Review of Phytopathology* 29:15–26. [Ch. 10]

Hull, E. S. 1869. Cryptogamous diseases. *Transactions of the Illinois State Horticultural Society* 2:35–36. [Ch. 5]

Humphrey, D. C. 1976. *From King's College to Columbia, 1746–1800.* New York: Columbia University Press. [Ch. 1]

Hurt, R. D. 1994. *American agriculture: A brief history.* Ames, Iowa: Iowa State University Press. [Ch. 8, 9, 10 ; Pts. 1, 3, 4 intro.]

Institute of Agriculture, Director's Office, Papers, 1837–1969. University Archives, University of Minnesota, Minneapolis, Minn. [Ch. 11]

J.J.T. 1844. Cherry black wart. *The Cultivator* 1:131. [Ch. 3]

Jenkins, J. W. 1991. *A centennial history: A history of the College of Agricultural and Life Sciences at the University of Wisconsin—Madison. College of Agricultural and Life Sciences.* Madison: University of Wisconsin. [Ch. 11]

Johnson, A. G. Topical Files, National Fungus Collections. U.S. Department of Agriculture. Systematic Botany and Mycological Laboratory, Beltsville, Md. [Epilogue]

Johnston, J.F.W. 1850. Lecture third: The relations of botany, vegetable physiology, and zoology, to practical agriculture. *Transactions of the N.Y. State Agricultural Society* 9:188–97. [Ch. 3]

Jones, L. R. 1891. *Potato blight and rot.* State Agricultural Experiment Station Bulletin 24. Burlington, Vt. [Ch. 8]

———. 1893. Report of the botanist. In *Sixth annual report of the State Agricultural Experiment Station, 1892,* 57–60. Burlington, Vt.: Free Press Association. [Ch. 8]

———. 1896a. *Potato blights and fungicides.* Vermont Agricultural Experiment Station Bulletin 49. Burlington, Vt. [Ch. 8]

———. 1896b. Report of the botanist. In *Ninth annual report of the Vermont Agricultural Experiment Station, 1895,* 88–98. Montpelier, Vt.: Argus and Patriot Printing House. [Ch. 8]

———. 1899. *Certain potato diseases and their remedies.* Vermont Agricultural Experiment Station Bulletin 72. Burlington, Vt. [Ch. 8]

———. 1901a. A soft rot of carrot and other vegetables. In *Thirteenth annual report of the*

Vermont Agricultural Experiment Station, 1899–1900, 299–332. Burlington, Vt.: Free Press Association. [Ch. 8]

———. 1901b. Bacillus carotovorus n. sp., die Ursache einer Fäulnis der Möhre. *Centralblatt* 7:12–21. [Ch. 8]

———. 1910. *The bacterial soft rots of certain vegetables.* Part II: *Pectinase, the Cytolytic Enzym (sic) produced by Bacillus carotovorus and certain other soft-rot organisms.* Vermont Agricultural Experiment Station Bulletin 147. Burlington, Vt. [Ch. 8]

———. 1914. Problems and progress in plant pathology. *American Journal of Botany* 1: 97–111. [Ch. 2, 10]

———. 1919. Our journal, Phytopathology. *Phytopathology* 9:159–64. [Ch. 12]

———. 1922. Whetzel resigns headship—deplores present system of administrative tenure. *Phytopathology* 12:499–500. [Ch. 11]

———. 1931. William Allen Orton: 1877–1930. *Phytopathology* 21:1–11. [Ch. 10]

———. 1934. Arthur Bliss Seymour, 1859–1933. *Mycologia* 26:284. [Ch. 11]

———. 1939. *Biographical memoir of Erwin Frink Smith, 1854–1927.* National Academy of Sciences Biographical Memoirs 21. Washington, D.C.: National Academy of Sciences. [Ch. 7, 9]

Jones, L. R. Papers. 1892–1943. The University of Wisconsin Division of Archives, University of Wisconsin, Madison, Wis. [Ch. 8]

Jones, L. R. Papers. 1896–1909. University Archives, University of Vermont, Burlington, Vt. [Ch. 8]

Jones, L. R., J. Johnson, and J. G. Dickson. 1926. *Wisconsin studies upon the relation of soil temperature to plant disease.* Agricultural Experiment Station of the University of Wisconsin—Madison Research Bulletin 71. Madison, Wis. [Ch. 11]

Jones, R. L. 1983. *History of agriculture in Ohio to 1880.* Kent, Ohio: The Kent State University Press. [Ch. 3]

Josselyn, J. 1674. *An account of two voyages to New-England, wherein you have the setting out of a ship, with the charges; the prices of all necessaries for furnishing a planter and his family at his first coming; a description of the countrey, natives and creatures, with their mercantile and physical use; the government of the countrey as it is now possessed by the English, &c. A large chronological table of the most remarkable passages, from the first discovering of the continent of America, to the year 1673.* London: G. Widdows. [Ch. 1]

Kastner, J. 1977. *A species of eternity.* New York: Alfred A. Knopf. [Ch. 4]

Keitt, G. W., and F. V. Rand. 1946. Lewis Ralph Jones, 1864–1945. *Phytopathology* 36:1–17. [Ch. 8]

Kellerman, W. A. 1891a. *Experiments with corn smut.* Experiment Station of the Kansas State Agricultural College Bulletin 23. Manhattan, Kan. [Ch. 8]

———. 1891b. *Spraying to prevent wheat rust.* Experiment Station of the Kansas State Agricultural College Bulletin 22. Manhattan, Kan. [Ch. 8]

Kellerman, W. A., and W. T. Swingle. 1890a. *Additional experiments and observations on oat smut, made in 1890.* Experiment Station Kansas State Agricultural College Bulletin 15. Manhattan, Kan. [Ch. 8]

Kellerman, W. A., and W. T. Swingle. 1890b. *Preliminary experiments with fungicides for stinking smut of wheat.* Experiment Station Kansas State Agricultural College Bulletin 12. Manhattan, Kan. [Ch. 8]

Kelly, H. A. 1977. *Some American medical botanists.* Portland, Maine: Longwood Press. [Ch. 4]

Kennedy, B. F., W. D. Widkin, and K. F. Baker. 1979. Bacteria as the cause of disease in plants: A historical perspective. *ASM News* 45:1–4. [Ch. 5]

Kenrick, W. 1841. *The New American orchardist: or an account of the most valuable varieties of fruits, of all climates, adapted to cultivation in the United States; with their history, modes of culture, management, uses, etc. with an Appendix on vegetables, ornamental trees, shrubs, and flowers, the agricultural resources of America, and on silk, etc.* 3d ed. Boston: Otis, Broaders, and Co. [Ch. 1]

Kent, G. C., and A. G. Newhall. 1983. Plant pathology at Cornell-Ithaca: Historical highlights. Unpublished paper. Cornell University, Department of Plant Pathology. [Ch. 11]

Kern, F. D. 1942. Obituary: Joseph Charles Arthur. *Science* 95:617. [Ch. 5]

Kerr, N. A. 1987. *The legacy: A centennial history of the state agricultural experiment stations: 1887–1987.* Columbia: Missouri Agricultural Experiment Station, University of Missouri. [Ch. 8]

Kimmelman, B. 1983. The American Breeders' Association: Genetics and eugenics in an agricultural context, 1903–13. *Social Studies of Science* 13:163–204. [Ch. 10]

Kirkendall, R. S. 1986. The agricultural colleges: Between tradition and modernization. *Agricultural History* 60:3–21. [Ch. 4]

Kirtland, J. 1855. Premature decay of the plum. *The Florist and Horticultural Journal* 4:34–36. [Ch. 3]

Klein, A. J. 1930. *Survey of land-grant colleges and universities.* U.S. Department of the Interior Office of Education Bulletin 9. Office of Education. Washington, D. C.: GPO. [Ch. 4]

Klippart, J. H. 1858. An essay on the origin, growth, diseases, varieties, etc., of the wheat plant. In *Twelfth annual report of the Ohio State Board of Agriculture with an abstract of the proceedings of the county agricultural societies, to the General Assembly of Ohio for the year 1857*, 768–90. Columbus, Ohio: Richard Nevins. [Ch. 3]

"Klippart, John Hancock." 1933. *Dictionary of American Biography*, vol. 10, s.v. New York: Charles Scribner's Sons. [Ch. 3]

Knoblauch, H. C., E. M. Law, and W. P. Meyer. 1962. *State agricultural experiment stations: A history of research policy and procedure*. USDA Misc. Pub. 904. Washington, D.C.: GPO. [Ch. 5, 6, 8]

Kommedahl, T., J. J. Christensen, and R. A. Frederiksen. 1970. *A half century of research in Minnesota on flax wilt caused by Fusarium oxysporum*. Minnesota Agricultural Experiment Station Technical Bulletin 273. University of Minnesota. [Ch. 8]

Korf, R. P. 1991. An historical perspective: mycology in the departments of Botany and of Plant Pathology at Cornell University and the Geneva Agricultural Experiment Station. *Mycotaxon* 40:108–28. [Ch. 11]

Kühn, J. 1858. *Die Krankheiten der Kulturgewächse, ihre Ursachen und ihre Verhütung*. Berlin: Besselman. [Ch. 4]

Large, E. C. 1940. *The Advance of the fungi*. New York: Henry Holt. [Ch. 7, 8]

Lazell, J. A. 1840. Observations upon the blight in fruit trees, with an account of a method of preventing the disease. *The Magazine of Horticulture* 6:414–16. [Ch. 1]

LeDuc, W. G. 1881. Grape culture and wine making. In *Annual report of the Commissioner of Agriculture, 1880*, 25–26. Washington, D.C.: GPO. [Ch. 7]

Lee, D. 1850. Agricultural education. In *Report of the Commissioner of Patents, for the year 1849*, 15–18. Washington, D.C.: Office of Printers to the Senate. [Ch. 3]

Lee, F. S. 1904. The School of Medicine. In *A history of Columbia University, 1754–1904*, 307–34. New York: Columbia University Press. [Ch. 1]

Lindley, G., and M. Floy. 1846. *A guide to the orchard and fruit garden; or, an account of the most valuable fruits cultivated in Great Britain, with additions of all the most valuable fruits cultivated in America*. New York: J. C. Riker. [Ch. 1]

Lodeman, E. G. 1916. The *spraying of plants: A succinct account of the history, principles and practice of the application of liquids and powders to plants for the purpose of destroying insects and fungi*. New York: Macmillan Co. [Ch. 1, 7, 11]

Lodge, W. C. 1866. Fruits and fruit trees of the middle states; their propagation, influences of stocks, diseases, and enemies. In *Report of the Commissioner of Agriculture for the year 1865*, 199–207. Washington, D.C.: GPO. [Ch. 6]

Loring, G. 1882. Microscopical Division. In *Report of the Commissioner of Agriculture for the Years 1881 and 1882*, 10–11. Washington, D.C.: GPO. [Ch. 6]

Lowell, J. 1826. Insect in pear trees. *New England Farmer* 5:17–18. [Ch. 1]

Lugger, O. 1890. *A treatise on flax culture.* University of Minnesota Agricultural Experiment Station Bulletin 13. Minneapolis, Minn. [Ch. 11]

Lyman, G. R. 1918a. Minutes of the War Emergency Board of American Plant Pathologists. *Science* 47:210–13. [Epilogue]

———. 1918b. The relation of phytopathologists to plant disease survey work. *Phytopathology* 8:219–28. [Epilogue]

M. W. 1857. The cherry tree-remedy for black wart. *The Cultivator* 5:285. [Ch. 3]

Marcus, A. I. 1985. *Agricultural science and the quest for legitimacy: Farmers, agricultural colleges, and experiment stations, 1870–1890.* Ames: Iowa University Press. [Ch. 4, 6, 8; Pt. 2 intro.]

Marti, D. B. 1979. To improve the soil and the mind: agricultural societies, journals, and schools in the northeastern states, 1791–1865. University Microfilms International, Ann Arbor, Mich. [Ch. 1]

———. 1980. Agricultural journalism and the diffusion of knowledge: The first half-century in America. *Agricultural History* 54:28–37. [Ch. 1]

Mason, E. 1857. Fruit trees. *Transactions of the Michigan State Agricultural Society: with reports of county agricultural societies, for 1856* 8:221–31. [Ch. 3]

Mason, W. W. 1848. Mildew on wheat. *Southern Cultivator* 6:73. [Ch. 3]

Massachusetts Board of Agriculture, Committee on Diseases of Vegetation. 1859. Report of the Committee on Diseases of Vegetation. In *Sixth annual report of the Secretary of the Massachusetts Board of Agriculture, together with reports of committees appointed to visit the county societies, with an appendix, containing an abstract of the finances of the county societies,* 33–38. Boston: William White. [Ch. 3]

———. 1860. Report of the Committee on Diseases of Vegetation. In *Seventh annual report of the secretary of the Massachusetts Board of Agriculture, together with reports of committees appointed to visit the county societies, with an appendix containing an abstract of the finances of the county societies for 1859,* 49–61. Boston: William White. [Ch. 3]

Mayo, L. S., ed. 1936. *The history of the colony and province of Massachusetts-Bay by Thomas Hutchinson,* vol. 1. Cambridge, Mass.: Harvard University Press. [Ch. 1]

McAllister, H. N., G. Buchanan, J. Alexander, J. K. Shoemaker, and W. J. Waring. 1854. Fruits of Centre County, Pennsylvania. In *Report of the Commissioner of Patents, for the year 1853,* 257–59. Washington: A.O.P. Nicholson. [Ch. 3]

McCallan, S.E.A. 1959. The American Phytopathological Society—The first fifty years. In *Plant Pathology: Problems and Progress, 1908–1958,* ed. by C. S. Holton, G. W. Fischer, R.

W. Fulton, H. Hart and S.E.A. McCallan, 24–31. Madison: University of Wisconsin Press. [Ch. 12]

McClellan, J. E. 1985. *Science reorganized: Scientific societies in the eighteenth century.* New York: Columbia University Press. [Ch. 1]

McCubbin, W. A. 1954. *The plant quarantine problem: A general review of the biological, legal, administrative and public relations of plant quarantines with special reference to the United States situation.* Copenhagen, Denmark: Ejnar Munksgaard. [Ch. 10]

McMahon, B. 1806. The *American gardener's calendar; adapted to the climates and seasons of the United States.* Philadelphia: B. Graves. [Ch. 1]

McNall, N. A. 1976. *An agricultural history of the Genesee Valley, 1790–1860.* Westport, Conn.: Greenwood Press. [Ch. 3]

McNew, G. L. 1963. The ever-expanding concepts behind 75 years of plant pathology. *Perspectives of biochemical plant pathology,* ed. S. Rich. Bulletin 663, Connecticut Agricultural Experimentation Station, 163–83. [Ch. 2]

Mease, J. 1814. On the disease in wheat, mentioned in the Agricultural Memoirs, vol. 1. *Memoirs of the Philadelphia Society for Promoting Agriculture* 3:422–26. [Ch. 2]

———. 1818. Account of the means to prevent the must and mildew of wheat, adopted with success in Flanders. *Memoirs of the Philadelphia Society for Promoting Agriculture* 4:76–81. [Ch. 2]

Meehan, T. 1868. Diseases of the pear. *Transactions of the Illinois State Horticultural Society for 1867* 1:48–53. [Ch. 5]

Meinecke, E. P. 1941. Haven Metcalf (1875–1940). *Phytopathology.* 31:289–95. [Ch. 10]

"Memorial from the agriculturists of the United States to their respective state legislatures." 1846. *Monthly Journal of Agriculture* 2:260–62. [Ch. 3]

Metcalf, H. 1904. A soft rot of the sugar-beet. In *The University of Nebraska: Seventeenth annual report of the Agricultural Experiment Station of Nebraska,* 69–112. Lincoln, Neb. [Ch. 10]

———. 1906. *A preliminary report on the blast of rice.* South Carolina Agricultural Experiment Station of Clemson Agricultural College Bulletin 121. Columbia, S.C.: R. L. Bryan Co. [Ch. 10]

———. 1908. *The immunity of the Japanese chestnut to the bark disease.* Part 6 of USDA Bureau of Plant Industry Bulletin 121. Washington, D.C.: GPO. [Ch. 10]

Metcalf, H., and J. F. Collins. 1909. *The present status of the chestnut bark disease.* Part 5 of USDA Bureau of Plant Industry Bulletin 141. Washington, D.C.: GPO. [Ch. 10]

Metcalf, H., and J. F. Collins. 1911. *The control of the chestnut bark disease.* USDA Farmers' Bulletin 467. Washington, D.C.: GPO. [Ch. 10]

Micheli, P. A. 1729. *Nova plantarum genera.* Florence: B. Paperinii. [Ch. 2]

Miller, H. S. 1970. *Dollars for research: Science and its patrons in nineteenth-century America.* Seattle: University of Washington Press. [Ch. 2, 6]

Miner, P. 1856. On the cultivation of tobacco. *The Arator* 1:335–37. [Ch. 3]

Mitchill, S. L. 1792. *Outline of the doctrines in natural history, chemistry, and economics which, under the patronage of the state, are now delivering in the College of New York.* New York: Childs and Swaine. [Ch. 1]

———. 1794. *The present state of learning in the College of New York.* New York: T. and J. Swords. [Ch. 1]

———. 1826. *Address pronounced before the N.Y. Horticultural Society, August 29, 1826.* New York: E. Conrad. [Ch. 1]

Moore, V. A. 1894. An inquiry into the alleged relation existing between the Burrill disease of corn and the so-called corn-stalk disease of cattle. *Agricultural Science* 8:368–85. [Ch. 5]

Morton, A. G. 1981. *History of botanical science: An account of the development of botany from ancient times to the present day.* London: Academic Press. [Ch. 4]

Morton, J. S. 1894. *Report of the Secretary of Agriculture, 1893.* Washington, D.C.: GPO. [Ch. 9]

Murrill, W. A. 1906a. A serious chestnut disease. *Journal of the New York Botanical Garden* 7: 143–53. [Ch. 10]

———. 1906b. Further remarks on a serious chestnut disease. *Journal of the New York Botanical Garden* 7: 203–11. [Ch. 10]

Neal, J. C. 1889. *The root-knot disease. Peach, orange, and other plants in Florida, due to the work of Anguillula.* USDA Division of Entomology Bulletin 20. Washington, D.C.: GPO. [Ch. 10]

"New remedy for mildew of grapes." 1852. *The Ohio Cultivator* 8:225–26. [Ch. 3]

Newhall, A. G. 1980. Herbert Hice Whetzel: Pioneer American plant pathologist. *Annual Review of Phytopathology* 18:27–36. [Ch. 11]

Newton, I. 1863. Report. In *Report of the Commissioner of Agriculture for the year 1862,* 3–25. Washington, D.C.: GPO. [Ch. 6]

Nichols, E. 1858. Scab or rot on apples. *The Ohio Cultivator* 14:60. [Ch. 3]

Niederhauser, J. S., and W. C. Cobb 1959. The late blight of potatoes. *Scientific American* 200:100–12. [Ch. 2]

"Notes and news." 1887. *Botanical Gazette* 12:170. [Ch. 7]

"Obituary [J. E. Teschemacher]." 1854. *The American Journal of Science and Arts,* series 2. 17:150, 292–95. [Ch. 2]

Ohio State Board of Agriculture. 1858. *Twelfth annual report of the Ohio State Board of Agriculture with an abstract of the proceedings of the county agricultural societies, to the General Assembly of Ohio: for the year 1857.* Columbus, Ohio: Richard Nevins. [Ch. 3]

"On the vine and rot in grapes." 1833. *Southern Agriculturist* 6:348. [Ch. 3]

Orton, W. A. 1900. *The wilt disease of cotton and its control.* USDA Division of Vegetable Physiology and Pathology Bulletin 27. Washington, D.C.: GPO. [Ch. 10]

————. 1902. *The wilt disease of the cowpea and its control.* Part I of *Some diseases of the cowpea.* USDA Bureau of Plant Industry Bulletin 17. Washington, D.C.: GPO. [Ch. 10]

————. 1909. The development of farm crops resistant to disease. In *Yearbook of the United States Department of Agriculture 1908,* 463–64. Washington, D.C.: GPO. [Ch. 10]

Orton, W. A., and W. W. Gilbert. 1912. *The control of cotton wilt and root-knot.* USDA Bureau of Plant Industry Circular 92. Washington, D.C.: GPO. [Ch. 10]

Osborne, E. A., ed. 1913. *From the letter-files of S. W. Johnson.* New Haven, Conn.: Yale University Press. [Ch. 4, 5, 8]

Overfield, R. A. 1993. *Science with practice: Charles E. Bessey and the maturing of American botany.* Ames: Iowa State University Press. [Ch. 4]

Overman, C. R. 1855. Remarks on the pear and its culture. *Transactions of the Illinois State Agricultural Society: With the proceedings of the county societies, and kindred associations* 1:447–51. [Ch. 3]

P.W.J. 1846. Smut, or blast in oats. *The Southern Cultivator* 4:107. [Ch. 3]

Paarlberg, D. 1988. The land grant trio. In *Toward a well fed world,* 60–71. Ames: Iowa State University Press. [Ch. 3]

Pammel, L. H. 1889. Root rot of cotton, or "cotton blight." In *First annual report of the Texas Agricultural Experiment Station for the year 1888,* 50–65. College Station, Texas: J. J. Pastoriza, Printer and Stationer. [Ch. 8]

————. 1891. *Fungus diseases. Treatment of fungus diseases.* Iowa Agricultural Experiment Station Bulletin 13, 31–70. Ames, Iowa. [Ch. 8]

————. 1892a. *Experiments with fungicides.* Iowa Agricultural Experiment Station Bulletin 16, 314–29. Ames, Iowa. [Ch. 8]

————. 1892b. *Some diseases of plants common to Iowa cereals.* Iowa Agricultural Experiment Station Bulletin 18. Ames, Iowa. [Ch. 8]

————. 1892c. *Treatment of some fungus diseases. Experiments made in 1891.* Iowa Agricultural Experiment Station Bulletin 17. Ames, Iowa. [Ch. 8]

————. 1894a. Diseases of foliage and fruit. In *Report of the Iowa State Horticultural Society, for the year 1893,* 28:467–74. Des Moines, Iowa: G. H. Ragsdale, State Printer. [Ch. 8]

————. 1894b. *Experiments with fungicides.* Iowa Agricultural College Experiment Station Bulletin 24. Ames, Iowa. [Ch. 8]

————. 1894c. The most important factor in the development of rust. *Agricultural Science* 8:287–91. [Ch. 8]

————. 1894d. Powdery mildew of the apple. In *Proceedings of the Iowa Academy of Sciences for 1893,* 1:92. Des Moines, Iowa : G. H. Ragsdale. [Ch. 8]

————. 1895. *Bacteriosis of rutabaga (Bacillus campestris n. sp.).* Iowa Agricultural College Experiment Station Bulletin 27. Ames, Iowa. [Ch. 9]

————. 1900. Powdery mildew of the apple. In *Proceedings of the Iowa Academy of Sciences for 1899,* 7:177–82. Des Moines, Iowa: F. R. Conaway. [Ch. 8]

————. 1919. In memoriam. Dr. Byron D. Halsted. In *Proceedings of the Iowa Academy of Science for 1919,* 26:31–33. Des Moines: State of Iowa. [Ch. 8]

Pammel, L. H. Papers. 1856–1967. Iowa State University Archives, Special Collections, Ames, Iowa. [Ch. 4, 8]

Pammel, L. H., and G. W. Carver. 1895. *Treatment of currants and cherries to prevent spot diseases.* Iowa Agricultural College Experiment Station Bulletin 30. Ames, Iowa. [Ch. 8]

"Papers read, Section G at the 42nd meeting in Madison, Wisconsin in August 1893." 1894. In *Proceedings of the American Association for the Advancement of Science, for the 42nd meeting held at Madison, Wis., August 1893,* 42:253–64. Salem, Mass.: Published by the Permanent Secretary. [Ch. 12]

[Pascalis, F.] 1814–1815. Peach trees becoming diseased, and dying from a disordered blossom and morbid staminal dust. *The Medical Repository* 17:394–95. [Ch. 1]

"The Patent Office report and the potato rot." 1847. *The Farmer's Library and Monthly Journal of Agriculture* 3:11–15. [Ch. 2]

"Peach prospects." 1838. *The Farmers' Cabinet; devoted to agriculture, horticulture and rural economy* 2:297. [Ch. 1]

Peck, C. H. 1872. Report of the second class, in the second department (Botany), February 6, 1872. *Transactions of the Albany Institute* 7:186–204. [Ch. 4]

Peffer, G. 1882. Apple tree blossoms-blight. *Transactions of the Wisconsin State Horticultural Society* 12:191–202. [Ch. 5]

Peters, R. 1808a. On peach trees, read February 11, 1806. *Memoirs of the Philadelphia Society for Promoting Agriculture* 1:15–24. [Ch. 1]

————. 1808b. On peach trees, with a letter from Dr. James Tilton of Bellevue, near Wilmington, Delaware. *Memoirs of the Philadelphia Society for Promoting Agriculture* 1:183–92. [Ch. 1]

Peterson, P. D. 1999. William Gilson Farlow. In *American National Biography*. New York: Oxford University Press. [Ch. 4]

Peterson, P. D., and C. L. Campbell. 1997. Beverly T. Galloway—Visionary administrator. *Annual Review of Phytopathology* 35:29–43. [Pt. 4 intro.]

Peterson, P. D., C. L. Campbell, and C. S. Griffith. 1992. James E. Teschemacher and the cause and management of potato blight in the United States. *Plant Disease* 76:754–56. [Pt. 1 intro.]

Peterson, P. D., C. S. Griffith, and C. L. Campbell. 1996. Frank Lamson-Scribner and American plant pathology, 1885–1888. *Agricultural History* 70:33–56. [Pt. 3 intro.]

Philadelphia Society for the Promotion of Agriculture. 1854. *Minutes of the Philadelphia Society for the Promotion of Agriculture, from its Institution in February, 1785, to March, 1810*. Philadelphia: John C. Clark & Son, Printers. [Ch. 1]

Pickering, T. 1811. On mildew. Read March 13th, 1810. *Memoirs of the Philadelphia Society for Promoting Agriculture* 2:164–66. [Ch. 2]

Pierce, J. 1859. *Condensed reports of the American Pomological Society, for the year 1858*. In *Report of the Commissioner of Patents, for the year 1858*. Washington, D.C.: James B. Steedman. [Ch. 3]

Pierce, N. B. 1890. The mysterious vine disease. In *Annual report of the State Board of Horticulture of the State of California, for 1890*, 169–77. Sacramento, Calif.: State Printers. [Ch. 7]

———. 1892. *The California vine disease. A preliminary report of investigations*. USDA Division of Vegetable Pathology Bulletin 2. Washington, D.C.: GPO. [Ch. 9]

Pinney, T. 1989. *A history of wine in America: From the beginnings to prohibition*. Berkeley: University of California Press. [Ch. 7]

Pohl, M. C. 1985. Louis H. Pammel: Pioneer botanist. A biography. In *Proceedings of the Iowa Academy of Science for 1985*, vol. 92, 5–9. Cedar Falls: University of Northern Iowa. [Ch. 8]

Pool, R. J. 1915. A brief sketch of the life and work of Charles Edwin Bessey. *American Journal of Botany* 2:505–18. [Ch. 4]

Prescott, O. 1813. *A dissertation on the natural history and medicinal effects of the secale cornutum or ergot*. Boston: Cummings and Hilliard. [Ch. 2]

Prévost, B. [1807] 1939. *Memoir on the immediate cause of bunt or smut of wheat, and of several other diseases of plants, and on preventives of bunt*. Reprint, Phytopathological Classics 6, trans. G. W. Keitt. Menasha, Wis.: American Phytopathological Society. [Ch. 2]

Prince, W. R. 1828. *A short treatise on horticulture: Embracing descriptions of a great variety of fruit and ornamental trees and shrubs, grape vines, bulbous flowers, green-house trees and plants, etc., nearly all of which are at present comprised in the collection of the Linnaean botanic garden, at Flushing, near New-York. With directions for their culture, management, etc.* New York: T. and J. Swords. [Ch. 1]

———. 1847. Cure of the yellows in peach trees. *The Horticulturist and Journal of Rural Art and Rural Taste* 1:318–19. [Ch. 3]

"Proceedings of Section G., AAAS, Brooklyn Meeting, 1894." 1894. *Botanical Gazette* 19:366–68. [Ch. 12]

"Proceedings of the Botanical Club of the AAAS" 1892. *Botanical Gazette* 17:285–90. [Ch. 12]

"Proceedings of the Madison Botanical Congress." 1893. *Botanical Gazette* 18:350–56. [Ch. 12]

Proctor, J. W. 1857. Smut upon the onion. *The Cultivator* 5:236. [Ch. 3]

R.R.S. 1858. Diseases of the pear and apple. *The Country Gentleman* 11:110. [Ch. 3]

Rand, F. V. 1949. Obituaries: [Albert Fred Woods]. *Journal of the Washington Academy of Sciences* 39:313–14. [Ch. 9]

Rasmussen, W. D. 1989. *Taking the university to the people: Seventy-five years of cooperative extension.* Ames: Iowa State University Press. [Ch. 11]

Ravenel, H. W. 1850. A catalogue of the natural orders of plants, inhabiting the vicinity of Santee Canal, S.C., as represented by genera and species; with observations on the meteorological and topographical conditions of that section of the country. In *Proceedings of the American Association for the Advancement of Science, 3rd meeting held at Charleston, SC., March 1850* 3:2–17. Charleston, S.C.: Steam-Power Press of Walker and James. [Ch. 12]

Read, W. H. 1858. Grapes and mildew. *The Horticulturist and Journal of Rural Art and Rural Taste* 8:134–35. [Ch. 3]

Reddick, D. 1915. Department of Plant Pathology. In *Twenty-seventh annual report of the New York State College of Agriculture at Cornell University and the Agricultural Experiment Station established under the Direction of Cornell University, Ithaca, N.Y., 1914,* xlix–lv. Albany, N.Y.: J. B. Lyon Co., Printers. [Ch. 11]

Reed, G. 1943. Phytopathology 1867–1942. *Torreya* 43:155–69. [Ch. 4, 5]

"Remedy for black knot in plum and cherry trees." 1858. *The Valley Farmer* 10:57–58. [Ch. 3]

"Review of *The essentials of botany.*" 1884. *Botanical Gazette* 9:184. [Ch. 4]

Richards, M. 1851. Grape rot. *Southern Cultivator* 9:110. [Ch. 3]

Riley, C. V. 1886. The mildews of the grape-vine—an effectual remedy for peronospora. In *Proceedings of the twentieth session of the American Pomological Society held in Grand Rapids, Michigan, September 9th, 10th, and 11th, 1885,* 49–54. Cleveland, Ohio: American Pomological Society. [Ch. 7]

Rodgers, A. D. 1944. *American botany, 1873–1892: Decades of transition.* Princeton, N.J.: Princeton University Press. [Ch. 4, 5, 8; Pt. 2 intro.]

————. 1949. *Liberty Hyde Bailey: A story of American plant sciences.* Princeton, N.J.: Princeton University Press. [Ch. 12]

————. 1952. *Erwin Frink Smith: A story of North American plant pathology.* Memoirs of the American Philosophical Society held at Philadelphia for promoting useful knowledge 31. Philadelphia: American Philosophical Society. [Ch. 4, 5, 7, 8, 9, 10, 11]

Rogers, D. P. 1977. *A brief history of mycology in North America.* Amherst, Mass.: Mycological Society of America. [Ch. 2, 5]

Rosenberg, C. E 1971. Science, technology, and economic growth: The case of the agricultural experiment station scientist, 1875–1914. *Agricultural History* 45:1–20. [Ch. 5]

Rosenberg, D. Y. 1989. The interaction of state and federal quarantines in the U.S. Chapter 3 of *Plant Protection and Quarantine,* ed. by R. P. Kahn, 59–74. Boca Raton, Fla.: CRC Press. [Ch. 10]

Ross, E. D. 1942. *Democracy's college: The land-grant movement in the formative stage.* Ames: Iowa State College Press. [Ch. 4]

————. 1946. The United States Department of Agriculture during the commissionership: A study in politics, administration, and technology, 1862–1889. *Agricultural History* 20:129–43. [Ch. 6]

Rossiter, M. W. 1975. *The emergence of agricultural science. Justus Liebig and the Americans, 1840–1880.* New Haven, Conn.: Yale University Press. [Ch. 2, 3, 4, 5]

————. 1976. The organization of agricultural improvement in the United States, 1785–1865. In *The pursuit of knowledge in the early American republic: American scientific and learned societies from colonial times to the Civil War,* ed. by A. Oleson and S. C. Brown, 279–98. Baltimore: The Johns Hopkins University Press [Ch. 1, 3, 4; Pt. 1 intro]

Rothrock, J. T. 1884. Review of *Briefer course in botany. The American Naturalist* 18:1247–48. [Ch. 4]

Rowell, J. B. 1983. Cereal rust laboratory. In *Aurora Sporealis,* 75th anniversary ed., ed. by C. E. Windels, and C. J. Eide, 78–84. St. Paul, Minn: The Department of Plant Pathology, University of Minnesota. [Ch. 11]

Rumbold, C. T. 1920. Effect on chestnuts of substances injected into their trunks. *American Journal of Botany* 7:45–56. [Ch. 10]

Rusk, J. M. 1890. Division of Vegetable Pathology. In *Report of the Secretary of Agriculture, 1890,* 35–36. Washington, D.C.: GPO. [Ch. 7]

[Rusk, J. M.] 1892. *Report of the Secretary of Agriculture, 1891.* Washington, D.C.: GPO. [Ch. 9]

Russell, H. L. 1898. *A bacterial rot of cabbage and allied plants.* The University of Wisconsin Agricultural Experiment Station Bulletin 65. Madison: University of Wisconsin. [Ch. 11]

———. 1911 *Report of the director, 1910–1911.* The University of Wisconsin Agricultural Experiment Station Bulletin 218. Madison: University of Wisconsin. [Ch. 11]

———. 1913. *Report of the director, 1911–1912.* The University of Wisconsin Agricultural Experiment Station Bulletin 228. Madison: University of Wisconsin. [Ch. 11]

———. 1916. Work done by the Experiment Station in 1915. In *Director's annual report.* Agricultural Experiment Station of the University of Wisconsin Bulletin 268. Madison: University of Wisconsin. [Ch. 11]

Russell, H. S. 1976. *A long, deep furrow: Three centuries of farming in New England.* Hanover, N.H.: University Press of New England. [Ch. 1]

"Rust on oats." 1858. *The Valley Farmer* 10:329–30. [Ch. 3]

S. S. 1844. The peach-tree. *The American Agriculturist* 3:43–45. [Ch. 3]

Salisbury, J. H., and C. B. Salisbury. 1864. Microscopic researches: Resulting in the discovery of what appears to be the cause of the so-called blight in apple, pear and quince trees; and the decay in their fruit, &c. In *Eighteenth annual report of the Ohio State Board of Agriculture,* 450–79. Columbus, Ohio: Richard Nevins, State Printer. [Ch. 5]

Sander, E. 1878–1886. Memorial volume to George Engelmann, M.D. *Transactions of the Academy of Science, St. Louis* 4:1–18. [Ch. 4]

Saunders, W. 1855. Grape mildew. *The Horticulturist and Journal of Rural Art and Rural Taste* 5:129–30. [Ch. 3]

———. 1862. Report of the superintendent of the garden attached to the Department of Agriculture. In *Report of the Commissioner of Agriculture for the year 1862,* 540–45. Washington, D.C.: GPO. [Ch. 6]

———. 1866. Report of the superintendent of garden. In *Report of the Commissioner of Agriculture for the year 1865,* 13–25. Washington, D.C.: GPO. [Ch. 6]

———. 1885. Report of the Superintendent of gardens and grounds. In *Report of the Commissioner of Agriculture. 1885,* 33–46. Washington, D.C.: GPO. [Ch. 6]

Sayre, R. M. 1994. *Art in phytopathology.* St. Paul, Minn.: APS Press. [Ch. 10]

Schaeffer, F. G. 1816–1817. Cultivation of the peach tree. *The Medical Repository* 18:199–200. [Ch. 1]

"Science the Loser." 1910. *The Sunday Star,* [Washington, D.C.], January 23. [Ch. 6]

Scott, R. V. 1970. *The reluctant farmer: The rise of agricultural extension to 1914.* Urbana, Ill.: University of Illinois Press. [Ch. 11]

Scott, W. M. 1908. *Self-boiled lime-sulphur mixture as a promising fungicide.* USDA Bureau of Plant Industry Circular 1. Washington, D.C.: GPO. [Ch. 10]

Scribner, F. L. 1885. Fungous diseases of plants. In *Report of the Commissioner of Agriculture. 1885,* 76–88. Washington, D.C.: GPO. [Ch. 6, 7, 11]

———. 1886. *Report on the fungous diseases of the grape vine.* USDA Division of Botany, Section of Plant Pathology Bulletin 2. Washington, D.C.: GPO. [Ch. 7]

———. 1887. Report of the Mycological Section. *Report of the Commissioner of Agriculture, 1886,* 95–138. Washington, D.C.: GPO. [Ch. 7]

———. 1888a. New observations on the fungus of black rot of grapes. *Proceedings of the ninth annual meeting of the Society for the promotion of agricultural science held at Cleveland, Ohio,* 68–72. Columbus, Ohio: Gazette Printing House. [Ch. 7]

———. 1888b. Observations the past season with grape rot and mildew. In *Proceedings of the New Jersey State Horticultural Society, at its thirteenth annual meeting, held at Trenton, N. J., December 14th and 15th, 1887,* 133–47. Newark, N.J.: Starbuck and Dunham. [Ch. 7]

———. 1888c. Report of the Section of Vegetable Pathology. In *Report of the Commissioner of Agriculture, 1887,* 323–97. Washington, D.C.: GPO. [Ch. 7]

———. 1888d. *Report on the experiments made in 1887 in the treatment of the downy mildew and the black rot of the grape vine, with a chapter on the apparatus for applying remedies for these diseases.* USDA Botanical Division, Section of Vegetable Pathology Bulletin 5. Washington, D.C.: GPO. [Ch. 7]

———. 1888e. Successful treatment of black rot. In *Proceedings of the ninth annual meeting of the Society for the Promotion of Agricultural Science held at Cleveland, Ohio,* 72–73. Columbus, Ohio: Gazette Printing House. [Ch. 7]

Scribner, F. L. Papers. 1885–1938. Manuscript Division, Library of Congress. [Ch. 7]

Scribner, F. L. Topical Files, National Fungus Collections. U.S. Department of Agriculture Systematic Botany and Mycological Laboratory, Beltsville, Md. [Ch. 6, 7]

Scribner, F. L., and Viala, P. 1888. *Black Rot (Laestadia bidwellii).* USDA Botanical Division, Section of Vegetable Pathology Bulletin 7. Washington, D.C.: GPO. [Ch. 7]

Selby, A. D. Files. Department of Plant Pathology, Ohio State University, Wooster, Ohio. [Ch. 12]

"September 16, 1861." 1868. *The Transactions of the Academy of Science of St. Louis,* vol. 2 *1861–1868,* 165–66. St. Louis: George Knapp and Co. [Ch. 4]

Setchell, W. A. 1927. Biographical memoir: William Gilson Farlow, 1844–1919. *Memoirs of the National Academy of Sciences* 21:1–22. [Ch. 4]

Shamel, A. D., and W. W. Cobey. 1907. *Tobacco breeding*. USDA Bureau of Plant Industry Bulletin 96. Washington, D.C.: GPO. [Ch. 10]

Shaw, E. A. 1986. Changing botany in North America: 1835–1860. The role of George Engelmann. *Annals of the Missouri Botanical Garden* 73:508–19. [Ch. 4]

Shear, C. L. 1906a. American Mycological Society, New Orleans Meeting, January 1, 1906. *Journal of Mycology* 12:85–86. [Ch. 12]

———. 1906b. Societies and academies: The American Mycological Society. *Science* 23:186. [Ch. 12]

———. 1910a. The American Phytopathological Society, Part 1. *Science* 31:746–57. [Ch. 12]

———. 1910b. The American Phytopathological Society, Part 2. *Science* 31:790–99. [Ch. 12]

———. 1919. First decade of the American Phytopathological Society. *Phytopathology* 9:165–70. [Ch. 12]

Sidar, J. W. 1976. *George Hammell Cook: A life in agriculture and geology*. New Brunswick, N.J.: Rutgers University Press. [Ch. 4]

"Sketch of Samuel Latham Mitchill." 1890–1891. *The Popular Science Monthly* 38:577, 691–98. [Ch. 1]

Smith, E. F. 1888. *Peach yellows: A preliminary report*. USDA Botanical Division Bulletin 9. Washington, D.C.: U.S. GPO. [Ch. 1, 7]

———. 1891. The chemistry of peach yellows II. In *Proceedings of the twenty-third session of the American Pomological Society held in Washington, D.C., September 22–24, 1891, 21–26*. Cleveland, Ohio: American Pomological Society. [Ch. 7]

———. 1894. *Peach yellows and peach rosette*. USDA Division of Vegetable Pathology Farmers' Bulletin 17. Washington, D.C.: GPO. [Ch. 7]

———. 1896. The bacterial diseases of plants: A critical review of the present state of our knowledge. *American Naturalist* 30:626–43. [Ch. 9]

———. 1897a. A bacterial disease of cruciferous plants. *Science* 5:963. [Ch. 11]

———. 1897b. *Pseudomonas campestris* (Pammel), the cause of a brown-rot in cruciferous plants. In *Centralblatt für Bakteriologie*, part 2, vol. 3: 284–91; 408–15; 478–86. [Ch. 11]

———. 1898. The first annual meeting of the society for plant morphology and physiology. *American Naturalist* 32:96–110. [Ch. 12]

———. 1899. *Wilt disease of cotton, watermelon, and cowpea (Neocosmospora nov. gen.)*. USDA Division of Vegetable Physiology and Pathology Bulletin 17. Washington, D.C.: GPO. [Ch. 9]

———. 1902. Plant pathology: Retrospect and prospect. *Science* 15:601–12. [Ch. 10]

————. 1911. *Bacteria in relation to plant diseases.* Vol. 2. of *Bacteria in relation to plant diseases.* Washington, D.C.: Carnegie Institution. [Ch. 5, 9]

————. 1916. In memoriam Thomas J. Burrill. *Journal of Bacteriology* 1:269–71. [Ch. 5]

————. 1929. Fifty years of pathology. In *Proceedings of the International Congress of Plant Sciences* 1:13–46. [Ch. 7]

Smith, E. F. Papers. 1865–1940. American Philosophical Society, Philadelphia, Pa. [Ch. 7, 9]

Smith, E. F. Papers. 1885–1927. Special Collections, National Agricultural Library, Beltsville, Md. [Ch. 7]

Smith, E. F., and C. O. Townsend. 1907. A plant-tumor of bacterial origin. *Science* 25:671–73. [Ch. 10]

Smith, E. F., N. A. Brown, and L. McCulloch. 1912. *The structure and development of crown gall: A plant cancer.* USDA Bureau of Plant Industry Bulletin 255. Washington, D.C.: GPO. [Ch. 10]

Smith, R. E. 1961a. Grape mildew as viewed in the early agricultural press of California. *Plant Disease Reporter* 45:700–702. [Ch. 7]

————. 1961b. Wheat smut as viewed in the early agricultural press of California. *Plant Disease Reporter* 45:712–16. [Ch. 3]

————. 1962. Early plant pathology in America. *Phytopathology* 52:119–22. [Ch. 4]

"Smut in wheat." 1845. *American Agriculturist* 3:371. [Ch. 3]

Solberg, W. U. 1968. *The University of Illinois 1867–1894: An intellectual and cultural history.* Urbana: University of Illinois Press. [Ch. 5]

Southworth, E. A. 1890. Anthracnose of cotton. *Journal of Mycology* 6:100–105. [Ch. 7]

Spalding, T. 1835. Rust in cotton. *Southern Agriculturist* 8:174. [Ch. 3]

Spaulding, P. 1909. A biographical history of botany at St. Louis, Missouri, part 3. *Popular Science Monthly.* 74:124–33. [Ch. 4]

Spaulding, P., and E. Field. 1912. *Two dangerous imported plant diseases.* Farmers' Bulletin 489. Washington, D.C.: GPO. [Ch. 10]

Stakman, E. C., and J. J. Christensen. 1954. Edward Monroe Freeman, pioneer plant pathologist. *Science* 120:285. [Ch. 11]

Stearns, R. P. 1970. *Science in the British Colonies of America.* Urbana, Ill.: University of Illinois Press. [Ch. 1]

Stevens, F. L. 1919. *The fungi which cause plant disease.* New York: Macmillan. [Appendix]

Stevens, F. L., L. H. Pammel, and M. T. Cook. 1920. Byron David Halsted, June 7, 1852–August 28, 1918. *American Journal of Botany* 7:305–6. [Ch. 8]

Stevens, N. E. 1933. Phytopathology: The dark ages in plant pathology in America:

1830–1870. *Journal of the Washington Academy of Sciences* 23:435–46. [Ch. 2; Pt. 1 intro.]

Stevenson, J. A. 1938. Beverly Thomas Galloway. *Mycologia* 30:597. [Ch. 7]

———. 1954. Plants, problems and personalities: The genesis of the Bureau of Plant Industry. *Agricultural History* 28:155–62. [Ch. 10]

———. 1957a. Cornelius Lott Shear, 1865–1956. *Phytopathology* 47:321–22. [Ch. 12]

———. 1957b. Cornelius Lott Shear, 1865–1956. *Taxon* 6:7–10. [Ch. 12]

———. 1958. Galloway, Beverly Thomas. *Dictionary of American Biography*, supplement 2, s.v. New York: Charles Scribner's Sons. [Ch. 7]

———. 1959. The beginnings of plant pathology in North America. In *Plant pathology: Problems and progress, 1908–1958*, ed. by C. S. Holton et al., 14–23. Madison, Wis.: The University of Wisconsin Press. [Ch. 1; Pt. 1 intro.]

Stone, G., and R. Smith. 1898. *Nematode worms in the greenhouse.* Hatch Experiment Station of the Massachusetts Agricultural College Bulletin 55. Amherst, Mass.: Carpenter & Morehouse. [Ch. 10]

Stover, W. G. 1925. Augustine Dawson Selby, 1859–1924. *Phytopathology* 15:1–10. [Ch. 12]

Sturgis, W. C. 1894. Report of the mycologist. In *Seventeenth annual report of the Connecticut Agricultural Experiment Station for 1893*, 72–111. New Haven, Conn.: Tuttle, Morehouse & Taylor. [Ch. 8]

———. 1896. Report of the mycologist. In *Nineteenth annual report of the Connecticut Agricultural Experiment Station for 1895*, 176–82. New Haven, Conn.: Tuttle, Morehouse & Taylor. [Ch. 8]

Swingle, W. T. 1892. Orange blight. In *Proceedings of the fifth annual meeting of the Florida State Horticultural Society held at Ormond, Florida, May 3d, 4th and 5th, 1892.* Florida State Horticultural Society. [Ch. 9]

———. 1894. Sulphur solution-blight-lemon scab. In *Proceedings of the seventh annual meeting of the Florida State Horticultural Society held at Jacksonville, Florida, April 10th, 11th, 12th, and 13th, 1894.* Tallahassee, Fla: Floridian Printing Company. [Ch. 9]

Swingle, W. T., and H. J. Webber. 1896. *The principal diseases of citrous fruits in Florida.* USDA Division of Vegetable Physiology and Pathology Bulletin 8. Washington, D.C.: GPO. [Ch. 9, 10]

Targioni-Tozzetti, G. [1767] 1952. *True nature, causes and sad effects of the rust, the bunt, the smut, and other maladies of wheat, and of oats in the field.* Reprint, Alimurgia, part 5. Phytopathological Classics 9, trans. L. R. Tehon. Ithaca, N.Y.: American Phytopathological Society. [Ch. 2]

Taylor, T. 1872. Report on fungoid diseases of plants. In *Report of the Commissioner of Agriculture for the year 1871*, 110–22. Washington, D.C.: GPO. [Ch. 6]

———. 1873. Microscopic investigations: The potato blight and rot. In *Twenty-seventh Annual Report of the Ohio State Board of Agriculture*, 32–37. Columbus, Ohio: Nevins & Myers. [Ch. 6]

———. 1874. Microscopic investigations. In *Report of the Commissioner of Agriculture for the year 1873*, 183–210. Washington, D.C.: GPO. [Ch. 6]

———. 1875. Microscopic observations. In *Report of the Commissioner of Agriculture for the year 1874*, 161–78. Washington, D.C.: GPO. [Ch. 6]

———. 1876. Microscopic observations. In *Report of the Commissioner of Agriculture for the Year 1875*, 187–206. Washington, D.C.: GPO. [Ch. 6]

———. 1877. Microscopic investigations. In *Report of the Commissioner of Agriculture for the Year 1876*, 74–86. Washington, D.C.: GPO. [Ch. 6]

Taylor, T. Topical Files, National Fungus Collections. U.S. Department of Agriculture Systematic Botany and Mycological Laboratory, Beltsville, Md. [Ch. 6]

Teschemacher, J. E. 1844a. The disease in potatoes. *New England Farmer and Horticultural Register* 23:125. [Ch. 2]

———. 1844b. The disease in potatoes. *New England Farmer and Horticultural Register* 23:158. [Ch. 2]

———. 1845a. Potato disease. *The Gardener's Chronicle and Agricultural Gazette for 1845* 8:125. [Ch. 2]

———. 1845b. The potato disease—salt as a remedy. *New England Farmer and Horticultural Register* 24:166. [Ch. 2]

———. 1845c. Potato rot. *New England Farmer and Horticultural* Register 24:114. [Ch. 2]

Thacher, J. 1822. *The American orchardist; or a practical treatise on the culture and management of apple and other fruit trees, with observations on the diseases to which they are liable, and their remedies. To which is added the most approved method of manufacturing and preserving cider. Compiled from the latest and most approved authorities, and adapted to the use of American farmers.* Boston: Joseph W. Ingraham. [Ch. 1]

Thaxter, R. 1890. Report of Roland Thaxter, mycologist. In *Annual report of the Connecticut Agricultural Experiment Station for 1889*, 129–53. New Haven, Conn.: Tuttle, Morehouse & Taylor. [Ch. 8]

———. 1891. Report of the mycologist, Dr. Roland Thaxter. In *Annual report of the Connecticut Agricultural Experiment Station for 1890*, 81–95. New Haven, Conn.: Tuttle, Morehouse & Taylor. [Ch. 8]

———. 1920. William Gilson Farlow. December 17, 1844–June 3, 1919. *American Journal of Science* 49:87–95. [Ch. 4]

Thomas, J. J. 1844. The diseases and insects injurious to the wheat crop. *Transactions of the New York State Agricultural Society* 3:201–16. [Ch. 2]

———. 1848. The pear blight. *The Cultivator* 5:20–21. [Ch. 3]

Thomson, S. V., D. C. Hildebrand, and M. N. Schroth. 1981. Identification and nutritional differentiation of the Erwinia sugar beet pathogen from members of *Erwinia carotovora* and *Erwinia chrysanthemi. Phytopathology* 71:1037–42. [Appendix]

Thornberry, H. H. Papers, 1929–1994. University of Illinois Archives, Urbana. [Ch. 5]

Thorne, G. 1961. *Principles of nematology.* New York: McGraw-Hill. [Ch. 10]

Thornton, T. P. 1989. *Cultivating gentlemen: The meaning of country life among the Boston elite, 1785–1860.* New Haven, Conn.: Yale University Press. [Ch. 1]

Thwaites, R. G., ed. 1900. *The University of Wisconsin. Its history and its alumni with historical and descriptive sketches of Madison.* Madison, Wis.: J. N. Purcell. [Ch. 5]

Tillet, M. [1755] 1970. *Dissertation on the cause of the corruption and smutting of the kernels of wheat in the head and on the means of preventing these untoward circumstances.* Reprint, Phytopathological Classics 5, trans. H. B. Humphrey. Ithaca, N.Y.: American Phytopathological Society. [Ch. 2]

"To prevent mildew on gooseberries." 1848. *Ohio Cultivator* 4:107–8. [Ch. 3]

Tobey, R. C. 1981. *Saving the prairies: The life cycle of the founding school of American plant ecology, 1895–1955.* Berkeley: University of California Press. [Ch. 4]

Tomlinson, D. 1852. Black knot on plum trees. *The Cultivator* 9:189. [Ch. 3]

Toumey, J. W. 1900. *An inquiry into the cause and nature of crown-gall.* The University of Arizona Agricultural Experiment Station Bulletin 33. Tucson, Ariz. [Ch. 10]

Townley, J. 1851. The leaf blight of the pear. *The Horticulturist and Journal of Rural Art and Rural Taste* 6:354–56. [Ch. 3]

Trelease, W. 1916. Thomas Jonathan Burrill, April 25, 1839–April 14, 1916. *Botanical Gazette* 62:153–55. [Ch. 5]

———. 1925. The evolution of a botanist and of a department of botany. *Proceedings of Indiana Academy of Science* 34:59–62. [Ch. 5]

True, A. C. 1928. *A history of agricultural extension work in the United States 1785–1923.* USDA Misc. Pub. 15. Washington, D.C.: GPO. [Ch. 11]

———. 1929. *A history of agricultural education in the United States, 1785–1925.* USDA Misc. Pub. 36. Washington, D.C.: GPO. [Ch. 1, 3]

———. 1937. *A history of agricultural experimentation and research in the United States,*

1607–1925. USDA Misc. Pub. 251. Washington, D.C.: U.S. GPO. [Ch. 1, 2, 3, 4, 6, 7, 8, 10]

Turner, J. B. 1852a. Notes on pear blight in Illinois. *The Horticulturist and Journal of Rural Art and Rural Taste* 7:163–66. [Ch. 5]

———. 1852b. Plan for an industrial university. In *Report of the Commissioner of Patents, for the year 1851,* 37–44. Washington, D.C.: Robert Armstrong. [Ch. 3]

U.S. Congress, Committee on Agriculture. 1856. *Department of Agriculture Report.* H. Rep. 321, 34th Cong., 1st Sess., 1–10. [Ch. 3]

U.S. Patent Office. 1844. *Report of the Commissioner of Patents, for the year 1843.* Washington, D.C.: Gales and Seaton. [Ch. 2, 3]

———. 1845. *Report of the Commissioner of Patents, for the year 1844.* Washington, D.C.: Gales and Seaton. [Ch. 2, 3]

———. 1846. *Report of the Commissioner of Patents, for the year 1845.* Washington, D.C.: Ritchie and Heiss. [Ch. 2]

———. 1848. Maize or Indian corn. *Report of the Commissioner of Patents, for the year 1847.* Washington, D.C.: Wendell and Van Benthuysen. [Ch. 3]

———. 1848. *Report of the Commissioner of Patents, for the year 1847.* Washington, D.C.: Wendell and Van Benthuysen. [Ch. 2]

University of Minnesota Bulletin, The College of Agriculture, 1908–1909. 1908. St. Paul: University of Minnesota. [Ch. 11]

University of Minnesota Bulletin, The College of Agriculture, 1909–1910. 1909. St. Paul: University of Minnesota. [Ch. 11]

University of Minnesota Bulletin, The College of Agriculture, 1913–1914. 1913. St. Paul: University of Minnesota. [Ch. 11]

University of Wisconsin Catalogue, 1908–1909. 1909. Madison: University of Wisconsin. [Ch. 11]

University of Wisconsin Catalogue, 1910–1911. 1911. Madison: University of Wisconsin. [Ch. 11]

University of Wisconsin Catalogue, 1914–1915. 1915. Madison: University of Wisconsin. [Ch. 11]

University of Wisconsin College of Agriculture, Administration Records. Archives, University of Wisconsin Memorial Library. Madison, Wis. [Ch. 11]

USDA Division of Botany, Section of Mycology. 1886. *Treatment of downy grape mildew (Peronospora viticola) and black rot (Phoma uvicola).* Circular 1. Washington, D.C.: GPO. [Ch. 7]

USDA Division of Botany, Section of Vegetable Pathology. 1886. *Grape mildew and black rot.* Circular 2. Washington, D.C.: GPO. [Ch. 7]

USDA Division of Vegetable Pathological and Physiological Investigations. Letters Sent 1885–1888. Records of the Division of Vegetable Pathological and Physiological Investigations. Records of the Bureau of Plant Industry, Soils, and Agricultural Engineering. Record Group 54. National Archives, Washington, D.C. [Ch. 7]

USDA Correspondence of the Office of the Chief (General) 1900–1908; 1908–1939. Records of the Bureau of Plant Industry, Soils, and Agricultural Engineering, Record Group 54, National Archives, Washington, D.C. [Ch. 10]

USDA Letters Received (General) 1891–1900; 1901–1902; 1903–1908. Records of the Division of Vegetable Pathological and Physiological Investigations. Records of the Bureau of Plant Industry, Soils, and Agricultural Engineering, Record Group 54, National Archives, Washington, D.C. [Ch. 8, 9, 10]

USDA Letters Sent (General). 1882–1897. Records of the Immediate Offices of the Commissioner and Secretary of Agriculture. Records of the Office of the Secretary of Agriculture, Record Group 16. National Archives, Washington, D.C. [Ch. 6]

Van Buren, J. 1854. On the black rot in apples. *Southern Cultivator* 12:92. [Ch. 3]

———. 1859. The diseases of fruit trees. *The Country Gentleman* 13:142. [Ch. 3]

Van Der Zwet, T., and H. L. Keil. 1979. *Fire blight: a bacterial disease of rosaceous plants.* USDA Agriculture Handbook 510. Washington, D.C.: U.S. GPO. [Ch. 1]

Venning, F. D. 1977. Walter Tennyson Swingle, 1871–1952. *The Carrell* 18: 1–32. [Ch. 9]

Veysey, L. R. 1965. *The emergence of the American university.* Chicago: University of Chicago Press. [Ch. 11]

Von Schrenk, H., and G. G. Hedgcock. 1907. *The wrapping of apple grafts and its relation to the crown-gall disease.* Part 2 of USDA Bureau of Plant Industry Bulletin 100. Miscellaneous Papers. Washington, D.C.: GPO. [Ch. 10]

Von Schrenk, H., and P. Spaulding. 1903. *The bitter rot of apples.* USDA Bureau of Plant Industry Bulletin 44. Washington, D.C.: GPO. [Ch. 10]

Wadsworth, W. 1858. Sowing smutty wheat. *The California Culturist* 1:125–26. [Ch. 3]

Waggoner, P. E., and P. Gough. 1975. The first two stations—Connecticut, California. *USDA Yearbook of Agriculture,* 2–8. Washington, D.C.: GPO. [Ch. 8]

Walker, J. C. 1957. *Plant pathology.* 2d ed. New York: McGraw-Hill. [Ch. 8]

———. 1986. Beginnings, L. R. Jones, 1910–1925. In *With one foot in the furrow: A history of the first seventy-five years of the Department of Plant Pathology at the University of*

Wisconsin—Madison, ed. by P. H. Williams and M. Marosy, 3–17. Dubuque, Iowa: Kendall/Hunt Publishing Co. [Ch. 11]

Walsh, T. R. 1971. Charles E. Bessey and the transformation of the industrial college. *Nebraska History* 52:383–409. [Ch. 4]

War Emergency Board, American Plant Pathologists. 1918a. Final report for the year ending December 31, 1918. American Phytopathological Society Records, Special Collections, Iowa State University, Ames. [Epilogue]

———. 1918b. Midyear report, June 1918. American Phytopathological Society Records, Special Collections, Iowa State University, Ames. [Epilogue]

———. 1918c. Notes on the conference meeting of the War Emergency Board, American Plant Pathologists, held at LaFayette, Indiana, March 15. In Whetzel, H. H. Papers. [Epilogue]

Ward, H. M. 1902. On the relations between hosts and parasite in the bromes and their brown rust, *Puccinia dispersa* (Erikss.). *Annals of Botany* 16:233–315. [Ch. 10]

———. 1905. Recent researches on the parasitism of fungi. *Annals of Botany* 19:1–54. [Ch. 10]

Warder, J. A. 1862. The pear orchard. In *Report of the Commissioner of Patents, for the year 1861*, 232–51. Washington, D.C.: GPO. [Ch. 3]

Warren, J. C. 1853. Notice of J. E. Teschemacher. *Proceedings of the Boston Society of Natural History* 4:393–94. [Ch. 2]

Waters, J. 1839. Blight in fruit trees. *Western Farmer* 1:37. [Ch. 1]

Watts, F. 1876. Report. In *Report of the Commissioner of Agriculture for the year 1875*, 7–16. Washington, D.C.: GPO. [Ch. 6]

Webber, H. J. 1897. *Sooty mold of the orange and its treatment.* USDA Division of Vegetable Physiology and Pathology Bulletin 13. Washington, D.C.: GPO. [Ch. 9]

Webber, H. J., and E. A. Bessey. 1900. Progress of plant breeding in the United States. In *Yearbook of the USDA 1899*, 465–90. Washington, D.C.: GPO. [Ch. 10]

Webber, H. J., and W. A. Orton. 1902. *A cowpea resistant to root knot (Heterodera radicicola).* Part 2 of *Some Diseases of the Cowpea.* USDA Bureau of Plant Industry Bulletin 17. Washington, D. C.: GPO. [Ch. 10]

"Webber, Herbert John." 1927. *The National Cyclopaedia of American Biography*, s.v. New York: James T. White & Co. [Ch. 9]

Weber, G. A. 1930. *The plant quarantine and control administration: Its history, activities and organization.* Washington, D.C.: The Brookings Institution. [Ch. 10]

Webster, N. 1819. Address. *Rural Magazine and Farmer's Monthly Museum* 1:83–86. [Ch. 3]

Wendell, H. 1850. Report of Dr. Herman Wendell, of Albany: Pears. In *Report of the Com-*

missioner of Patents for the year 1849. Part 2. Agriculture, 446–48. Washington, D.C.: GPO. [Ch. 5]

Weston, W. H. 1933. Roland Thaxter. *Mycologia* 25:70. [Ch. 8]

Wetmore, C. J. 1891. Report to the Viticultural Commission. *Pacific Wine and Spirit Review* 27:9–10. [Ch. 9]

"Wheat culture-uredo (or smut), etc." 1848. *Southern Cultivator* 6:115. [Ch. 3]

"Wheat." 1848. *Report of the Commissioner of Patents, for the year 1847*, 101–11. Washington, D.C.: Wendell and Van Benthuysen. [Ch. 3]

Whetzel, H. H. 1909. Department of Plant Pathology. In *Twenty-first annual report of the Cornell University Agricultural Experiment Station, Ithaca, N.Y. 1908*, 39–45. Albany, N.Y.: J. B. Lyon Co., State Printers. [Ch. 11]

———. 1910. Department of Plant Pathology. In *Twenty-second annual report of the Cornell University Agricultural Experiment Station, Ithaca, N.Y., 1909*, lxiii–lxxi. Albany, N.Y.: J. B. Lyon Co., State Printers. [Ch. 11]

———. 1911. Department of Plant Pathology. In *Twenty-third annual report of the New York State College of Agriculture at Cornell University and the Agricultural Experiment Station established under the Direction of Cornell University, Ithaca, N.Y., 1910*, 59–64. Albany, N.Y.: J. B. Lyon Co., State Printers. [Ch. 11]

———. 1914. Department of Plant Pathology. In *Twenty-sixth annual report of the New York State College of Agriculture at Cornell University and the Agricultural Experiment Station established under the direction of Cornell University*, lii–lviii. Albany, N.Y.: J. B. Lyon Company, Printers. [Ch. 11]

———. 1916. Department of Plant Pathology. In *Twenty-eighth annual report of the New York State College of Agriculture at Cornell University and of the Agricultural Experiment Station established under the direction of Cornell University, Ithaca, N.Y., 1915*, xxxv–xxxviii. Albany, N.Y.: J. B. Lyon Co., State Printers. [Ch. 11]

———. 1918. A call to the colors. American Phytopathological Society Records. Special Collections. Iowa State University, Ames. January 8, 1918. [Epilogue]

———. 1918. *An outline of the history of phytopathology*. Philadelphia: W. B. Saunders Co. [Ch. 11, 12]

———. 1919. George Francis Atkinson. *The Botanical Gazette* 67:366. [Ch. 8]

———. 1945. The history of industrial fellowships in the department of plant pathology at Cornell University. *Agricultural History* 19:99–104. [Ch. 11]

Whetzel, H. H. Papers. Rare and Manuscript Collection, Department of Manuscript and University Archives, Cornell University Library, Ithaca, N.Y. [Ch. 11, Epilogue]

Whetzel, H. H., and S.E.A. McCallan. 1930. Studies on fungicides. Part 1, Concepts and terminology. *Cornell University Agricultural Experiment Station Memoir* 128:3–7. [Ch. 11]

White, L. D. 1958. *The Republican era, 1869–1901, A Study in Administrative History.* New York: Macmillan. [Ch. 6, 8]

Wiebe, R. H. 1967. *The search for order, 1877–1920.* New York: Hill and Wang. [Pt. 4 intro.]

Wilcoxson, R. D., and T. Kommedahl. 1992. Elvin Charles Stakman: American champion of plant pathology, 1885–1979. *Review of Tropical Plant Pathology* 7:223–36. [Ch. 11]

Wilson, J. T. 1857. To raise gooseberries without mildew. *The Michigan Farmer* 15:147. [Ch. 3]

Wood, F. A. 1975. Those Minnesota plant doctors. *Minnesota Science* 31:34–35. [Ch. 11]

Woods, A. F. 1898. Report of the Acting Chief of the Division of Vegetable Physiology and Pathology. In *Annual Reports of the Department of Agriculture, 1898,* 27–36. Washington, D.C.: GPO. [Ch. 10]

———. 1905. The relation of plant physiology to the development of agriculture. In *USDA Yearbook of Agriculture, 1904,* 119–32. Washington, D.C.: GPO. [Ch. 7]

———. 1938. Beverly Thomas Galloway. *Science* 88:6. [Ch. 7]

Woodward, C. R., and I. N. Waller. 1932. *Botany: Byron David Halsted.* New Jersey's Agricultural Experiment Station, 1880–1930. New Brunswick, N.J.: New Jersey's Agricultural Experiment Station. [Ch. 8]

Woronin, M. 1934. *Plasmodiophora brassicae: The cause of cabbage hernia.* Phytopathological Classics 4. Ithaca, N.Y.: American Phytopathological Society. [Ch. 8]

Wyman, L. 1847. Salt as applied to the plum tree. *The Horticulturist and Journal of Rural Art and Rural Taste* 1:472–73. [Ch. 3]

Young, J. M., G. S. Saddler, Y. Takikawa, S. H. De Boer, L. Vauterin, L. Gardan, R. I. Gvozdyak, and D. E. Stead. 1996. Names of plant pathogenic bacteria 1864–1995. *Review of Plant Pathology* 75:721–63. [Appendix]

Younger, J. 1850. Rust in wheat. *The Southern Planter* 10:309. [Ch. 3]

Wheat:

    bunt, 30, 59, 162, 186, 194, 196

    mildew, 31

    rusts, 3, 14, 31, 34, 39, 53, 57, 59, 63, 87, 128, 195, 198, 218–219, 257, 266, 314, 337–338

    smut, 3, 34, 39, 53, 59, 128,

    stem rust, 293–294, 297, 300, 336

Whetzel, H. H., *239–240*, 286–291, 305, 309, 321, 325–328, 330, 334–335, 338

White pine blister rust, 276–279, 325–326

Wilson, J., 251–252, 278

Wollenweber, 306

Woods, A. F., 104, *153*, 160, 217–218, 223, 252–253, 260, 298, 319, 325

Woronin, M. S., 114, 190

Xanthomonas sp.:

    campestris, 199

    malvacearum, 201

Xylella fastidiosa, 208

Yale University, 92, 105, 116

Young, A., 3

Zimmerman, A., 264

THE AUTHORS

C. Lee Campbell holds B.S. and M.S. degrees from Colorado State University and a Ph.D. from the Pennsylvania State University, all in plant pathology. He is professor of plant pathology at North Carolina State University. The author of several textbooks in plant pathology and plant disease epidemiology, he is a fellow of the American Association for the Advancement of Science and served as president of the American Phytopathological Society for 1997–1998.

Paul D. Peterson holds a B.A. degree in history and an M.A. in public history/archival management from North Carolina State University. He is a visiting lecturer/research assistant in the Department of Plant Pathology at North Carolina State University and is pursuing studies leading to the Ph.D. degree in plant pathology. He is also historian of the American Phytopathological Society.

Clay S. Griffith holds B.A. and M.A. degrees in history from North Carolina State University. He is a freelance research writer based in Raleigh, North Carolina. In addition to papers on the history of plant pathology in the United States, he has published a variety of nonfiction and fiction works.